416-327-71

DEVELOPMENTS IN
EARTH & ENVIRONMENTAL SCIENCES 5

GLOBAL WARMING AND GLOBAL COOLING: EVOLUTION OF CLIMATE ON EARTH

VOLUME 1 GEOSCIENCES, ENVIRONMENT AND MAN
         by H. Chamley

VOLUME 2 MEDICAL GEOLOGY
         EFFECTS OF GEOLOGICAL ENVIRONMENTS ON
         HUMAN HEALTH
         by M.M. Komatina

VOLUME 3 OIL POLLUTION AND ITS ENVIRONMENTAL
         IMPACT IN THE ARABIAN GULF REGION
         by M. Al-Azab, W. El-Shorbagy and S. Al-Ghais

VOLUME 4 MEDITERRANEAN CLIMATE VARIABILITY
         by P. Lionello, P. Malanotte-Rizzoli and R. Boscolo

Developments in Earth & Environmental Sciences, 5

GLOBAL WARMING AND GLOBAL COOLING: EVOLUTION OF CLIMATE ON EARTH

By

O.G. SOROKHTIN
Moscow, Russia

G.V. CHILINGAR
Los Angeles, CA, USA

L.F. KHILYUK
Los Angeles, CA, USA

ELSEVIER

Amsterdam – Boston – Heidelberg – London – New York – Oxford
Paris – San Diego – San Francisco – Singapore – Sydney – Tokyo

Elsevier
Radarweg 29, PO Box 211, 1000 AE Amsterdam, The Netherlands
Linacre, House, Jordon Hill, Oxford OX2 8DP, UK

First edition 2007

Copyright © 2007 Elsevier B.V. All rights reserved

No part of this publication may be reproduced, stored in a retrieval system
or transmitted in any form or by any means electronic, mechanical, photocopying,
recording or otherwise without the prior written permission of the publisher

Permissions may be sought directly from Elsevier's Science & Technology Rights
Department in Oxford, UK: phone (+44) (0) 1865 843830; fax (+44) (0) 1865 853333;
email: permissions@elsevier.com. Alternatively you can submit your request online by
visiting the Elsevier web site at http://elsevier.com/locate/permissions, and selecting
Obtaining permission to use Elsevier material

Notice
No responsibility is assumed by the publisher for any injury and/or damage to persons
or property as a matter of products liability, negligence or otherwise, or from any use
or operation of any methods, products, instructions or ideas contained in the material
herein. Because of rapid advances in the medical sciences, in particular, independent
verification of diagnoses and drug dosages should be made

Library of Congress Cataloguing-in-Publication Data
A catalog record for this book is available from the Library of Congress

British Library Cataloguing in Publication Data
A catalogue record for this book is available from the British Library

ISBN: 978-0-444-52815-5
ISSN: 1571-9197

For information on all Elsevier publications
visit our website at books.elsevier.com

Printed and bound in The Netherlands

07 08 09 10 11   10 9 8 7 6 5 4 3 2 1

Working together to grow
libraries in developing countries

www.elsevier.com | www.bookaid.org | www.sabre.org

ELSEVIER   BOOK AID International   Sabre Foundation

This book is dedicated to

Dr. Jemmy T. M. You

and

Dr. and Mrs. Shaw Yung Li

for their vision of the problem and
continuous support of research on global warming
at the University of Southern California, Los Angeles, California

CONTENTS

Foreword   xi

Preface   xiii

CHAPTER 1. Introduction to the Global Evolution of the Earth   1
   1.1. Origin of the Earth and its Composition   1
   1.2. Formation of the Earth-Moon Double Planet   11
   1.3. Beginning of Tectonomagmatic Activity on Earth   21
   1.4. Separation of Earth's Core (Main Factor Driving the Earth's Global Evolution)   26
   1.5. Energy Balance of the Earth   41
   1.6. Tectonic Activity of Earth   50
   1.7. Evolution of the Chemical Composition of the Mantle   55
   1.8. Formation of Oceanic and Continental Lithospheric Plates   61

CHAPTER 2. Degassing of Earth and Formation of Hydrosphere and Atmosphere   83
   2.1. Formation of the Earth's Hydrosphere   83
   2.2. Formation of Iron-bearing Deposits   92
   2.3. Nature of Global Transgressions   99
   2.4. Formation of the Earth's Atmosphere   103
   2.5. Evolution of Partial Pressure of Nitrogen in the Atmosphere   105
   2.6. Evolution of Carbon Dioxide Partial Pressure   108
   2.7. Oxygen Accumulation in the Earth's Atmosphere   116
   2.8. Generation of Abiogenic Methane   120
   2.9. Origin of the "Ozone Holes"   130

CHAPTER 3. The Adiabatic Theory of Greenhouse Effect   133
   3.1. Introduction   133
   3.2. Geenhouse Effect   133
   3.3. Principal Characteristics of Present-day Atmosphere   138
   3.4. Key Points of the Adiabatic Theory of Greenhouse Effect   140

viii   Contents

3.5. Possible Directions of Generalizing the Adiabatic Theory	148
3.6. Prognostic Atmospheric Temperature Estimates	156
3.7. Impact of Anthropogenic Factor on the Earth's Climate	160
3.8. Influence of Oceans on the Atmospheric Content of Carbon Dioxide	162
3.9. Methane Emission Effect	165

CHAPTER 4. Evolution of Earth's Climates	167
4.1. Introduction	167
4.2. Climatic Cooling during Cenozoic Era	167
4.3. Bacterial Nature of the Earth's Glaciations	171
4.4. Precession Cycles and the Earth's Climate	180
4.5. Continental Drift and the Climates of Earth	190

CHAPTER 5. Influence of Climate on Formation of Mineral Deposits on Earth	195
5.1. Introduction	195
5.2. Plate Tectonics and Origins of Endogenic Mineral Deposits	195
5.3. Earth's Core Separation Determining the Geologic Evolution of Earth	205
5.4. The Influence of Ocean and Climates of Earth on the Formation of Sedimentary Mineral Deposits	211
5.5. Origin of Diamond-bearing Kimberlitic and Similar Rocks	219
5.6. Origin of Exogenic Deposits of Minerals	222
5.7. Plate Tectonics and Formation of Hydrocarbon Deposits	224

CHAPTER 6. Origin and Development of Life on Earth (Influence of Climate)	235
6.1. Uniqueness of Earth	235
6.2. The Earth's Climate and Origin of Life	236
6.3. Influence of Geologic Processes on the Evolution of Earth's Life	240
6.4. Influence of Continental Drift and Marine Transgressions on Ecology during Phanerozoic Eon	244
6.5. Future Development of Atmosphere and the Death of Earth	255

√ CHAPTER 7. Global Forces of Nature Driving the Earth's Climate	259
7.1. Introduction	259
7.2. Solar Irradiation Reaching the Earth	261

7.3.	Orbital Deviations and the Earth's Mass Redistribution	263
7.4.	The Earth's Outgassing	265
7.5	Global Climatic Cooling due to Increase in CO_2 Content	271
7.6.	Inner Sources of the Earth's Energy	273
7.7.	Smoothing Role of World Ocean	276
7.8.	Microbial Activity at Earth's Surface	278
7.9.	Glacial and Interglacial Periods	279
7.10.	Global Warming or Global Cooling?	281
7.11.	Conclusions	284

References	287
Further Reading	295
Subject Index	307

Foreword

Global climate change has become one of the most hotly contested science-based public policy issues in many years, but the debate suffers from lack of scientific breadth and data. The debate is one of data versus theoretical computer models. Computer models argue forcefully that human contributions to greenhouse gases are significantly changing the global climate for the warmer, whereas observations argue equally forcefully that natural processes dominate climate change. Temperature evolution is one of Earth's major dynamic processes, and the possibility that humankind can alter this process has alarmed many scientists. Media have taken sides in favor of human control of the process, as have many scientific societies. These learned societies have, however, taken their positions without full study of the issue; rather, they reflect the current "humanocentric" bias of an environmentally zealous social community.

The greatest lack in the debate is the understanding of climate history that only the geological professions can provide. Only a few scattered geological refereed papers and fewer books have addressed the history and natural causes of climate change over time. Prior to the industrial revolution, all climate change was naturally driven. For 4.5 billion years climate has changed continuously, in both directions, at many scales of intensity and duration. The geologic history of the Earth and the dynamism of the Earth are the rocks upon which the computer models of climate change must be driven to test their hypotheses. We have permitted the public debate to focus on the last thirty years, ignoring the climate change that has taken place over the last 250 years, or 1,500 years, or 1,000,000 years. We permitted an international body to tout around a model of historic temperature change that flew in the face of not only geologic history, but all recorded human history.

Separating whether climate is changing from what causes climate to change is a first step in restoring scientific integrity to the debate. Climate is changing all of the time, but unlike weather, its rate of change cannot be quantified over short time spans. Addressing then the causes of climate change, starting at the origin of climate on Earth, and progressing to modern times permits modern change to be placed in context in Earth's history and allows us to separate any human contribution from natural change.

There are a plethora of processes that can cause climate to change. Documenting these requires understanding of not only how the processes work, but what their scale of intensity of change is, and over what time span. Understanding greenhouse effects is part of that equation, but also understanding the correlation of temperature and solar intensity variations, orbital variations, and short timescale events such as meteorite impacts and major volcanic eruptions is also required.

All of this leads to the conclusion that there is need for study of climate change over time, and of the processes that have historically driven such change. Sorokhtin, Chilingar, and Khilyuk, in this volume, have provided the reader with a new approach to study of Earth history by combining a review of Earth's geological history with a review of Earth's climate history, emphasizing the dynamic nature of Earth processes. They theorize, model, and richly document coupling of climate change to other dynamic and compositional changes of Earth, bringing to the reader their new sound concepts and models, the literature and the conceptual views of European, Asian, and North American research scientists.

The authors have developed a sequence of studies starting with evolution of the Earth as a planet, followed by the formation of the hydrosphere and atmosphere. Once those physical attributes are understood, the transition to understanding the greenhouse effect and the evolution of climates follows. This discussion addresses human influences as well as the physics of the greenhouse.

Climate does not only influence development of life on Earth, but also influences the development of some of Earth's mineral deposits and energy resources.

A full discussion of the Earth processes that may contribute to climate change outlines many of the issues now in the public debate, and also includes a review of some lesser known processes that may well have a bearing on the extent of climate change.

It is time for the Earth sciences to contribute to the discussion and I commend these authors for this contribution. It might be possible to bring science back into the debate and to rebuff the "beliefs" that now dominate with interpretations of data and observations based on sound theory.

Lee C. Gerhard
Lawrence
Kansas

Preface

Understanding and explaining the past and present climate evolution and forecasting the future climatic changes in order to establish efficient global policies on environmental protection and avoid the future possible disasters are strong stimuli for the development of new advanced theories of evolution of Earth's climate. Climate is a set of long-term patterns of changes in the state of atmosphere. The state of atmosphere can be described by a certain collection of physical parameters, mainly the temperature, pressure, and humidity. A good theory should be able to reveal and explain the past and present climatic changes and predict their future trends. The scientific method dictates the construction of a sound theory based on links to underlying processes and phenomena, matching the proposed theory to all available observations. Examination of the problem allows one to conclude that the Earth's climate is determined by the (1) mass and composition of atmosphere and (2) energy brought to, contained in, and transferred from the atmosphere. A sound theory of climate evolution should adequately describe and explain all the established facts related to climatic evolution and provide means for prediction of future changes in the physical, chemical, and biological processes, which determine the mass–energy transformations in the atmosphere. Before developing a theory of climate evolution, one needs to understand the global evolution of Solar System and our planet Earth, and the processes, which form and change the Earth's atmosphere. This theory should not be self-contradictory and should be based on basic laws of physics. One should be able to verify the theory by direct or indirect experiments, enabling one to test quantitatively the predictions of the theory.

According to the general physical principles and the first law of thermodynamics (mass–energy conservation law), the most influential input to the Earth's evolution belonged to the processes that decreased the inner potential energy of Earth and the Earth–Moon system in the highest degree during the Earth's evolution. Inasmuch as the heat resulting from these processes finally dissipates into space, the process of evolution of Earth and the Earth–Moon system is irreversible. As the boundary conditions of evolutionary theory, one should use the data on structure and composition of contemporary Earth and the geologic history of its development.

According to the theory of evolution presented in Chapter 1, the process of chemical–density differentiation of Earth's matter (leading to the formation and gradual growth of a dense iron oxide core, emergence of chemical–density convection in the mantle, and formation of a lighter silicate crust) is the main planetary process driving the evolution of Earth. At the present time, this process accounts for 90% of endogenic energy, whereas the process of radioactive decay constitutes about 9% of inner energy. The tidal deformations add about 1% of energy generated inside the Earth's body. The inputs of these processes into the total inner Earth's energy varied, however, throughout the geologic history. Thus, at the time of Earth's formation, the gravitational energy of space matter accretion had been dominant, whereas in the juvenile Earth the tidal energy generation had been prevalent. Beginning with the Archaean time and until now, the energy of gravitational matter differentiation played a leading role in the tectonic and other evolutionary processes on the Earth. The energy generated by the radioactive decay did not influence the Earth's evolution significantly.

The laws of Earth's degassing and stages of formation of hydrosphere and atmosphere are described in Chapter 2. The juvenile Earth was completely deprived of hydrosphere and the original atmosphere consisted almost entirely of nitrogen. The sea basins were formed only during the Early Archaean time. At the same time, the degassed carbon dioxide started entering the atmosphere, which led to the formation of dense Archaean carbon dioxide–nitrogen atmosphere. The first oceans were formed during the Late Archaean time. After that, the oceanic waters began interacting with the rocks of oceanic crust. This process gave rise to the considerable changes in the composition and pressure of atmosphere and the formation of iron-bearing deposits of Precambrian time. The latter suppressed oxygenation of atmosphere and generation of abiogenic methane.

The adiabatic theory of greenhouse effect (Chapter 3) is the backbone of the book. It is shown that the process of heat transfer in the atmosphere occurred mostly by convection (67%) and possessed the adiabatic properties. The adiabatic model allows one to describe quantitatively the temperature regimes of planetary troposphere and evaluate the influence of atmospheric composition and pressure on climate. The influence of Earth's precession angle on the climate is also presented. It was shown that in the case of small precession angles the climate cools down, whereas with increasing precession angles the climate becomes warmer. The adiabatic theory allowed the writers to compare the relative effects of natural and anthropogenic influences on Earth's climate and to conclude that the anthropogenic influence is negligible in comparison with natural factors. In the light of present-day "hot" debates on the causes of currently observed global warming, the obtained upper (the worst-case scenario) estimate of 0.01 °C for the possible influence of anthropogenic emission of greenhouse gases (carbon dioxide, methane, etc.) on global atmospheric temperature shows that the traditional explanation of greenhouse effect as a result of anthropogenic emission of greenhouse gasses is a myth.

In Chapter 4, the writers discussed the evolution of Earth's climate throughout the geologic history. The computed temperature regimes under the most probable initial and boundary conditions are presented for all geologic periods. The predictions of future Earth's atmosphere temperature regimes are presented together with numerous illustrations. The nature and origin of ice ages in the Pleistocene Epoch are explained based on theory of precession cycles. In Chapter 5, the evolution of climate was used to explain the nonuniformity of distribution in time of periods of intense accumulation of mineral deposits. In particular, this theory allowed us to explain the formation of unique (largest) iron-ore deposits at the end of Archaean and Early Proterozoic time.

The origin of life on Earth (Chapter 6) is linked to the formation of reducing environment in the Earth's atmosphere due to intense generation of abiogenic methane at the very beginning of Early Archaean time. This, in turn, led to abiogenic formation of first organic compounds (formaldehyde and hydrogen cyanide, for example), which served as building materials for development of primitive life forms. The subsequent development of life on Earth occurred according to "biological laws", but under the strong influence of geochemical and climatic conditions. The main stages in the life transformations (beginning from the origin of life at the very beginning of Archaean time and to the time of emergence of highest life forms at the boundary between the Proterozoic and Phanerozoic time, with further life evolution during the Phanerozoic time) coincided with the main geotectonic breaks in the development of Earth. In turn, the development of life influenced significantly the nature of geologic processes. Thus, the emergence and vast proliferation of nitrogen-consuming bacteria possibly reduced the partial pressure of nitrogen and, consequently, the total atmospheric pressure. As a result, beginning with the mid-Proterozoic time, the phase of general climate cooling was observed.

The writers identified and described the global forces of nature driving the Earth's climate: (1) solar radiation as a dominant source of exogenic energy; (2) outgassing as a major endogenic supplier of gasses to World Ocean and atmosphere; and (3) microbial activities directly generating and consuming atmospheric gases at the interface between lithosphere and atmosphere (Chapter 7). The scope and extent of their influences on the global climate were quantitatively evaluated. It was shown that the influences of global forces of nature are at least 4–5 orders of magnitude greater than those due to human activities. Thus, the effect of anthropogenic influence on global climate is negligible.

In spite of the fact that the book integrates abundant information from a broad variety of scientific disciplines (physics, chemistry, geology, biology, and evolutionary theory) and uses a great deal of mathematical modeling for analyzing the underlying processes, the authors attempted to make the book understandable for readers with basic background in college physics and chemistry. Detailed theoretical constructions and cumbersome mathematical computations are not presented. The authors, however, tried to (1) state the basic system modeling assumptions; (2) provide the description (mostly mathematical) of developed models; (3) present the final results in graphic form, comparing them (if available) with the corresponding experimental data and previous theories; (4) provide their interpretations; and (5) describe the possible practical applications.

The book is addressed to the graduate and postgraduate students in geosciences and the practicing engineers and scientists in the fields of geology, geophysics, climate studies, environment, geography, and biology. Scientists, engineers, and decision makers will obtain the guiding theory and methodology, clear graphical presentation of global climate evolution, explanation of main causes of changes in the parameters of atmosphere, ready quantitative estimates of resulting effects of various influences on the global characteristics of Earth's atmosphere.

The book is also addressed to the policy makers and concerned general public, whose awareness on the subject of climate evolution rose immensely due to broad international debates on currently observed global warming. It presents important factual information and sound theory of global climate evolution in a systematic and easy-to-understand form and allows the readers to develop the conceptual basis necessary for understanding the real and imaginary threats to their well-being and existence.

Chapter 1

Introduction to the Global Evolution of the Earth

1.1. Origin of the Earth and its Composition

One can start this book with the "Big Bang", which created the universe. The "Big Bang" occurred about 13.7 billion years ago. At the moment of occurrence, its location could not be identified because no object existed in the "universe" that could serve as a benchmark. After occurrence of this event, matter (in the form of elementary particles) and light, which were emitted from the "starting point", began rapidly expanding according to the Hubble's law (Alpher and Herman, 1988) filling-in the space of universe. The early universe was filled homogeneously with a high-density energy, and the matter was subjected to extremely high temperatures and pressures. It expanded and cooled, going through phase transitions related to elementary particles. The first atoms (mostly hydrogen) possibly formed about 300 thousand years after the Big Bang (Gamow, 1948). Over time, the matter of slightly denser regions gravitationally attracted nearby matter creating new formations such as gas clouds, stars, and galaxies. Over billions of years, traveling through the universe, these objects experienced various metamorphoses forming, in particular, clouds of interstellar gas and dust, which served as original building material for our Solar System. One of the sources of matter for these clouds is explosion of "supernovas", which are large stars with a mass several times that of the Sun.[1] After the first protogalaxies were formed, the first primitive hydrogen–helium protostars emerged. Gravitational compaction of these stars resulted in rapid temperature increase within them. That, in turn, led to initiation of the nuclear synthesis. As a result, heavier elements (up to iron) began gradually and consistently forming in deep areas of such stars.

Large stars are unstable structures and their evolution ends up in huge explosions. The greater the star's initial mass, the faster such fatal evolution

---

[1] Nova is a variable star that suddenly increases in brightness by thousands to hundreds of thousands of times up to 14 magnitudes and then decreases in brightness over a period of months to years. Supernova is an extremely bright nova that suddenly increases up to a billion times in brightness and reaches an absolute magnitude of $c.$-16.

unfolds. The supernova explosion occurs when the light elements supply is totally exhausted in their internal regions, and a core consists only of iron and nickel (i.e. elements with the lowermost nuclear attraction potential energy between the protons and neutrons in the nuclei). As a result, all nuclear reactions stop in the cores of the large stars after completion of their evolution. The heat energy to prevent their unlimited contraction is no longer generated. There is no more counterforce to gravity and the stars collapse. Under a tremendous pressure in the center of the star (building as a result of unlimited gravitational attraction during the star's collapse) the electrons are "squeezed into" the iron nuclei. The protons turn into neutrons and the star core itself becomes a coagulum of neutrons or even a "black hole".

Gravitational collapse of the star nucleus is accompanied by a violent collapse of the star's outer gaseous layer. There is an emergence of shock waves in this layer with huge gas temperature and pressure increase. As the iron core of the star collapses, fusion reaction occurs in its outer layer (shell) if it contains sufficient quantities of hydrogen, helium, and other light elements (C, O, Mg, Si, etc.). The shell's matter at this time must be subjected to intense irradiation by neutron flows from the collapsing core of the star. As a result, the drastic increase in the pressure, temperature, and neutron flow in the star's shell cause drastic, avalanche-type increase in the scale of nuclear fusion reactions. Thus, a huge amount of energy is released in a very short time. In minutes, or maybe seconds, as much energy is released in the collapsed shell as would take millions of years of a quiescent evolution. This leads to drastic acceleration of various nuclear synthesis reactions and formation of the entire gamut of isotopes, including those of elements heavier than iron. Release of huge energy in the lower part of the star's shell results in the nuclear explosion and shedding of the shell matter into space. Thus, in place of the former "normal" star remains a small, but very dense ($\geqslant 10^{14}$ g/cm^3), neutron star (pulsar) or even a "black hole".

The dispersed matter of many exploded stars gradually forms gas–dust clouds within the galaxies. As the mass of such clouds reaches a certain critical level, the cloud begins to self-gravitate, becomes denser, heats up, and condenses into a new star. Over the lifetime of a universe, several generations of stars dispersed their matter into the interstellar space. The source material for the formation of each new generation of stars was the matter shed by the previous generation of supernovae.

The Solar System, which is approximately 4.7 billion years old, is a relatively newcomer in this lengthy chain of cosmic transformations. Therefore, the composition of the matter that formed should bear the footprints of the long evolution of universe. The isotope analysis of meteorites (short-life isotopes, such as ^{244}Pu, ^{129}I, and ^{26}Al) indicates that shortly before the formation of Solar System at least two supernovae exploded. The latter of these explosions (the one that enriched the protoplanetary matter with ^{129}I and ^{26}Al isotopes) most likely was an impetus to the formation of the Sun and Solar System.

Concepts of the origins of Solar System and its planets were discussed extensively by Safronov (1969, 1972), Ruscol (1975), Vityasev et al. (1990), Jones and Keynes (1984), Dormand and Woolfson (1989), Galant (1990), and Newsom and Jones (1990). Consequently, we will give only a general description of the formation of Solar System. We will be more specific in describing the origin and evolution of Earth–Moon system, which affected considerably the development of the Earth's systems.

According to contemporary concepts, the planets of the Solar System formed as a result of accretion (coalescence and consequent growth) of solid particles from the gas–dust protoplanetary cloud (Safronov, 1972). Usually, the initial density of the interstellar cloud is insufficient to achieve a gravitational compaction and development of spontaneous star and planet formation. However, the supernovae explosions are accompanied by the generation of shock waves in the interstellar medium. On encountering the gas–dust cloud, the pressure and density of matter at the wave front drastically increase. As a result, blobs of matter can emerge, which are capable of further compaction due to gravity. Thus, the supernovae explosions not only supply new matter to the cosmic space, but also serve as a compaction mechanism in the formation of new stars and their surrounding planetary systems.

It is plausible that this occurred about 4.7 billion years ago in the vicinity of our Solar System. Having received an initial impulse of compaction and rotation, and being supplied with a new matter, the cloud began to contract under its own gravitational field. As the compression proceeded, the pressure and temperature in the central area of the cloud started to increase rapidly. Gradually, a gigantic gas agglomeration (Proto-Sun) formed in that zone. At the beginning, prior to initiation of nuclear reactions and the main development route of star, the Proto-Sun temperature was relatively low (no higher than 900–1,000 °C), and its radiation was mostly within the infrared and red spectra.

Simultaneously, with the compression of the Proto-Sun cloud by the centrifugal and gravity forces, its peripheral areas gradually shrank to the equatorial plane and turned into a flat lenticular disc of a protoplanetary cloud. The density of matter in the cloud rapidly increased, especially in the vicinity of equatorial plane. Particle trajectories affected by ever more frequent collisions and turbulent deceleration gradually approached the elliptic Keplerian orbits.

As a rule, interstellar clouds comprise the mixture of gas and micron-sized dust particles. Gases are composed mainly of hydrogen and helium but also include H_2O, CO, CO_2, CH_4, NH_3, and N_2. Dust particles were most likely a mixture of refractory metal oxides and silicates with metals, sulfides, and, to a smaller extent, hydrosilicates and carbonates. In open space, the dust particles can grow only by sorption at the surface of metal atoms and oxides or sulfides from the gas phase. This process, however, was progressing slowly because of a rarefied state of the interstellar matter.

When a protoplanetary cloud began contracting, the situation within the cloud became different. As the density of the cloud grew, the probability of particles collision and adhesion increased significantly. The adhesion resulted in initially small and loose clots of matter (similar to dirty snow balls). Further compaction of swarms of these primordial clots accelerated their growth. Gradually, they turned into larger bodies, embryos of future planets (planetesimals), with sizes up to several kilometers. The largest planetesimals (primordial protoplanetary bodies) were up to a few hundred kilometers in size. They possessed their own appreciable gravitational fields, which increased their capacity to capture smaller bodies. Thus, the smaller bodies dropped onto the large ones increasing further their masses. Because of that, the larger planetesimals grew faster than the smaller ones. One of the largest planetesimal bodies positioned within the internal belt of the protoplanetary cloud disk eventually became the embryo of Earth.

Formation of the Sun from the compacting primordial gas and dust cloud occurred much faster than the formation of planets. This process took just a few millions of years or tens of millions of years. Eventually, helium-synthesis nuclear reactions "flared up" within the juvenile Sun. After that, it joined the main course of the stellar evolution. Prior to that, at the beginning of "flare up", the Sun must have gone through a brief T-Taurus stage, i.e. rapid revolution, strong magnetic field, and stellar wind of high intensity.

These particulars in the evolution of juvenile Sun should have affected the matter accretion in the protoplanetary cloud disc. Initially, the strong solar wind (high energy flow of charged particles), typical of T-Taurus stars, must have swept away all gas and volatile components from the near-solar space to the distant periphery of the Solar System. After that, the ionizing effect of the solar wind on the surrounding matter must have resulted in strong interaction between the Sun's magnetic field and protoplanetary disc matter. As a result of such "engagement" between the fast-revolving juvenile Sun and the surrounding matter and the tidal interaction of the Sun and juvenile planets, the kinetic momentum was redistributed from the central celestial body to the protoplanetary disc periphery. After that, the rotational speed of the Sun decreased, whereas the orbital velocity of planet revolution around the Sun increased. Perhaps the same mechanism led to a noticeable matter separation within the protoplanetary cloud as all ionized elements were swept away from the near-solar space by the magnetic field force directed to the protoplanetary disc periphery. Next, the chemical differentiation in the protoplanetary cloud must have been significantly affected by a greater heating (from the Sun) of the central areas of the disc at the stage of its compaction and, especially, after the "flare up" of nuclear reactions within the Sun. For that reason, many volatile elements and the compounds (such as sulfur and its compounds, carbon dioxide, etc.) were evaporated and removed by the pressure of the Sun's radiation from this area to a distant periphery of the Solar System.

As a result, mostly refractory elements and compounds (refractory metals including Fe and Ni, and oxides Al_2O_3, CaO, MgO, FeO, Ti_2O_3, SiO_2, Cr_2O_3, etc.) were condensed in central areas of the protoplanetary disc. Average concentration of low-melting and easily ionized elements (Li, Na, Ka, Rb, Cs, In, Ba, rare Earth elements, Hg, Pb, Rn, etc.) was significantly lowered in that part of the protoplanetary cloud. The Earth planet matter was somehow less depleted in sulfur, zinc, tin, and some other elements. The gaseous components (H_2, He, and other noble gases; H_2O, CO_2, CO, CH_4, NH_3, H_2S, SO_2, SO_3, HCl, HF) were almost entirely swept away from the internal areas of protoplanetary cloud and concentrated at the periphery where giant planets with massive dense gaseous atmosphere formed later. Apparently, the cloud's internal area was also depleted in hydrosilicates and carbonates that dissociated under the Sun's radiation and lost the volatile components.

Under these conditions, the primordial protoplanetary cloud was considerably differentiated even before the planet formation process began. Probably, that is the reason why the planetary density clearly correlates with the distance from the Sun (Mercury, $5.54\,g/cm^3$; Venus, $5.24\,g/cm^3$; Earth together with Moon, $5.44\,g/cm^3$; Mars, $3.94\,g/cm^3$; Jupiter, $1.33\,g/cm^3$; Saturn, $0.67\,g/cm^3$; Uranus, $1.3\,g/cm^3$; Neptune, $1.67\,g/cm^3$).

The composition and relatively small masses of the Earth atmosphere and hydrosphere (total of 2.4×10^{-4} of the Earth's mass) as well as those of other planets of Earth's group show that these planets formed out of the matter that almost totally lost its gas components. The Earth's atmosphere contains very small amounts of the primordial noble gases. Besides, the Earth matter is strongly depleted in hydrosilicates, carbonates, sulfur and its compounds, and to a smaller degree, in alkali metals and other low-melting metals.

A good presentation of the contemporary accretion theory of the planet formation process is given in the publications by Safronov (1969, 1972). He showed that the Earth's growth lasted for about 100 million years. At the beginning, it proceeded at the accelerating accretion rate. Later on, when the solid matter supply in the near-Earth planetesimal swarm depleted, the accretion process slowed down. In the process of Earth's accretion, a huge amount of gravitational energy was released (about 23.2×10^{38} erg). That energy would be more than enough not only to melt the entire Earth matter, but also to totally evaporate it. However, most of the accretion energy was released within the very near-surface shell of the growing Proto-Earth and was lost with the heat radiation. The slower was the Earth's growth, the greater was the heat loss.

From the accretion theory, one can deduce that: (1) the juvenile Earth was relatively a cold planet (the temperature of its interior never exceeded the melting temperature of the Earth's matter) and (2) composition of the matter of the juvenile Earth was uniform in the Earth's body. From these two important conclusions one can deduce very important for geology consequences, which can be verified using the geologic history of the Earth: (1) the juvenile Earth was a tectonically passive planet with absence of magmatic activity and (2) the juvenile Earth had no dense core and no crust.

The Earth's core contains 32% of the total mass of the Earth. Because of this, the juvenile Earth must have been almost totally melted for rapid separation of the core. Simple physical considerations show that this did not happen. Under the pressures of lower mantle, the melting temperature of silicates reaches 4,000–8,000 K (Boehler, 1993, 1996), and the melting temperature of the core matter ranges from 2,200 to 3,100 K. If the Earth's core separated at the beginning of Earth's evolution, then the Earth's average temperature (taking into account the amount of energy, which is necessary for the mantle melting and the core separation) should have risen by 4,800 K. This would have led to the complete differentiation of the Earth's matter and the formation of light (about 2.7 g/cm^3 density) and thick (up to 80–100 km) anorthosite-type crust, which would contain all the lithophylic elements including uranium, thorium, and potassium. At the same time, the deep layers of the Earth would have been completely deprived of all the sources of endogenic energy. As a result, the Earth (similar to the Moon) would have become a tectonically passive planet.

It is impossible to destroy such a thick and, which is even more important, light crust by any external bombardment by planetesimals (as many geologists are prone to think). Even if such a bombardment did occur, then the light rocks of the primordial crust would have floated to the surface and be conserved until the present time. In spite of an active search, however, no traces of such ancient anorthosite crust had been discovered. The most ancient rocks of the Earth's crust have an age of about 3.8 billion years, which is a direct evidence that the formation of the Earth crust started only about 3.8 (at most 3.9) billion years ago.

The planetesimals colliding with the Earth were subject to intense degassing. In this intense process, when the plenetesimals hit the Earth, they experienced the heat explosions, and their matter could be totally evaporated. The gaseous matter did not accumulate in the atmosphere, however, because all the chemically active gasses (CO_2, H_2O) and other volatiles immediately reacted with porous ultrabasic regolith covering the growing Earth and were rapidly removed from the primordeal Earth's atmosphere. In the primordeal atmosphere, only the noble gases and chemically inactive nitrogen accumulated. It is clear that this intense planetesimal degassing in no way could characterize the internal thermal conditions of the juvenile Earth and, therefore, could not serve as an indicator of its early differentiation. In addition, if the early Earth were melted completely, this would be accompanied by the separation of the Earth's core and would lead to complete degassing of the Earth's interior. In this rapidly developing process, in a relatively short time interval, about 5×10^{23} g of carbon dioxide (presently bound in carbonate rocks) and 2.5×10^{24} g of water would enter the Earth's atmosphere. Emergence of such a dense carbon dioxide atmosphere with a pressure of about 100 atm would lead to emergence of a strong greenhouse effect with the surface temperature increasing above the critical temperature of water (374 °C). After that, the oceans would evaporate and the atmospheric pressure would rise by an additional 500 atm. This will give rise to an *irreversible* greenhouse effect in the Earth's atmosphere (like the one on Venus), with the temperatures exceeding 550–600 °C. In this scenario, water in the liquid phase would not exist on the Earth and possibly even the most primitive organisms would not emerge. We are lucky that this chain of events did not occur, and the Earth remained a "live" planet.

In addition to the numerous indirect evidences, there are also convincing direct proof that the juvenile Earth was never molten and did not have a dense metallic core. The most persuasive irrefutable evidence is the isotopic lead ratios on the Earth and the Moon. In the Moon rocks, which were formed after complete melting of its body, the ratios of radioactive lead isotopes with the atomic weights of 206, 207, and 208, formed after radioactive decay of uranium 238 and 235 and also thorium 232 to the stable primary isotope 204, are extremely high ($^{206}Pb/^{204}Pb \approx 207$; $^{207}Pb/^{204}Pb \approx 100$; $^{208}Pb/^{204}Pb \approx 226$, and higher). Meanwhile, for the Earth's rocks, in the pelagic sediments, these ratios are about one order of magnitude lower ($^{206}Pb/^{204}Pb \approx 19.04$; $^{207}Pb/^{204}Pb \approx 15.68$; $^{208}Pb/^{204}Pb \approx 39.07$). They are even lower for the primary lead, ranging from 9.5 to 29.45 (judging from the isotopic composition of the iron meteorite "Diablo Canyon", Arizona, USA; Handbook of Geochemistry, 1990).

Thus, one can conclude that in the process of melting and destruction of the parental body of the Moon (Proto-Moon) at the Roche limit (Sorokhtin, 1988, 1999, 2002), the Moon matter lost from 96% to 98% of primary (not radioactive) lead. At first, this lead was transferred to the Moon's core, and after that fell on the Earth's surface. In the meantime, the radioactive lead accumulated mostly in the Moon's crust and basalts. There is no other explanation of the considerable losses of primary lead, except total melting of the Proto-Moon matter with subsequent transition of lead and its sulfides into the core of the Moon. The iron meteorite "Diablo Canyon", in which the contents of the lead isotopes are close to their primary ratios, should be considered as a fragment of the core of a certain satellite, which (like the Proto-Moon) went through the stages of tidal melting, differentiation, and destruction at the time when the Solar System planets were forming. The attempts to explain the differences in isotopic ratios stating that the Moon went through differentiation already in the process of formation, whereas the Earth, because of its larger mass, did not, are not satisfactory. Both planets (Moon and Earth) had the same source of building material (from the same gas–dust protoplanetary disc) and formed through accretion of uniform (in composition) planetesimals.

As opposed to the Moon, the Earth's matter was never subject to rapid and complete differentiation. The Earth's core formed gradually without melting of the silicates owing to zonal (in Archaean) and later to baric diffusion (in Proterozoic and Phanerozoic) mechanisms of the mantle matter differentiation. The rate of progression of these processes of differentiation was always limited by low magnitudes of the thermal conductivity and diffusion coefficients of the mantle silicates. Thus, the Earth was never completely molten and did not experience radical (complete) rapidly progressing differentiation. In spite of the overwhelming factual evidence of this situation, many geologists continue to promote the 18th–19th century Kant–Laplace hypothesis on "hot" origin of the Earth. According to this hypothesis, the planets of the Solar System formed as a result of condensation and compression of the protoplanetary gas–dust cloud that led to melting of the protoplanetary matter. Later, the molten portions of matter originated the planets of the Solar System. Under this idea, the Earth was originally a hot and completely molten planet, which after its creation experienced steady gradual cooling.

For deciphering the Earth's evolution, one needs to know its original composition. As a rule, the original composition of Earth is estimated based on the meteorite composition. However, the meteorites themselves, most probably, are the fragments (which have undergone differentiation) of the planetary bodies (destroyed by the tidal forces) that formed in the other regions of the Solar System (Sorokhtin and Ushakov, 1989a, b, 1991, 2002). Because of this, the "meteorite" estimate is not acceptable.

Independent evaluation of the average composition of the Earth's matter can be obtained if one theoretically combines the compositions of main geospheres of the Earth: mantle, core, and crust. The composition of the Earth's crust is well studied. The mantle composition can be satisfactorily estimated based on the composition of xenolites in volcanic matter and oceanic crust (which, without water, actually represents the mantle's upper layer), and velocities of the seismic waves propagation in the mantle. Estimation of the composition of the core is a more difficult problem. Obviously, one does not have any direct data on the composition of the Earth's core. In this situation, one can use data obtained as a result of specially designed laboratory experiment. The experimental data of the shock-compression of the metals and their compounds (Al'tshuller et al., 1968; Sharipjanov, 1971) and of the static compression tests (Boehler, 1996; Lin et al., 2003) indicate that the outer core contains about 88–90% of iron, whereas the inner core contains mostly iron with admixture of nickel.

Most probable light additives to the iron of the core could be sulfur and oxygen. However, the sulfur content in the Earth's and cosmic matter is not sufficient for this purpose. Because of that, the hypothesis of the liquid iron oxide outer core, consisting mostly of Fe_2O, was suggested (Sorokhtin, 1971). This hypothesis explains best the density of the outer core. For the inner solid core the iron–nickel composition is commonly accepted. This composition of the outer core can be explained by the fact that iron is a transitional metal with free electron energy levels at the inner electron orbits. Under high pressures at great depths, the electrons in iron atoms transfer from outer to inner orbits, and chemical properties of iron change – it becomes univalent. This iron oxide hypothesis was further supported and developed by the seismologist Bullen (1973) and petrologist Ringwood (1977). Later, it was experimentally shown (Ohtani and Ringwood, 1984; Boehler, 1993, 1996) that under high pressures (exceeding 200 kbar) iron oxide dissolves in the molten iron almost completely forming eutectic alloy $Fe \cdot FeO$, stoichiometry of which is close to that of FeO.

Having determined the chemical composition of the core, one can estimate the original composition of the juvenile Earth theoretically mixing together the chemical compositions of the crust, mantle, and core. Table 1.1 contains the results of the corresponding computations.

The information presented in Table 1.1 shows that, in general, chemical composition of the juvenile Earth material is close to that of meteorites. It is considerably different, however, in the content of SiO_2 and heavy fraction (FeO, FeS, Fe, and Ni).

The matter density distribution in the mantle and the Earth's core can be estimated based on the data on the velocity of seismic waves propagation

Table 1.1: Chemical Composition of the Contemporary Earth and Original Earth's Matter*.

Oxides	Continental crust composition	Model composition of the Earth's mantle	Model composition of the Earth's core	Composition of the juvenile Earth's matter (calculated)	Average composition of chondrites	Average composition of coal chondrites
SiO_2	59.3	45.5	—	30.78	38.04	33.0
TiO_2	0.7	0.6	—	0.41	0.11	0.11
Al_2O_3	15.0	3.67	—	2.52	2.50	2.53
Fe_2O_3	2.4	4.15	—	—	—	—
FeO	5.6	4.37	49.34	22.76	12.45	22.0
MnO	0.1	0.13	—	0.09	0.25	0.24
MgO	4.9	38.35	—	25.77	23.84	23.0
CaO	7.2	2.28	—	1.56	1.95	2.32
Na_2O	2.5	0.43	—	0.3	0.95	0.72
Ka_2O	2.1	0.012	—	0.016	0.17	—
Cr_2O_3	—	0.41	—	0.28	0.36	0.49
P_2O_5	—	—	—	—	—	0.38
NiO	—	0.1	—	0.07	—	—
FeS	—	—	6.69	2.17	5.76	13.6
Fe	—	—	43.41	13.1	11.76	—
Ni	—	—	0.56	0.18	1.34	—
Total	100.0	100.0	100.0	100.0	99.48	98.39

*Computed based on the data of Ronov and Yaroshevsky (1978), Dmitriev (1973), Ringwood (1966), Urey and Craig (1953), and Florenskiy et al. (1981).

in these geospheres. The density distribution of the juvenile Earth was computed using its chemical composition given in Table 1.1 and the experimental data on the shock compression of chemical compounds (Naimark and Sorokhtin, 1977a, b). The computed density distributions for (1) original and (2) contemporary Earth compositions is shown in Fig. 1.1.

1.2. Formation of the Earth–Moon Double Planet

The Earth and Moon actually form a double-planet system. Their current interaction is not significant although noticeable. At the early stages of planetary development, this interaction was very strong and resulted in radical changes in the evolution of both planets. That is a reason for joint consideration of the origins of the Earth and the Moon. The Moon, as the Earth's satellite, served as a trigger that started and substantially amplified tectonic evolution of the juvenile Earth at the very beginning of the Archaean time. Besides, the Moon gave a jump-start to the Earth's rotation and determined by its capture orbit the inclination of the Earth's axis of rotation. This phenomenon caused the Earth's climatic zoning and originated magnetic field of the Earth. The Moon accelerated the Earth's evolution and indirectly facilitated the emergence on its surface of a highly organized organic life. At this point, we are going to discuss a new model of the origin of Moon (Sorokhtin, 1988), for which we assume that the contemporary Moon formed as a result of tidal destruction of a more massive Proto-Moon planet at the Roche limit. The model was constructed using the universal principle

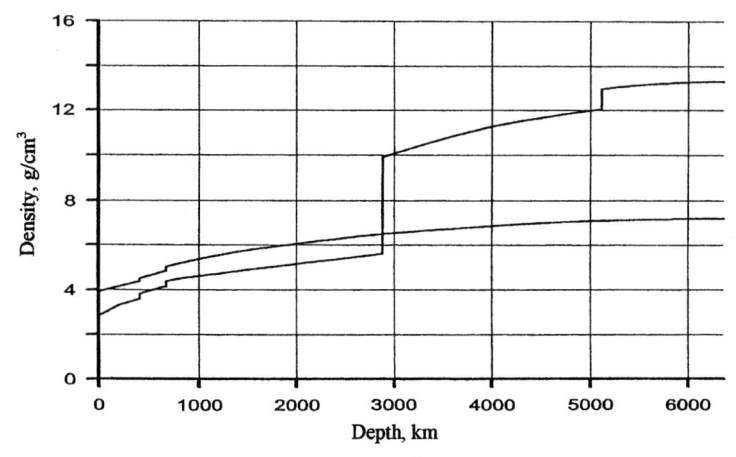

Figure 1.1: Distribution of Density (g/cm^3) of the Primordeal Earth's Matter with Depth (km).

of minimum potential energy: the Earth's evolution was mostly influenced by the processes that were reducing the potential energy of the Proto-Moon system to the greatest degree. It appears that this model explains better than the others the discovered facts on the composition, structure, and history of evolution of our natural satellite. It explains as well the origin of the Earth's axial rotation and the kinetic momentum distribution between the Earth and the Moon.

One of the major difficulties in the construction of an adequate Moon formation theory is Moon's sharp impoverishment in iron and siderophilic and chalcophilic elements. Based on the Moon's average density (3.34 g/cm^3), it contains no more than 5% of the iron–nickel phase (Ringwood, 1982). Taking into account the average FeO concentration in its mantle, it is only 13–14% of the heavy fraction, i.e. FeO + FeS + Fe + Ni. That is much lower than average iron compound contents in the undifferentiated matter of carbon chondrites (28.6%) and, even more so, in the Earth's matter (about 37%). Besides, as evidenced by the lead isotopes, the Moon lost almost completely all of its primordial lead. The lead of the contemporary lunar rocks is almost entirely of radiogenic origin (i.e. formed in the process of uranium and thorium decay).

Besides the iron anomaly, the lunar basalt composition is surprisingly similar to those of the Earth's mid-oceanic ridges. In addition, the oxygen isotopic data indicate the affinity of the Earth and Moon origins and a different origin of the carbon chondrite and regular chondrite meteorites. Based on this observation, Ringwood (1982) convincingly showed the geochemical affinity between the lunar matter and the Earth's mantle matter. On these grounds, however, Ringwood deduced an exotic conclusion that the Earth, after its formation and a dense core separation, began rapidly spinning that led to a rotational instability and detachment of a large chunk of the mantle matter that later turned into the Moon. This idea is not new. It was proposed about a hundred years ago by G. Darwin (1965), son of C. Darwin. Unfortunately, it contradicts the basic laws of mechanics.

Considering the Moon's origin, one needs to take into account an extreme differentiation of its matter, which resulted in separation of the silicates from iron, and the substantial scarcity of siderophilic elements. This kind of differentiation could occur only within a large and molten planet. That is important evidence that cannot be discarded easily. In favor of the theory of the formation of Moon from an originally molten matter is the composition of its thick anorthosite (calcium feldspar) crust. The Moon could have separated from a totally molten celestial body larger than the Moon (3–4 times larger). The age of lunar anorthosites is estimated at about 4.4–4.6 billion years, which is close to the time of formation of the Earth–Moon system. Thus, it seems that the Moon underwent the total planetary melting and differentiation stages at the time of its formation.

Another important fact to consider is the fact that the total kinetic momentum of the Earth–Moon system corresponds to the situation when both planets were positioned at the Roche limit distance and possessed a synchronized angular speed of rotation and revolution. This cannot be just a random coincidence. On the contrary, it is the strong evidence that the Moon at the time of its formation was positioned at the Roche limit and could have been subjected to destruction.

The aforementioned facts support the hypothesis that the Moon formed as a result of the destruction of a larger planet "Proto-Moon". The Proto-Moon was captured by the growing Earth from the nearby closest orbit (or possibly formed near to the Earth from the protoplanetary swarm of planetesimals) and was destroyed by the Earth's gravitational field at the Roche limit. Similar ideas on the two-stage formation of Moon as a result of tidal destruction of a larger planet with subsequent capture of its fragments were published by Wood and Miller (1974) and Opic (1961). We propose the hypothesis that the Moon is a surviving outer tidal hump of the previously molten and completely differentiated Proto-Moon, after its destruction at the Roche limit.

This theory of the Moon's origin (Sorokhtin, 1988) is not commonly accepted. There are at least four other proposed major theories of the Moon origin (DeYoung, 2000):

The fission theory – a long time ago the Moon split off from the spinning Earth. Some scientists state that the Pacific Ocean basin is the scar that remains as an evidence of this loss of the Earth's material. This theory contains at least four inherent contradictions: (1) the today's momentum of the Earth–Moon system is absolutely insufficient for fission to have ever occurred; (2) the Moon should be positioned directly above the equator (actually, the Moon's orbit is always tilted (between 18° and 28°) to the Earth's equator); (3) when the Moon after fission was moving outward, the Earth's gravity would have dispersed its matter into the Saturn-type rings; and (4) the most important evidence (presented in our book) is that the Moon's rocks are chemically different from the analogous rocks on the Earth.

The capture theory – gravity forces pulled the Moon into a permanent Earth's orbit when it was wandering close to the Earth. The capturing mechanism, however, is problematic. For example, all the space fly-by probes sent to the other planets were not captured but instead were thrown outward at a high speed.

The condensation (contraction, accretion) theory – concurrent formation of both Earth and the Moon from the gas–dust space material. In this scenario, the chemical composition of both celestial bodies must be very similar. Yet, as we showed before, the compositions of analogous rock's on the Earth and the Moon are quite different.

The lunar-origin theory (very popular today) – collision between the early Earth and another large planetary body, which resulted in an orbiting cloud of debris. Eventually, on coalescing this debris formed the Moon. One can estimate, however, that the Earth's crust would be totally melted (which did not occur) on impact with a celestial body large enough for its debris to form a body with a mass of contemporary Moon.

1.2.1. Tidal Interaction between the Planets

The phenomenon of tides was studied by Newton, Bernoulli, Euler, Laplace (who in 17th century formulated the present-day theory on tidal interaction between planets) (see G. Darwin, 1898; and later MacDonald, 1964; Goldreich, 1966; Ruskol, 1975).

The tidal interaction between the Earth and the Moon is about twice as strong as that between the Sun and the Earth. In order to describe the evolution in the Moon–Earth mutual position, one needs to apply the laws of celestial mechanics (Kepler's third law) and the momentum conservation law of the system. It is also necessary to take into account the dissipation of the tidal energy in the planets. If (for simplicity) one assumes that the Moon revolves in the Earth's equatorial plane, then the momentum conservation law can be written in the following form (Ruskol, 1975):

$$I\Omega^3 \frac{Mm}{M+m} L^2 \omega = K = \text{const} \qquad (1.1)$$

where $I = 8.03 \times 10^{44}\,\text{g cm}^2$ is the contemporary Earth's momentum of inertia; Ω is the angular velocity of the Earth spinning around its axis; ω is the angular velocity of the Moon's orbital revolution around the Earth; $M = 5.977 \times 10^{27}$ g = Earth's mass; $m = 7.35 \times 10^{25}$ g = lunar mass; and L the distance between the centers of gravity of Earth and Moon ($L = 3.844 \times 10^{10}$ cm).

Third law of Kepler can be rewritten in the following form:

$$\omega^2 \times L^3 = \gamma(M+m) = \mu = \text{const} \qquad (1.2)$$

where γ is the gravitational acceleration. The energy of the self-rotation of Earth E_Ω and the total orbital energy of the Moon E_ω may be determined from the following simple equations:

$$E_\Omega = \frac{I\Omega^2}{2} \qquad (1.3)$$

$$E_\omega = \frac{-\gamma m M}{2\pi L} \qquad (1.4)$$

Current values of energy of the Earth rotation and Moon revolution are $E_\Omega = 2.12 \times 10^{36}$ erg and $E_\omega = -3.80 \times 10^{35}$ erg, respectively (the Moon's orbital energy is in essence a potential energy and is negative).

The tidal interactions redistribute kinetic momentum between the planets, but the total kinetic momentum remains constant. Transfer of energy from one planet to another is caused by the same interactions. On the other hand, the energy of rotation in the system does not remain constant because it is converted into heat due to the tidal deformation of the planetary bodies and is dissipated into the outer space. Currently, the Earth's rotational energy is being transferred to the Moon. This results in gradual deceleration of the Earth's axial rotation (spinning) and simultaneous increase in the distance between the Earth and the Moon.

From these observations, one can derive a very important conclusion. If a satellite during its formation or capture had its own rotation with an angular velocity that was not equal to the velocity of its revolution around a massive central planet, then the satellite was subjected to a pair of tidal forces slowing down its axial rotation. This resulted in synchronous rotation, i.e. the angular velocities of orbital and axial rotations became equal. After synchronization, the same side of a satellite will always face the central planet (as one can observe now in the process of revolution of the Moon around the Earth).

1.2.2. Hypotheses of Formation of Proto-Moon

When the formation of the Solar System planets as a result of planetesimal accretion was almost completed (about 4.6 billion years ago), one can consider the classical double-planet formation mechanism: after formation of the Proto-Earth, Proto-Moon formed from the remaining Proto-Earth planetesimal swarm (Ruscol, 1975). According to another popular theory, the Proto-Moon formed as a result of tangential mega impact of a Martian-size planet on the Proto-Earth. These popular hypotheses, however, cannot explain chemical composition of Moon and geochemistry of its rocks, as well as the perfect match of the real system kinetic momentum to the Earth–Moon system momentum with the Moon positioned at the Roche limit from the Earth.

At the Roche limit, attraction of the central planet is equal to the self-gravitation of satellite and, consequently, the force of gravity at the satellite surface facing the central planet becomes equal to zero. If the satellite is molten, then the matter from its surface starts flowing toward the central planet. This kind of situation occurred about 4.6 billion years ago when a greater part of the Proto-Moon matter fell on the Proto-Earth surface, including the molten iron of its core. This, in part, explains a pronounced deficiency of iron in the Moon. The Moon itself formed from a silicate matter (with a small content of iron) of an outer tidal hump of destroyed proto-satellite. That is why an average content of FeO in the Moon basalts is 21.6% (Handbook, 1990), which is close to FeO content in the original matter of the Earth (see Table 1.1).

For these (and some other) reasons, we believe that the most probable scenario is the gravitational capture from a nearby orbit of growing and already large enough Proto-Moon by the growing Proto-Earth. It was shown that the probability of a satellite capture from a distant orbit is next to zero. On the other hand, the probability of a satellite transfer from a nearby heliocentric orbit to the orbit around the growing planet is greater (Aliven and Arrhenius, 1972, 1979; Zinger, 1972). Kaula and Harris (1973) believe that the collisions of a fly-by satellite with the planetesimal swarm of bodies (still revolving around the growing planet) could trigger the capturing process. Because of these collisions, the satellite could noticeably slow down, which could cause a trajectory conversion from hyperbolic (with respect to the central planet) to elliptic and, later, circular.

Direct mathematical modeling of the gravitational interaction between several bodies within a nonuniform protoplanetary plasma disc in the vicinity of a growing planet showed a possibility of capture of cosmic bodies (Otsuki and Ida, 1998). According to Sorokhtin and Ushakov (1989a, b), the satellite capture is possible only in forward direction and is the main cause of planetary spin-up. Unfortunately, their model did no include the destruction of captured massive bodies at the Roche limit and the spin-up of planet to the angular velocity of synchronous rotation with the satellite revolution. Because of that, Otsuki and Ida (1998) explained the origin of Moon by subscribing to either a tangential impact of a larger protoplanetary body or to the accretion of the Proto-Earth disk fragment. Then, how could one explain the origin of numerous satellites of external planets? Were all of them created through tangential impact of protoplanetary bodies on the gaseous atmospheres of planets? At the present time, the influence of Moon on the development of tectonic processes of Earth is insignificant, and the energy of this influence does not exceed 1% of the energy driving the matter–heat transformations in the mantle. However, at the early stages of planetary evolution, when the Moon revolved close to the Earth, this influence dominated all other sources of the inner energy of Earth. Consequently, it is not possible to analyze the global evolution of Earth without taking into consideration the history of its interaction with the Moon. Due to strong tidal interactions of a central planet with a satellite, the distance between them should rapidly decrease, which would lead to the intense overheating of

Proto-Moon by almost 2,000 K (without considering surface cooling). As a result, it was melted and went through practically a complete cycle of gravitational matter differentiation. The latter conclusion is strongly supported by the lead isotope ratios in the Moon and Earth rocks: in the Moon rocks the isotopic ratios range from 100 to 220, whereas in the Earth rocks these ratios range from 15 to 39. In the process of tidal interaction, the Earth also gained additional heat. This additional heating, however, was considerably lower (about 180 °C) because of significantly larger mass of the body.

High isotopic ratios in the Moon rocks are explained by the fact that in the process of the Proto-Moon melting and differentiation the original lead (together with iron) was transferred to its core. Consequently, only the radiogenic lead could accumulate in the Proto-Moon mantle and core. The loss of original lead at the stage of planet formation could not be explained by the higher volatility of lead. Otherwise, analogous ratios of lead isotopes must be observed in the Earth rocks (which is not the case), because the Earth and the Moon formed from the planetesimals of the same material with the same lead isotopic ratios.

After the capture of Proto-Moon about 4.6 billion years ago, the Earth started spinning in the forward direction with the period of rotation of about 6 h. At the Roche limit, the molten Proto-Moon began disintegrating (Fig. 1.2). The closest to the Earth hump fell on the Earth's surface, bringing the molten iron of the Moon's core. This, in particular, explains the extreme scarcity of iron in the Moon rocks. The Moon itself formed from a silicate matter (with a low iron content) of the outer tidal hump of the destroyed Proto-Moon.

This scenario of the Moon formation is also corroborated by the fact that contemporary total kinetic rotational momentum of the Earth–Moon system is very close to the momentum of this system when the Moon was at the Roche limit. Our calculations give estimates of 3.451×10^{41} g cm^2/s for the contemporary system and 3.447×10^{41} g cm^2/s for the system with the Moon at the Roche limit. This coincidence could not be random. Most probably, it indicates that the formation of Moon occurred as a result of destruction of a larger celestial body at the Roche limit.

Due to tidal interactions with the Proto-Moon, the Earth's rotation was spun off to the maximum angular velocity of the satellite at Roche limit. After the formation of Moon and its transfer to a stationary orbit (with the period of revolution equal to the period of proper axial rotation of the Earth), the tidal interaction became quasi-stable, and the tidal retardation forces were reduced to zero. In the meantime, under the influence of solar perturbations retarding revolution of the Moon around the Earth in even greater degree than the proper rotation of the Earth, the tidal forces of

Figure 1.2: Destruction of Proto-Moon and Formation of Moon at the Roche Limit of Earth about 4.6 BY Ago.

interaction between the central planet and satellite changed the sign to opposite. Thus, the Moon began moving rapidly away from the Earth. This saved the Moon from total destruction and the proper rotation of the Earth began slowing down. Spinning off of the planets as a result of destruction of their satellites at the Roche limit should be a common mechanism in the Universe. This conclusion is supported by the distribution of specific kinetic rotational momentum for the Solar System planets (unretarded) and some classes of rapidly revolving stars (Fig. 1.3).

What were the conditions on the surface of juvenile Earth? There are numerous unsubstantiated myths derived from science fiction literature during the time when the contraction hypothesis ("hot" origin) of the creation of Earth prevailed. As a typical image of "hot" origin, one can visualize scenes of violent volcanic activity on the black surface of Earth covered by a dense, thick atmosphere. Actually, the landscape of the juvenile Earth was that of a bleak and cold desert under black skies studded with bright but not twinkling stars. The yellow, hardly warming Sun had luminance 26–30% lower than the current one. The lunar disk was enormously large without the usual present-day "maria". The Earth's terrain was similar to the crater-speckled modern lunar surface. However, due to strong and practically continuous tidal earthquakes, the Earth topography was substantially leveled down and contained only the monotonous dark-gray original matter covered with a

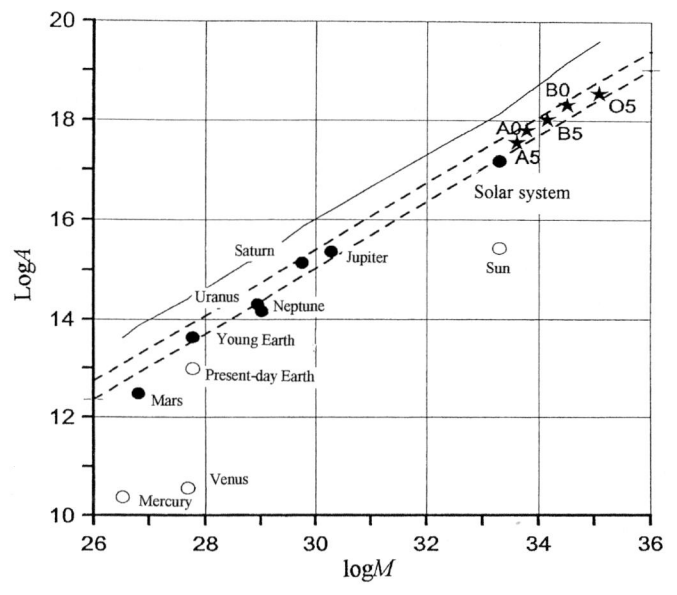

Figure 1.3: Relationship between Specific Axial Rotational Momentum (A) for the Planets of Solar System (and Some Classes of Rapidly Rotating Stars) and Their Mass (M). Solid Dots Indicate Unretarded Objects; Open Dots Indicate Retarded Objects; and Asterisks Indicate Stars. The Dashed Line Shows the Value of Specific Rotational Momentum at the Spinning Off of the Planets and Stars to the Angular Velocity of the Satellite Rotation at the Roche Limit as a Function of the Dimensionless Inertia Momentum of a Planet (or a Star), the Planet's Density, and the Density of Satellites Disintegrating at the Roche Limit. The Solid Line shows the Limits of Specific Rotational Momentum (A) in the Case of Rotational Instability of the Same Planets and Stars.

thick dark-gray regolith layer. More differentiated rocks, like basalts, peridotites, anorthosites, or granites, simply did not exist.

From time to time, a desolate landscape of the juvenile Earth was disturbed and shaken by explosions from the remaining planetesimals falling on the Earth's surface. The frequency of these impacts rapidly declined in time. Only in the equatorial zone, the satellite swarms of fragments (rocks and iron) of recently destroyed Proto-Moon continued to shed prolific, torrential showers on the Earth's surface. That is why the Earth's surface in the equatorial zone remained incandescent some time after the formation of Moon (for at least several hundred or a few thousand years). There was no volcanic activity (eruptions of lava, gas, and water vapor) because there were no hydrosphere and no dense atmosphere. Insignificant amounts of gases and water vapor emitted as a result of explosions from dropping planetesimals and the Proto-Moon fragments were rapidly absorbed by the porous ultrabasic regolith that formed a thick layer all over the Earth's surface.

The Sun rushed over the skies, and in only 3 h crossed the expanse of heaven and rose again in the east over the lifeless horizon of the primordial Earth. Although, the year's duration was almost the same as now, it comprised almost 1,500 days. The motion of the Moon was slower as it was rapidly orbiting the Earth in the same direction. The lunar phases, however, alternated frequently running through all the stages in 6–8 h. At the very beginning of the Earth–Moon system evolution, a month lasted about 6 h, but its duration at that time was rapidly increasing. The Moon's visible size was amazingly huge: 300–500 times the visible area of the present-day moon disk.

At the beginning of its existence, the Moon was still a hot planet and radiated energy in the red (and infrared) spectral range. It reflected the solar rays during the day and night and also radiated its own dark-red light. Being of a huge visible size, the Moon noticeably warmed the Earth's surface. Sometimes (most often on the eastern side of the lunar disc), bright orange spots flashed as magma erupted onto the Moon's surface from underneath a thin crust of solidified rocks upon impacts of meteorites on the Moon's surface. Besides, from time to time, the entire huge Moon's disc was cut with zigzag bright arrows that indicated the fractures in a thin lunar crust when the radius of curvature of the tidal bulges of Moon sharply changed. (The Moon was rapidly moving away from the Earth.) A disk of small particles was revolving around the Earth at close orbits in equatorial plane. Individual particle trajectories merged and created illusion of solid translucent disks resembling the Saturn's rings. At a certain angle of sight, these disks jointly reflected the light of Sun and may have appeared solid. In reality, they were ephemeral (short-lived) and shortly all their matter dropped onto the Earth.

Besides the Moon, the Earth had several smaller satellites (micro-Moons). They were revolving at greater distances than the Moon, but were clearly visible in the reflected light of the Sun as small disks or bright "starlets". Apparently, there were at least ten such satellites. Eventually all of them dropped on the Moon's surface and formed large basalt-filled craters ("lunar maria").

The most impressive events (for a hypothetical observer) were the series of devastating continuous earthquakes caused by the intense and continuous tidal deformation of the Earth's body. Following the Moon's motion and tidal bulges, these quakes literally shook the primordial face of the Earth. Periodicity of the tidal quakes, which was determined by the difference between the periods of the Earth's axial rotation and the Moon orbital revolution, was continuously changing. At the beginning, when the Moon was positioned at the Roche limit and the angular velocity of Earth's rotation coincided with the angular revolution velocity of the Moon, the tidal humps of the Earth practically did not move around the Earth surface and the tidal earthquakes did not occur. As soon as the Moon left the Roche limit and began distancing itself from the Earth, intense tidal quakes were initiated, which shook the Earth surface twice per each Moon revolution around the Earth.

The tide height (caused by the Moon) on the Earth's surface varies inversely with the cubed distance between the gravity centers of Earth and Moon. At the Roche limit, this distance was 17,200 km (Moon surface hovered only around 7,000 km above the Earth surface). Consequently, the lunar tides were very high (about 1.5 km; whereas the "Earth" tides on the Moon may have reached 870 km!). As high tides moved on the Earth's surface, they were accompanied by swarms of numerous and strong earthquakes. The Moon experienced a rapid tidal repelling effect (the repelling force varies inversely with the 5.5 power of the distance between the planets). After one million years of the formation of planets, the tide height declined to 130 m; in 10 million years more, to 45 m; and after 100 million years, to just 15 m. About 4 billion years ago, when the distance between the Earth and Moon was about 168,000 km, the lunar tides declined to 7 m. The current distance of the Moon from the Earth is 384,400 km, and the tides on the surface of the present-day solid silicate crust of the Earth in the vicinity of "under-moon" point are about 46 cm high.

Disregarding the tidal earthquakes, the juvenile Earth, during the first 600 million years of its existence, was a tectonically passive cold planet (average temperatures ranging from $-10\,°C$ to $-5\,°C$) and the Earth's surface resembled that of the Moon – lifeless desert.

1.3. Beginning of Tectonomagmatic Activity on Earth

According to the accretion theory of planet formation (Safronov, 1969, 1972), partial melting of the Earth's matter (and the tectonomagmatic activity of the Earth) could start only after heating its inner layers to the melting point of the Earth's matter. Necessary heating was supplied by the radioactive decay inside the Earth and its tidal interaction with the Moon. This required a certain time for raising the temperature in the Earth at depth to the melting point. This event occurred at the very beginning of Archaean time about 4 billion years ago. The temperature distribution at different times of the Earth's evolution is presented in Fig. 1.4. The shift from curve 1

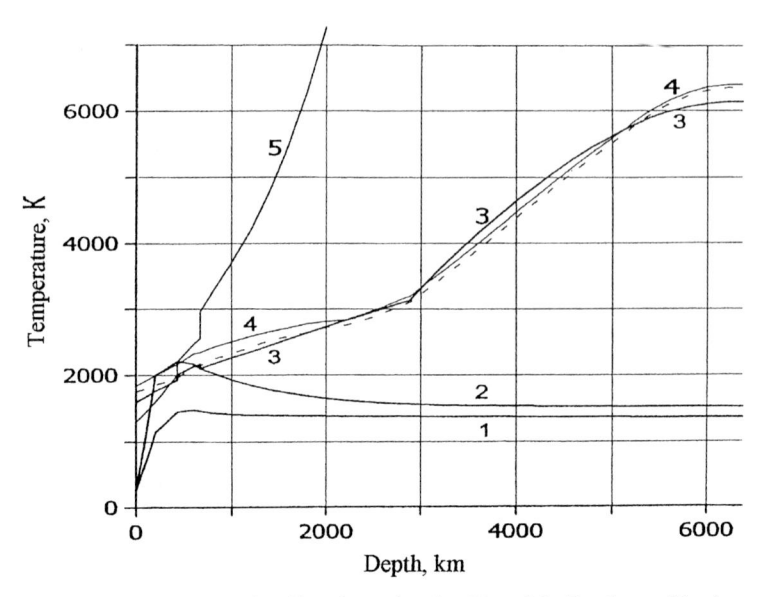

Figure 1.4: Temperature Distributions in the Earth's Body at Various Stages of its Evolution. Curve (1): at the Time of Earth's Formation about 4.6 Billion Years Ago; Curve (2): at the Boundary between Pre-Archaean and Archaean Eons about 4 Billion Years Ago; Curve (3): at the Present Time; Curve (4): Iron Melting Temperature (Boehler, 1993); Curve (5): Melting Temperature of Earth's Matter. Dashed Line Represents the Melting Temperature of Eutectic Composition Fe · FeO (the Outer Layer of the Earth's Core).

to curve 2 (Fig. 1.4) represents the increase in temperature from the initial one to the melting temperature of the Earth's matter. Because of the tidal interaction of Earth with the Moon, its tectonic evolution was accelerated by at least 2.5–3 billion years. The Earth's outgassing, which created the hydrosphere and the atmosphere, began between the Pre-Archaean and Archaean time. This theoretical conclusion is supported by the geological data, first of all by the ages of the most ancient magmatic rocks of the Earth.

In spite of the strong efforts of the World geologic community to find the Earth's oldest rocks, the ages of discovered samples do not exceed 3.75–3.80 billion years (Murbat, 1980; Taylor and McLennon, 1988), although the Earth (judging from the age of meteorites) formed about 4.6 billion years ago. Recently, the Australian geologists (Wilde et al., 2001) reported finding small zircon fragments with the ages of 4.2–4.4 billion years. These embryonic zircon fragments were found in the considerably younger sedimentary rocks (sandstones and conglomerates) of about 3.5–3.8 billion years old, which indicates that they originated in the shallow and short-lived melting zones. The latter could emerge as a result of meteorite bombardment of the juvenile Earth in the Pre-Archaean time (4.0–4.6 billion years ago). Therefore, the Pre-Archaean age of magmatic zircon fragments does not indicate that the Earth crust formed at this time as some of the geologists continue to believe.

Based on geologic data, the first mantle alloys formed at the beginning of Archaean time, i.e. 600–800 million years after formation of the Earth. The crust rocks are the products of gravitational differentiation of the Earth's matter and, consequently, they are always lighter than the original mantle material. Being elevated to the Earth's surface, such rocks cannot be completely destroyed by the succeeding tectonic processes, according to the Archimedes principle. Thus, during the entire Pre-Archaean time there were no magmatic processes in the bowels of the Earth (in accordance with accretion theory of the planet formation). Safronov (1969, 1972) estimated that this latent period has lasted from 0.5 to 1 billion years.

Among the evidences indicating the beginning of the tectonomagmatic processes on the Earth (besides appearance of the first erupted rocks about 3.8–3.9 billion years ago), an important place belongs to the beginning of basalt magmatism on the Moon (Sorokhtin, 1988). This, at first sight paradoxical, coincidence is determined by the tidal interactions of the Earth with the Moon and the other satellites, which existed in the near-Earth space at that time.

It is important to discuss here the tidal mechanism of interaction of the central planet and a satellite in detail. The rate of increase of the distance between a satellite and a central planet is determined by a relatively simple expression (Sorokhtin, 1988):

$$\frac{dL}{dt} = \frac{-0.903\sqrt{\gamma}R^5 m(M+m)^{0.5}}{MQL^{5.5}} \tag{1.5}$$

where γ is the specific gravity, m is the mass of satellite, M is the mass of planet, R is the radius of planet, L is the distance between the centers of gravity of planet and satellite, and Q is the mechanical merit factor of the planet (Earth).

The mechanical merit factor characterizes the proximity of rheologic properties of a real body to the state of ideal elasticity: the higher the mechanical merit factor, the closer the properties of the body to the ideally elastic materials; and, vice versa, the lower the mechanical merit factor, the higher the manifestation of the viscous properties. A classic example of a body with high merit factor is a long-sounding bronze bell. If the bell is made using plasticine, then it would not sound at all because the energy of blow transforms completely into plastic deformations. Equation (1.5) shows that the factor Q considerably influences the tidal interaction between the planet and the satellite.

Upon its emergence, the juvenile Earth was a cold celestial body. The factor Q for the juvenile Earth was very high – about 5,000. Due to the Earth heating caused by the tidal and radiogenic energy, the mechanical merit factor gradually decreased (especially at the equatorial plane) to less than 1,000 by the end of Pre-Archaean time. The tectonic activity of the Earth started about 4 billion years ago, at the beginning of Archaean time. As a result, the first basins of molten rocks emerged and degassing of the mantle started. After formation of the first shallow sea basins, the merit factor decreased sharply (to 16 by our estimates). Therefore, the rate at which the Moon moves away from the Earth increased dramatically (Fig. 1.5). According to Eq. (1.5), the rate of distancing is proportional to the mass of satellite. This led to "sweeping away" of lighter cosmic bodies, falling on the surface of more massive neighbors (when their orbits approached each other). As a largest satellite planet, the Moon gradually swept away all the bodies of the near-Earth satellite swarms. The most intense "sweeping" occurred during the fastest distancing (moving away) of Moon from the Earth in Pre-Archaean and Early Archaean time.

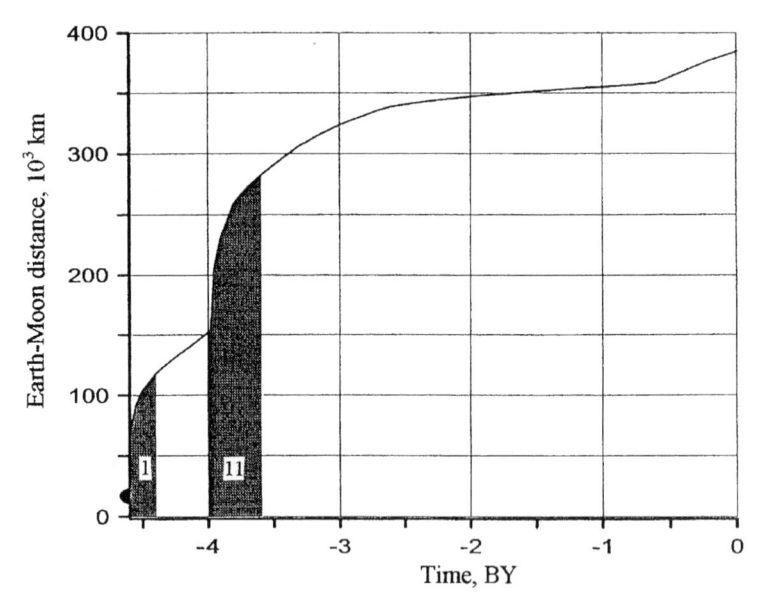

Figure 1.5: Evolution of the Distance between the Moon and the Earth. Interval I is the Time of Development of the Anorthosite Magmatism on the Moon; Interval II is the Time of Development of the Basalt Magmatism on the Moon.

Simultaneously with the Moon, the other cosmic bodies revolving around the Earth also enlarged their orbits with the rates proportional to their masses. Consequently, the Moon should collide at first (in Pre-Archaean time) with the small or medium-size bodies. Later, in the early Archaean time, the Moon collided with larger and more massive remaining satellites (which also grew at the expense of accretion of lighter bodies orbiting the Earth).

In the early Pre-Archaean time, the Moon did not have enough time to cool down completely after the tidal heating of the Proto-Moon; thus, there was an ocean of magma under the lithosphere. The magma was differentiated based on density: the light anorthosite magma on the top and the heavier basalt magma below. At the beginning, when the thickness of the lithosphere did not exceed 10–20 km, bombardment of the Moon's surface led to anorthosite eruptions only. After 600–800 million years of the Moon's formation (at the beginning of Archaean time), the thickness of lithosphere of the cooled planet increased to 100–200 km. Only the most massive satellites (with the diameters of tens to hundreds of kilometers) could penetrate this thick lithosphere. That was the time when the most massive satellites fell on the Moon's surface. They pierced the anorthosite lithosphere, creating the avenues for the outflow of basaltic magma.

Judging from the isotopic analyses of samples of Moon rocks made by "Apollo 13", at the first stage of lunar tectonic activity the magma consisted only of anorthosite. At the second stage, which started about 4 billion years ago and lasted to 3.6–3.2 billion years ago, the magma consisted of molten basalts (Jessberger et al., 1974; Tera and Wasserburg, 1974, 1975). Today, the traces of these large holes, filled with basalt lavas, can be observed at the Moon's surface as the "lunar maria" (Moon's seas). In the Moon's gravitational field, they stand out as positive anomalies linked to the so-called "lunar mascons" (large circular dense masses lying 40–50 miles below the centers of lunar maria).

The beginning of basalt magmatism on the Moon about 4 billion years ago corresponds to the time of profound decrease in the mechanical merit factor of Earth caused by the emergence of first basins of molten rocks in the bowels of the Earth. This event, in fact, marks the beginning of tectonomagmatic activity of the Earth.

Why did the older Earth's rocks disappear? Why the rocks ranging in age from 4.6 to 4 billion years ago (from the time of Earth's formation to the beginning of Archaean) were not found? How could this "amnesia" in geologic history be explained? The only possible explanation is that the juvenile Earth was originally (during the first ≈ 600 million years) a cold and passive planet. That is why the processes of matter differentiation, which lead to the melting out of the lighter rocks (anorthosites, basalts, and plagiogranitoids) did not develop in the bowels of the Earth. After beginning of the tectonomagmatic activity and emergence of the first basins of molten rocks in the bowels of the planet, the Earth matter was subjected to the gravitational differentiation mechanism. The process of differentiation involved floating upward of the lighter rocks (crust rocks) and sinking of denser primordeal Earth matter. Erupted rocks (extreme differentiates of the Earth's primordeal material) should be much lighter (2.9–3.0 g/cm^3) than original Earth matter (about 4.0 g/cm^3). That is why they were conserved until today having formed the most ancient small areas of the Earth's crust. Formation of the Earth's crust started during the Archaean time (the most ancient geologic era), which was clearly marked in the geologic chronicle.

1.4. Separation of Earth's Core (Main Factor Driving the Earth's Global Evolution)

An important issue in the planetary geophysics is the time and mode of core separation. These are crucial in deciphering the paths of planetary evolution of Earth. Inasmuch as one-third of the total Earth's mass is concentrated in its core, the process of radical differentiation of the Earth's matter should influence considerably the entire geologic history of our planet. The process of core separation was not short, because the rapid separation could have occurred only upon the total melting of the Earth including its central zones. As discussed before, however, based on lead isotopic ratios in the Earth and Moon, the Earth was never molten completely. This conclusion is corroborated by computations of the temperatures of juvenile and contemporary Earth, both of which are several hundred degrees below the expected melting temperature of terrestrial matter in deep layers (at high pressures) of the Earth. This fact practically excludes the possibility of total melting of the Earth (Fig. 1.4).

Indirect evidence on the time of core separation is presented by the paleomagnetic data, which date the emergence of dipole magnetic field at about 2.6 billion years ago (Hale, 1987), i.e. about 2 billion years after formation of the Earth (Fig. 1.6). As shown in Fig. 1.6, the Earth magnetic field did exist in the Archaean time, but most likely it was toroidal and not dipole. On analyzing the most probable scenario of the Earth's development in the Archaean time (Fig. 1.8), one can assume that the Earth's magnetic field emerged in the Archaean time upon appearance of the mantle molten matter. The intensity of this field grew jointly with the mass of convection

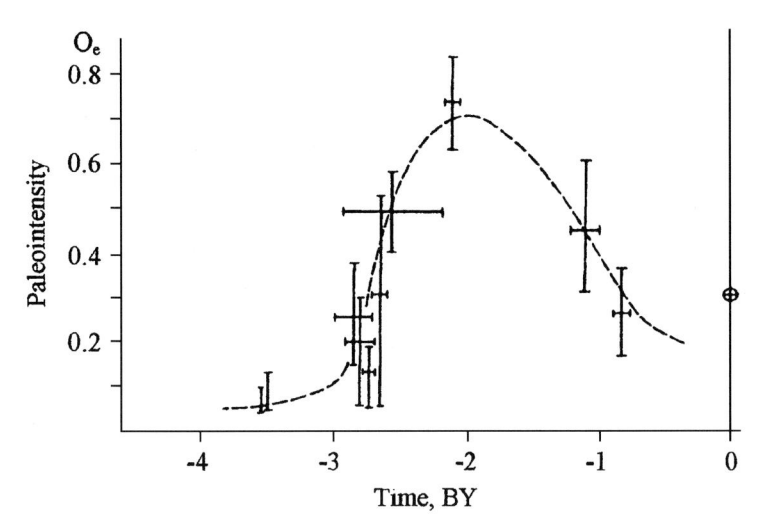

Figure 1.6: Changes in the Intensity of Earth's Magnetic Field Throughout the Geologic History based on the Paleomagnetic Data (Hale, 1987). Horizontal and Vertical Segments Indicate the Confidence Intervals of Measurements; the Open Circle with a Cross shows the Intensity of Contemporary Geomagnetic Field.

matter of the mantle. At the boundary of Archaean and Proterozoic eras, the intensity of Earth magnetic field attained its maximum, and after the core separation became dipole. The intensity of dipole geomagnetic field started to decrease in Proterozoic and Phanerozoic eras when the tectonic activity of the Earth began decreasing.

To explain the mechanisms of core separation, one does not need to assume the total melting of the Earth's matter. As pointed out by Monin and Sorokhtin (1981) and Sorokhtin and Ushakov (1991, 2002), the process of gravitational differentiation of matter could originate in the deep layers of Earth at temperatures considerably below the melting points of silicates. These mechanisms, however, are determined by the thermal conductivity of the Earth's matter and the low rate of diffusion of iron compounds out of silicate crystals, which excludes the possibility of rapid separation of the Earth's core. Consequently, the core separation process was extended for billions of years.

In the Pre-Archaean time, radioactive decay heated the juvenile Earth more or less uniformly. Heating of the Earth's matter by tidal deformation, however, occurred mostly in the upper layers of the equatorial belt of the Earth. Therefore, most likely, the primary asthenosphere emerged in the same equatorial belt at depths of 200–400 km at the beginning of Archaean time when the melting temperatures of iron and silicates were reached (Fig. 1.4).

The primordeal Earth's matter contained about 13% of metallic iron (Sorokhtin, 1974). Consequently, at the beginning, differentiation of the Earth's matter was linked to the separation of molten metallic iron from the silicates of the Earth's matter. Further differentiation progressed through the mechanism of zonal melting of the Earth's matter (Vinogradov and Yaroshevskiy, 1965, 1967). During the Archaean time, however, this process was sustained by the gravitational energy (not by radiogenic energy), which is one order of magnitude larger than radiogenic. The process of differentiation proceeded from the surface to the center (not from the bowels of the Earth to the surface). There are no sources of energy, which could continuously sustain zonal melting process in the contemporary Earth. On the other hand, the juvenile Earth possessed such a source of energy, i.e. the energy of the gravitational differentiation of matter. After the Earth's iron started to melt, the process of the Earth's matter differentiation could proceed upward and downward just due to release of gravitational energy. This energy was generated in the asthenosphere (molten iron layer) because of sinking of heavier molten iron and floating up of the lighter silicates. Studying the process of zonal melting in the juvenile Earth's mantle, N. Sorokhtin (2001) showed that the energy generated in the process of migration to the surface of unit mass Δm ($= \Delta V \rho$) of silicate matter (which underwent the gravitational matter differentiation) through the layer of molten iron with a thickness of h is equal to:

$$\Delta E = [1 - C(\text{Fe})] \Delta \rho h g \, \Delta V \tag{1.6}$$

where $C(\text{Fe})$ is the iron concentration in the primordeal matter of the Earth; $\Delta \rho = \rho(\text{Fe}) - \rho(\text{Si})$, where $\rho(\text{Fe})$ is the density of molten iron and $\rho(\text{Si})$ is the density of silicate phase; g is the gravitational acceleration; ΔV is the volume of unit of mass, and h is the thickness of molten iron. This energy was consumed for heating [$\Delta T = T_F(H) - T_0(H)$] of the relatively cold Earth's matter at depth H from the temperature $T_0(H)$ to the iron melting temperature $T_F(H)$ at depth H. Using notations $q (\approx 2.77 \times 10^9 \text{ erg/g})$ for the specific melting heat of iron and $\delta T(H)$ for overheating of molten iron and its oxides above their melting temperature at a depth H, one can write the energy balance equation as follows:

$$[1 - C(\text{Fe})]\Delta \rho h g \approx \rho_0 c_p \Delta T + C(\text{Fe})\rho(\text{Fe})q + \delta T(H)\rho(\text{Fe})c_p(\text{Fe}) \tag{1.7}$$

where ρ_0 is the density of the primordeal Earth's matter; $c_p \approx 8.9 \times 10^6$ erg/g °C — specific thermal capacity of primordeal Earth's matter; $c_p(\text{Fe}) \approx 5 \times 10^6$ erg/g °C — the specific thermal capacity of the molten iron.

From Eq. (1.7), the value of overheating of the molten iron in the zone of differentiation of Earth's matter is equal to

$$\delta T(H) \approx \frac{\{[1 - C(\text{Fe})]\Delta \rho h g - \rho_0 c_p \Delta T - C(\text{Fe})\rho(\text{Fe})q\}}{\rho(\text{Fe})c_p(\text{Fe})} \tag{1.8}$$

where

$$\frac{1}{\rho_0} = \frac{C(\text{Fe})}{\rho(\text{Fe})} + \frac{[1 - C(\text{Fe})]}{\rho(\text{Si})} \text{ and } g = \frac{\gamma m(r)}{r^2} \quad (1.9)$$

In the Eq. (1.9), $\gamma = 6.67 \times 10^{-8}\,\text{cm}^3/\text{g}\,\text{s}^2$ – gravitational constant; $m(r)$ is the mass of the Earth's matter contained in the sphere with radius $r(= R{-}H)$; and R is the radius of the Earth's globe.

The process of separation of molten iron, however, could not extend deeper than a certain limiting level, beginning with which the generated gravitational energy was not sufficient for heating the matter (lying below) to the point of iron melting. Besides iron, at depths exceeding 860 km, iron oxide (FeO) began melting. Its concentration in the primordeal matter of Earth reached 23% (Sorokhtin, 1974). This process facilitated the zonal differentiation in the Archaean time.

The thickness h of molten iron layer at a depth H can be determined from the mass balance equation for the Earth's matter, which underwent differentiation:

$$h = H + \left\{ [R^3 - (R-H)^3]\left(\frac{\rho}{\rho(\text{Fe})}\right) + (R-H)^3 \right\}^{1/3} - R \quad (1.10)$$

The mantle overheating from depth H to the Earth's surface can be recalculated using the following equation:

$$\delta Tm_0 = \frac{\delta T(H) \cdot T_{\text{Fe}}(O)}{T_{\text{Fe}}(H)} \quad (1.11)$$

In this case, the mantle temperature Tm_0 (°C) reduced to the surface is equal to

$$Tm_0 = 1530 + \delta Tm_0 \quad (1.12)$$

where 1,530 °C is the melting temperature of iron at normal conditions.

The mantle temperature reduced to the surface is shown in Fig. 1.7. As shown in this figure, the overheating of the upper mantle increased rapidly during the first 200 million years after the beginning of zonal differentiation of iron. At the front of zonal differentiation, the temperature reduction after the first maximum point can be explained by the gradual increase in difference between the melting temperature of iron and the temperature of the Earth's matter (Fig. 1.4). The second maximum of the mantle overheating occurred in the late Archaean time when iron oxides became involved in the process of core formation. Sharp decrease in the mantle temperature at the very end of Archaean time (2.7 billion years ago) was caused by squeezing upward of the cold primordeal matter from the inner layers of the Earth to the hot upper mantle (Fig. 1.8).

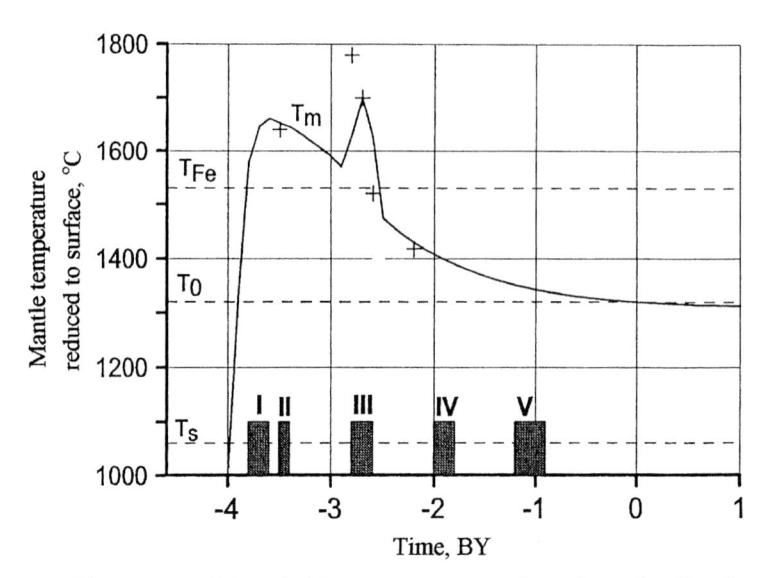

Figure 1.7: The Upper Mantle Temperature Reduced to the Surface in °C. $T_s = 1{,}600$ is the Solidification Temperature of the Mantle Matter; T_{Fe} ($= 1{,}530$) is the Iron Melting Temperature under Normal Conditions; and *I* through *V* are Intervals of Magmatic Tectonic Cycles in the Precambrian Time (Luts, 1985).

Because of relatively low temperature of the deep layers of juvenile Earth, the viscosity of its inner matter was very high. Below 1,000 km, the viscosity exceeded 10^{27} P with sharp increase to 10^{35}–10^{40} P in the central areas. At such high viscosity, any convection, including flows of molten iron through such a viscous medium to the Earth's center, as suggested by Elsasser (1963), is impossible. Consequently, rapid formation of the iron core immediately after the Earth's formation could not occur. Only because of the slow warming of the cold matter of inner layers (at the front of zonal differentiation) this process could gradually proceed toward the center of the Earth. This heating progressed slowly with the rates of about 0.2 cm/year. Consequently, the process of formation of the Earth's core lasted for about 1.6 billion years (from 4 to 2.6 billion years ago).

The early zonal differentiation of the Earth's matter during the Archaean time resulted in extreme gravitational instability of the planetary body: a high-density peripheral ring-shaped layer consisting of molten iron and its oxides was formed due to zonal melting and was positioned above the lighter matter of the core. This instability was resolved by squeezing out of the core matter through the equatorial belt to the Earth's peripherals and sinking of the heavy molten iron to the center of the planetary body from the opposite hemisphere, which gradually formed an iron core (Fig. 1.8). This process, which caused a pronounced asymmetry in the mass distribution of the Earth's body, ensured stable orientation of the Earth's main axis of inertia along the axis of the rotation and, consequently, stabilized rotation of the planet.

Thus, after accretion of the Proto-Moon and formation of the Moon, the huge tidal energy (about $(4-5) \times 10^{37}$ erg) was released mostly in the upper layers of Earth's equatorial belt. As a result, the Earth was heated at the high latitudes considerably more than in the polar areas. Consequently, the first asthenosphere and related to it the iron separation zone had to emerge under the equatorial belt of our planet. That is why, during entire Archaean Eon, zonal matter differentiation developed in the mantle only at low and medium latitudes of the Earth (Fig. 1.8). All the ancient Archaean shields and platforms emerged in the same equatorial belt.

After gradual warming of the upper layers of Earth at high latitudes (at the expense of dissipation of radiogenic and tidal energy), their viscosity was reduced to the level of $10^{24}-10^{25}$ P and below. At this "reduced" viscosity, the Earth's gravitational instability could be resolved by squeezing out of the lighter but firm core (density of about 2.7 g/cm^3) toward the Earth's surface (Fig. 1.8d, e). Due to large mass of the "core" matter accumulated in the ring-shaped differentiation zones (up to 12–13% of the Earth's mass) and pronounced difference between its density and that of the primordeal matter (about 4 g/cm^3), the process of squeezing of core matter should have accelerated at that time and attained catastrophic proportions.

The formation of the Earth's core was accompanied by a huge energy release (about 5.5×10^{37} erg). About half of this energy was spent to increase the inner heat of the Earth, i.e. for additional heating of the mantle and the matter of former core. The other half was transformed into the energy of convectional flows of the mantle matter and, finally, into the Earth's heat radiation. Additional heating of the bowels of the Earth has led to acceleration of the formation process of the core. As a result, the entire process of the Earth's core formation was accelerated and completed by the end of Archaean Eon, about 2.6 billion years ago with a catastrophic event of the dense core separation at the center of the Earth. Paleomagnetic data also indicate that the present-type dipole magnetic field emerged about 2.6 billion years ago at the boundary of the Archaean and Proterozoic eons (Hale, 1987). Prior to that time, the Earth must have had a magnetic field of the toroidal type.

Separation of the Earth's core was accompanied by emergence of intense convective currents in the Earth's mantle, completely changing the formerly existing tectonic layout of the lithospheric shell of Earth. Thus, it appears that at the end of Archaean eon, there was an emergence of a single-cell

32   Global Warming and Global Cooling: Evolution of Climate on Earth

convective structure with a rising flow above the former core and a descending flow above the sink of the "core" matter. Under these circumstances, one can suppose that the first supercontinent (called Monogaea by Sorokhtin and Ushakov, 1989a, b) formed above the descending flow about 2.6 billion years ago at the boundary between Archaean and Proterozoic eons. It is widely accepted, however, that the "first" supercontinent was Rodinia created about

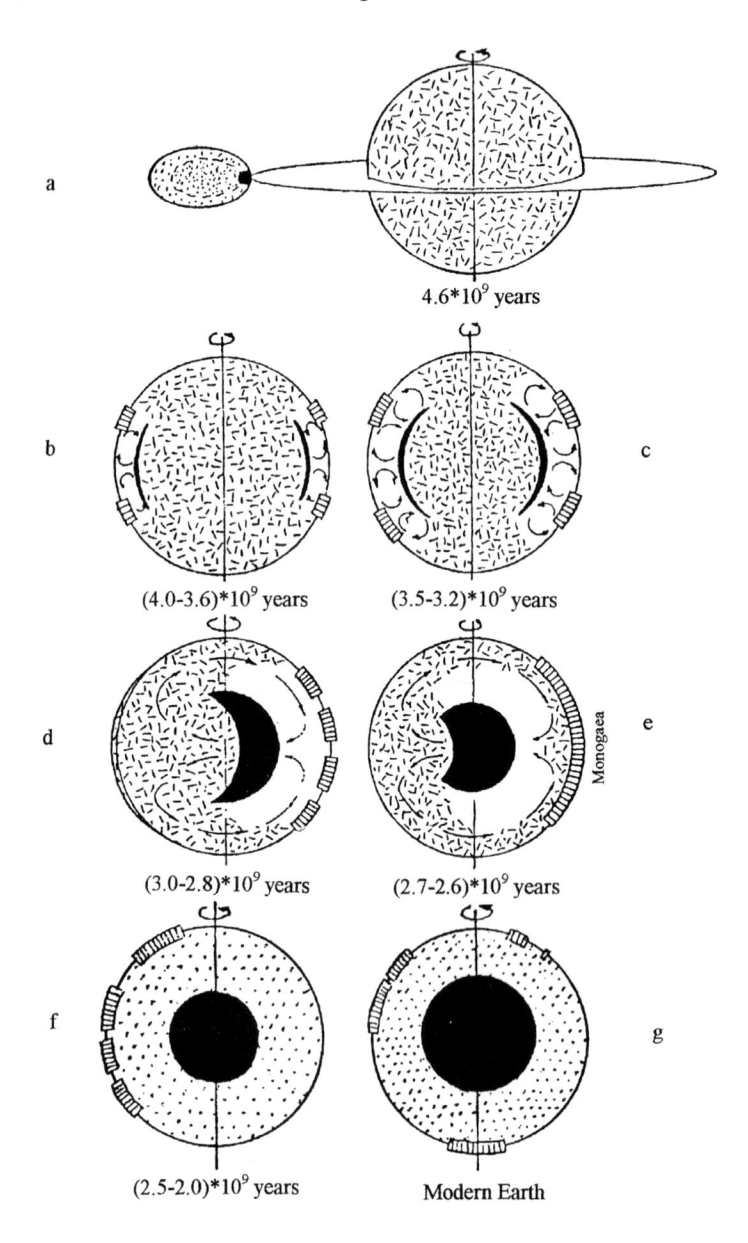

a

$4.6*10^9$ years

b

$(4.0-3.6)*10^9$ years

c

$(3.5-3.2)*10^9$ years

d

$(3.0-2.8)*10^9$ years

e

$(2.7-2.6)*10^9$ years

Monogaea

f

$(2.5-2.0)*10^9$ years

g

Modern Earth

1.1 billion years ago (Fig. 1.9). It is noteworthy that analogous supposition about formation of a supercontinent in early Proterozoic was made by Khain and Bozhko (1988) based upon independent geologic data. They called it Pangaea O.

The first stage of the Earth's core formation was completed by filling in the central area of the planet with heavy "core matter" (iron and iron oxides) and squeezing out of the lighter silicate matter to the periphery of the Earth. At the same time, the first stage of the Earth's matter differentiation through the mechanism of zonal separation of iron (and iron oxides) was completed. Further development of the Earth's metallic core progressed through the gradual barodiffusion mechanism, which will be discussed later.

After formation of the core at the boundary between the Archaean and the Proterozoic eons, the mantle temperature began decreasing gradually. The rate of cooling can be estimated using the contemporary Earth's heat flux of 2.5×10^{19} erg/s. Our estimates show that the contemporary mantle temperature reduced to the Earth's surface conditions is close to $1,320\,°C$ and is gradually decreasing at a rate of $1.5 \times 10^{-8}\,°C/year$.

1.4.1. The Earth's Core Growth after its Formation in the Archaean Eon

The mass of the core formed in the Archaean time was approximately 63% of its contemporary mass. The rest 37% of the core matter was separated during the last 2.6 billion years. What was the mechanism of accumulation of the core mass after the Archaean time? Development of a plausible theoretical model of the contemporary core differentiation mechanism of the Earth's matter, leading to the accumulation of dense core matter at the center of Earth is a very difficult task. Obviously, the geophysical process of the Earth's matter differentiation is not accessible for direct observations. Consequently, one needs to develop realistic hypotheses on the separation

---

Figure 1.8: Stages in the Earth's Core Differentiation. Legend: (a) Juvenile Earth; (b) and (c) Zonal Differentiation of the Earth's Matter in the Early- and Mid-Archaean Times; (d) and (e) Formation of Dense Earth's Core at the End of Archaean Time; (f) and (g) the Earth's Core Development in the Proterozoic and Phanerozoic Times. The Regions of Molten Iron and its Oxides are Represented in Black; the Archaean Mantle Impoverished in Iron, its Oxides, and Siderophilic Compounds are shown in White; Short Dashes Indicate the Regions of Primordeal Earth's Matter; Dots show the Regions of Normal Mantle in the Proterozoic and Phanerozoic Times; and Short Radial Dashes are used to show the Continental Massifs.

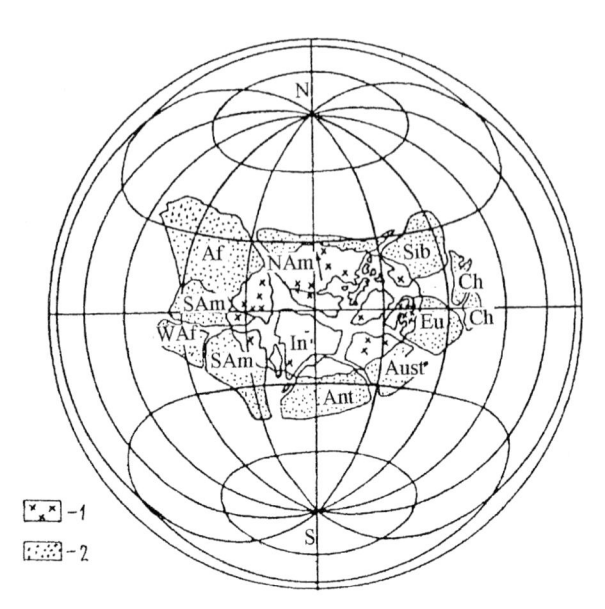

34 *Global Warming and Global Cooling: Evolution of Climate on Earth*

Figure 1.9: Reconstruction of Monogaea for the Time about 2.5–2.4 Billion Years Ago in the Lambert Projection: (**1**) – Tillites and Tilloids (Chumakov, 1978); (**2**) – Consolidated Continental Crust; Regions of Glaciations are shown in White. (1) Africa; (2) Antarctica; (3) Australia; (4) China; (5) Europe; (6) India; (7) North America; (8) South America; (9) Siberia; and (10) Western Africa.

mechanism of matter based on its indirect manifestations at the Earth's surface and analyzing the data of experimental and theoretical physics. One of the most difficult questions of the theory is a considerable difference between the melting temperature of silicates at depth and the adiabatic mantle temperature (Fig. 1.4). Most probable estimates show that the melting temperature of the mantle silicates at pressures existing at the bottom of the mantle exceeds 8,000 K, whereas the mantle temperature at the same depths is close to adiabatic, i.e. about 3,200 K. Existence of such a significant temperature difference of about 5,000 K rules out completely the assumption of separation of the Earth's core as a result of its melting out from the mantle.

Thus, one can suppose that, at the present time, separation of iron oxides from the mantle silicates occurs by diffusion. Under high pressures at the mantle bottom, solid mantle solutions decompose and the iron oxides diffuse from the silicate crystals into the intergranular pores of the mantle matter (Monin and Sorokhtin, 1981). This hypothesis is based on the Le Chatelier Principle and the experimentally established fact of decomposition of solid solutions under high pressures. This breaking-up occurs when the total mole volume of the solution exceeds the sum of the mole volumes of the solution components, i.e. when solution of a certain component leads to an increase in the relative volume of solution. A solid solution decomposes due to reduction in solubility of components under high pressure.

Experimental data for real solid solutions indicate that their decomposition can be described by the following equation:

$$C_i = C_{0i} \exp\left\{\frac{-\Delta V_i p}{RT}\right\} \tag{1.13}$$

where C_i is the limit concentration of the component i in the solid solution, C_{0i} is the concentration of the component i in saturated solution at a pressure $p = 0$, ΔV_i is the increase in the solution volume on dissolving in it one mole of the component i, R is the universal constant, and T is the absolute temperature in degrees Kelvin. Equation (1.13) allows using the theory of solid solution decomposition to describe the separation of iron oxides from the mantle silicates.

Decomposition of the mantle silicates alone is not sufficient for the differentiation of iron oxides because it is necessary to remove them also beyond the crystalline grid of silicates into the intergranular space. Such withdrawal mechanism is obviously linked only to diffusion. In contrast to usual thermal diffusion, which leads to more homogeneous matter, in this situation one should observe concentration of iron oxides in the intergranular space and in the intercrystalline films. This situation can occur only if there are energy barriers along the way of diffusing atoms that allow atoms to migrate out of the crystalline grid, but makes it difficult for them to return back.

Such natural energy barriers arise at the faces of crystals and surfaces of silicate grains. When the diffusion of iron oxides from the silicate crystal occurs, it leads to generation and dissipation of energy δE, which is proportional to decrease in the total volume of the matter δV at a given pressure p, $\delta E = p \delta V$. In order to reverse diffusion of oxides of iron into the crystal, the same amount of energy has to be spent to overcome the pressure and expansion of the participating matter by the same increment of volume δV. Because of reduction of the total volume of matter, the iron oxides, which diffused out of the crystals and grains of silicates, cannot return back to the silicate grid and gradually accumulate in the intercrystalline films and intergranular space.

Analyzing diffusion processes, one needs to remember that atoms and not molecules migrate through vacancies and defects of the crystalline grid (Reed-Hill and Abaschian, 1992). Consequently, one should consider the separate diffusion of atoms of iron and oxygen with corresponding partial diffusion coefficients. Inasmuch as these diffusion coefficients are not equal,

36 *Global Warming and Global Cooling: Evolution of Climate on Earth*

the rates of migration of atoms are different. This violates the stoichiometric ratios in stable compounds of diffusing elements. This, in turn, leads to emergence of, and increase in, chemical potentials in boundary layers, separating the unbound atoms. Injection of new electrochemical energy balances the rates of diffusion of atoms of iron and oxygen according to the composition of a stable (under given conditions) chemical compound or eutectic alloy.

One should keep in mind at all times that only atoms can diffuse in the solid medium. The process of matter differentiation is determined by pressure in deep layers of the Earth and diffusion of atoms caused by high pressures in these layers. The writers call this process the barodiffusion mechanism of the matter differentiation, i.e. diffusion of atoms in silicate crystalline grid.

The critical concentration of a component, which increases the relative volume of solution, decreases with pressure according to the exponential law (Eq. (1.13)). Fractional separation can occur only if the initial concentration of component exceeds its critical concentration at a given pressure and temperature (Fig. 1.10). Inasmuch as the critical concentration of oxides in their saturated silicate solution decreases according to the exponential law (Eq. (1.13)), one can conclude that there is a certain level below which the condition of disintegration of solid solution is realized. This level corresponds to a depth of about 2,000 km with a pressure of 0.89 Mbar and a temperature of about 2,730 K. One can expect that at this depth the critical concentration of core matter in the silicate solution is equal to the concentration of iron oxides in the mantle, which is 0.074 after recalculation of the concentration of oxides to Fe_2O. Thus, one can estimate that at the surface of Earth's core (at a pressure of 1.38 Mbar and a temperature of 3,130 K) the critical concentration of iron oxides in the silicate solution is about 0.027. Therefore, below a depth of 2,000 km the concentration of iron oxides

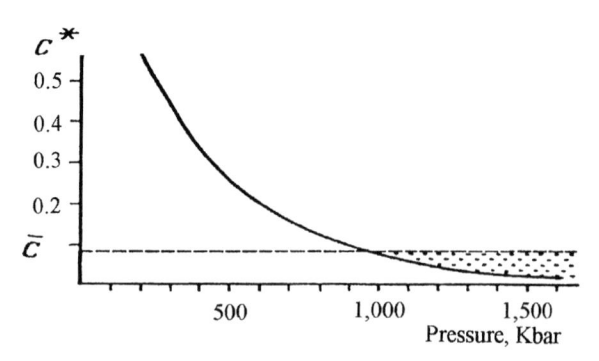

Figure 1.10: Saturation Concentration of Solid Solutions of Iron Oxides in the Mantle Silicates as a Function of Pressure. Area of Development of Barodiffusion of Iron Oxides from the Mantle Silicates is Marked by Specks.

(recalculated to Fe·FeO) in the mantle is always higher than their critical concentration in the silicate solutions as shown in Fig. 1.10. Thus, one can conclude that differentiation of iron oxides from the mantle matter occurs at a depth range of 2,000–2,886 km (depth of core surface). In the future, with decreasing concentration of iron oxides in the mantle (as a result of their transition to the core) the depth range of differentiation of the mantle matter will only decrease.

Supposition that the differentiation of mantle matter occurs only in the lower mantle is indirectly supported by the seismic data, which indicate significant changes in the mechanical merit factor at a depth range of 2,000–2,900 km. At this depth range, damping of seismic waves increases pronouncedly, and the mechanical merit factor for longitudinal waves drops from about 500 to 115 near the surface of the core (Teng, 1968). At the lower part of the mantle, there is a layer D'', in which the mechanical merit factor decreases by almost ten times (Zharkov et al., 1974). This anomaly is so distinct that Zharkov et al. suggested to call the layer D'' the second asthenosphere of the Earth. In the light of discussion on the mantle matter differentiation, the comparison of the lower mantle layer with the asthenosphere is justified because in the zone of separation of liquid phase (eutectic alloy Fe·FeO) one should also expect a sharp decrease in the effective viscosity of matter. Moreover, analyzing dynamic peculiarities of the seismic waves reflected from the Earth's core, Berson et al. (1968) identified a relatively thin (about 20 km) boundary layer at the very bottom of the mantle, in which the mechanical merit factor decreases gradually to almost zero (Berson et al., 1968; Berson and Pasechnik, 1972). The authors of this book call this boundary layer the Berson Layer. In the Berson Layer, the mantle matter gradually acquires the properties of a low-viscosity liquid, without the melting of silicates.

When the mantle mater of the downward convective flow approaches the core surface, a fraction of the iron oxides (diffused from the silicate crystals and grains into the intergranular space) gradually increases. The liquid state of the outer core testifies that, in the intergranular space, the core matter (i.e. Fe_2O) must also occur as a liquid phase. Experimental data of Boehler (1993) on melting of iron compounds under high pressures corroborate this conclusion. Therefore, the rigid bonds that exist between silicate crystals in the lower part of the mantle weaken on approaching the core. This phenomenon also causes a pronounced decrease in the seismic merit factor in D'' layer and, especially, in the Berson layer. When the separate films and inclusions of the liquid iron oxides merge in the vicinity of the Earth's core, the mantle matter begins to disintegrate.

In the case of downward convective flows (i.e. under heavier mantle matter) some mantle humps should form, which sink into the core matter,

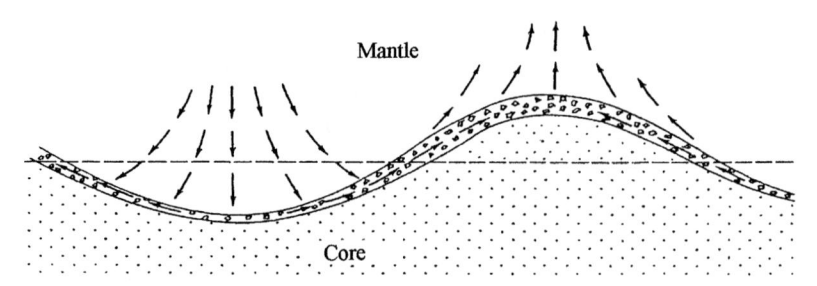

Figure 1.11: Schematic Diagram of Convective Mantle Currents at the Vicinity of Core–Mantle Border. Formation of Upward Flows in the Mantle from the Disintegrated Mantle Matter.

whereas as in the case of upward convective flows uplifts of the core should emerge, as shown in Fig. 1.11. This hypothesis was corroborated by the seismic thomography of the core (Morelli and Dziewonski, 1987) that showed the wavy topography of the core surface (mantle–core interface) with the level deviating ± 6 km from the average radius of the core.

Large variations in the level at the bottom of lower mantle and significant difference in density between the core and the mantle matter (about $4\,g/cm^3$) should lead to emergence of significant tensions of the order of 10 kbar (Sorokhtin, 1974) at the "roots" of downward convective flows. Owing to such strains, final disintegration of the polycrystalline mantle matter into separate crystals and granules (dispersed in $Fe \cdot FeO$ (Fe_2O) melt) occurs under downward flows at the bottom of mantle. In this process, liquid FeO, differentiated earlier from the silicate crystals of the mantle matter, transfers into the Earth's core leading to the gradual growth of its mass. This explains why in the transitional Berson Layer the mantle matter is gradually losing its rigidity in the process of its transition into liquid phase at the surface of the core.

At first sight, spreading of disintegrated mantle matter from under the roots of downward flows (toward upward convective flows) should level off the surface of Earth's core. This leveling, however, is continuously compensated by fresh portions of the settling mantle matter and by its uplifting after differentiation in the zones of development of upward currents. As a result, stable dynamic circulation of the disintegrated mantle matter in the form of closed convective cells occurs at the boundary between the mantle and the core. In the upward convective flows of the mantle matter, the diffusion of iron oxides progresses in the opposite direction – from intergranular space into the silicate crystals. Simultaneously, Fe_2O_3/FeO ratio increases in crystals, with the formation of magnetite (Fe_3O_4) molecules and zonal distribution of iron oxides in crystals.

Seismic data and the theoretical estimates indicate that the viscosity of core matter is relatively low. Gans (1972) estimated that the dynamic viscosity of core matter most probably did not exceed 0.01–0.1 P. At such low viscosity and significant differences in density between the core matter and suspended in it crystals and granules of silicates, the currents of mantle matter (disintegrated into granules) on the surface of core should be rapid and relatively narrow. Because of significant variations in the topography of the core surface, such flows should significantly erode the lower surface of the mantle (base of mantle). As a result, at the bottom of the mantle currents (mixture of the core and disintegrated mantle matter) should form separate streams flowing upward. The speed of flow of these streams can reach several meters per second.

At such high flow rates, these streams are subject to Coriolis force, which deflects these currents to equatorial plane. The resulting motion of the electro-conducting core matter induces very strong electric currents and magnetic fields. This leads to a highly plausible hypothesis that the Earth's geomagnetic field is induced by such currents at the core surface. This link of geomagnetic field to the convective streams in the mantle allows one to explain the existing interdependence of the frequency of the variation in magnetic polarity and the tectonic periodicity in the Earth's history.

It is important to note here that the Earth's core does not have any powerful sources of energy, and its heating occurs by the differentiating mantle matter at the surface only.

Therefore, the temperature distribution in the core should be below adiabatic. The latter implies the existence of a stable stratification of core matter that prevents emergence of the convective currents in the core's body, which supposedly originate magnetic field of the Earth. These *non-existing* core currents are usually used to justify the most popular hypotheses on the origin of geomagnetic field induction.

In conclusion, one can state that *iron is the main element that determines the global geologic evolution of the Earth.*

1.4.2. Formation of the Earth's Core

The rate of growth of Earth's core in the barodiffusion process of the core matter differentiation (separation of core iron oxides from mantle silicates) is proportional to the difference between the concentration of iron oxides in the mantle (c) and the critical concentration of iron oxides in the solid silicate solutions (C) at the base of the mantle (Fig. 1.10). On the other hand, the rate of separation of the core matter from the mantle matter (i.e. the rate of core growth) $\dot{M}_c$ should be proportional to the gravitational acceleration g at the core surface, area of the core surface, and the rate of barodiffusion k:

$$\dot{M}_c = k(c - C)g4\pi R_c^2 \tag{1.14}$$

where R_c is the radius of growing core.

To characterize quantitatively the process of core growth, Monin (1977) introduced the evolutionary Earth parameter χ as the ratio of core mass M_c to the total mass of the core matter in the Earth's body:

$$\chi = \frac{M_c}{MC_0} \tag{1.15}$$

where M is the Earth's mass and C_0 is the total concentration of core matter (iron, iron oxides, and other heavy elements and compounds in transition to the Earth's core) in the Earth's body. Contemporary Earth's core contains 1.94×10^{27} g of the core matter, whereas the contemporary mantle contains 1.97% Fe_2O_3 and 6.55% FeO. These contents recalculated as eutectic alloy $Fe \cdot FeO$ give 7.4% or 0.297×10^{27} g of the core matter remaining in the mantle. Therefore, the total concentration of core matter in the Earth $C_0 = 0.375$, and the contemporary value of the evolutionary parameter $\chi_0 = 0.865$.

Inasmuch as the gravitational acceleration at the center of Earth is equal to zero, barodiffusion could occur only after the formation of Earth's core embryo. A probable scenario of the core development was presented earlier (Fig. 1.8). Judging from geological data, this process lasted for about 1.4 billion years and was completed at the boundary of Archaean and Proterozoic eons about 2.6 billion years ago upon formation of the core embryo with the relative mass of 0.543. The further growth of core occurred due to barodiffusion mechanism of differentiation of the Earth's matter.

During the time interval when the embryo was forming (4–2.6 billion years ago, during the Archaean time) the differentiation mechanism was different. Therefore, evolution of the parameter χ in time should be evaluated based on the mass of core matter separated into the ring-like zone of differentiation (into the layers of iron alloys). Besides, one should take into account the gradual extension of the low-latitude ring-shaped belt, in which the processes of zonal differentiation took place. The evolution of the parameter χ, i.e. the curve of change in the dimensionless mass of the Earth's core, is presented in Fig. 1.12. This figure shows that during the first 1.4 billion years of geologic development (from 4 to 2.6 billion years ago), about 63% of the contemporary core mass was formed, whereas the remaining 37% were separated during the Proterozoic and Phanerozoic times (i.e. during the last 2.6 billion years). Thus, according to the model presented here, the Earth did not have a core in the Archaean time. A real core (smaller than the contemporary size)

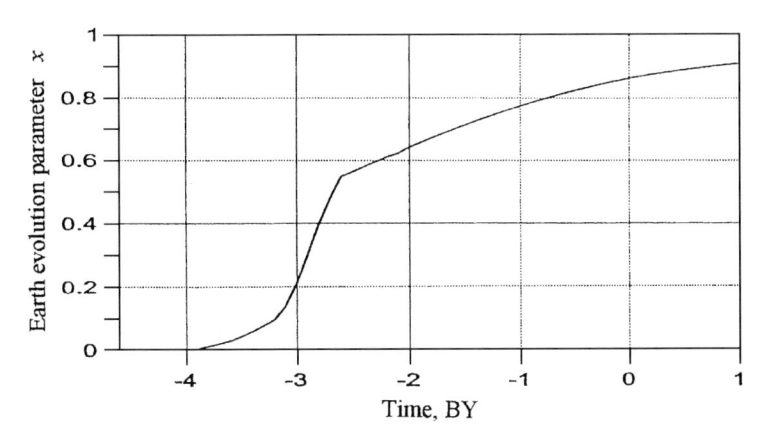

Figure 1.12: Relationship between the Evolutionary Parameter of the Earth (Dimensionless Mass of the Earth's Core) and Time (Billion of Years).

formed at the boundary of Archaean and Proterozoic times. At that time the toroidal geomagnetic field changed to the bipolar one of present-day type (Fig. 1.6).

The rate of change of the parameter χ at the present time is about $0.657 \times 10^{-10}\,\text{year}^{-1}$. Using this information, one can estimate that at the present time about $1.5 \times 10^{17}\,\text{g/year}$ (or 150 billion tons/year) of the core matter (Fe_2O) (i.e. 131 billion tons/year of metallic iron) is transferred from the mantle into core. During the late Archaean time, around 2.8 billion years ago, the mass of metallic iron being transferred into the core was about 1.65 trillion tons/year.

Galimov (1998) presented similar ideas about the role of Earth's core accumulation as a source of inner energy and an important factor in the evolution of redox potential of the mantle. He assumed, however, that the core originated at the time of Earth's formation, which appears to be unrealistic. In addition, according to Galimov, the core has grown by only 5% during the entire time of Earth's evolution, which contradicts the Earth's energy balance.

1.5. Energy Balance of the Earth

A fundamental question of planetary geophysics is the question about origins of endogenic sources of energy, which determine the thermal regime and the Earth's tectonic activity. One should address this question based on today's accumulated knowledge on the composition and evolution of the Earth and according to strict physical laws.

Sorokhtin (1974, 2004) and Sorokhtin and Ushakov (2002) emphasize that the main processes driving the Earth's tectonic activity are the endogenic energetic processes that reduce most considerably the potential (inner) energy of the Earth and of the Earth–Moon planetary system. The reduction of potential energy is due to its transformation into heat and kinetic energy of the moving Earth's masses: mantle convection, movement of lithospheric plates, continental drift, orogeny, etc. In turn, any displacement of the Earth's mass is accompanied by the dissipation of kinetic energy and heat generation, which leads to the partial melting of upper mantle matter or the rocks of continental crust. This contributes to the Earth's magmatic activity. Finally, all the generated heat, emanating through the Earth's surface, gradually dissipates into space. Therefore, the magnitude of this heat flux can be used as a measure of the Earth's tectonic activity. This is a very important theoretical conclusion, because one can determine the tectonic evolution of the planet on estimating the heat flux through the Earth's surface.

There are three major sources of endogenic energy in the Earth: (1) the gravitational matter differentiation resulting in stratification of the Earth's body (dense iron–iron oxide core, remaining silicate mantle, light aluminum-silicate crust, and hydrosphere and atmosphere); (2) the radioactive decay heating the crust and upper mantle; and (3) the tidal interaction of Earth and Moon. All the other endogenic sources are either considerably lower, or are completely reversible due to convective mass exchange in the mantle. Considerably larger heat flux delivered to the Earth's surface by the Sun is reflected mostly back into space, or through the chain of transformations in the atmosphere, hydrosphere, and the near-surface layers of the Earth's crust is partially conserved in the sediments forming fossil fuels. Taking all the aforementioned into account, one needs to consider the three major sources of energy: gravitational, radiogenic, and tidal.

In comparison with the present state, the initial heating of the juvenile Earth was not very high. Judging from the temperature distributions presented by Safronov (1969), the initial heat content of Earth's was about $(7-8) \times 10^{37}$ erg. Using a more precise estimate, based on the temperature distribution for the juvenile Earth (Fig. 1.4, curve 1), the initial heat content of juvenile Earth was estimated at 7.12×10^{37} erg (Sorokhtin and Ushakov, 2002). For comparison it should be noted that the present-day Earth's heat content is considerably higher – about 15.9×10^{37} erg. Thus, during its lifetime, the Earth was heated considerably (average of about 1,650 °C), i.e. the Earth's heat content was increased more than twice.

Figure 1.13 shows the graph of energy generation as a result of gravitational matter differentiation in the process of Earth's core separation. As shown in this figure, the total energy released in the process of gravitational

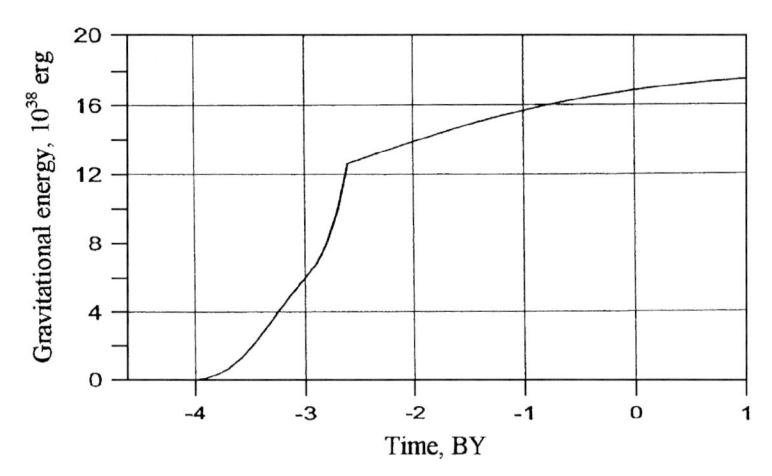

Figure 1.13: Evolution with Time of Energy Generated as a Result of Gravitational Differentiation of the Earth's Matter.

matter differentiation until the present time is about 16.84×10^{37} erg. From this total, 4.2×10^{37} erg of energy was spent for additional compression of the Earth's body, whereas 12.64×10^{37} erg were transformed into the kinetic energy of convective currents in the mantle and heat. To estimate the tectonic activity of the Earth, one needs to know the rate of energy release (Fig. 1.14). The graph of this rate is given in Fig. 1.14. Thus, the release of energy due to Earth's gravitational matter differentiation started at an accelerating rate about 4 billion years ago. In the early Archaean time, the rate of energy release was 1.8×10^{21} erg/s, which is about 6.2 times greater than the present-day rate of about 2.77×10^{20} erg/s. After deceleration in the middle of Archaean time, this process accelerated significantly during the late Archaean time after initiation of the Earth's core separation. During this time, the rate of energy release exceeded the present level by 14–18 times. This was a time of formation of dense iron core.

After separation of the Earth's core in the end of Archaean time, the rate of gravitational energy generation during the early Proterozoic time was reduced considerably to the level of about 7.6×10^{20} erg/s. There was further deceleration in the process of gravitational energy generation with the rate gradually decreasing to the present-day level of about 2.77×10^{20} erg/s. This deceleration will continue in the future.

In order to estimate generation of radiogenic energy, one needs to evaluate the contents of radioactive elements in the Earth's body. Radioactive isotopes concentrate mostly in the light aluminum silicates. In the process of Earth's matter differentiation, radioactive isotopes migrate to the zones

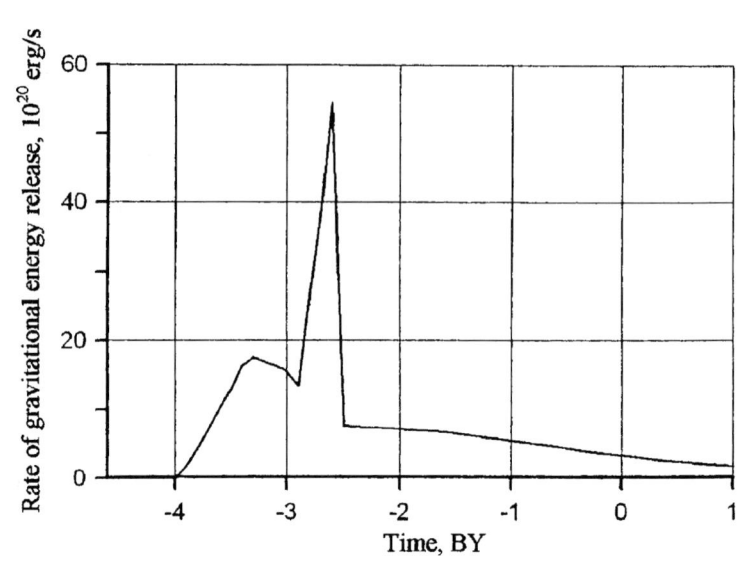

Figure 1.14: The Rate of Change in the Energy Generated as a Result of Gravitational Differentiation of the Earth's Matter.

having the highest concentrations of aluminum silicates, which have high contents of silicon dioxide (SiO_2), aluminum oxide (Al_2O_3), and alkalies (Li, Na, K, Rb, etc.), i.e. to the continental crust.

The concentrations of radioactive elements should be considerably lower in the dense ultrabasic rocks of mantle with low aluminum and high magnesium contents. They should be practically absent in the Earth's core, as evidenced by the contents of iron meteorites and their sulfide phase (troilite). On considering also K/U and K/Th ratios in the crust, mantle, and Moon rocks, and comparing the values of heat fluxes from the mantle and continental crust (Sorokhtin and Ushakov, 1991), the authors arrived at the contents of radioactive elements in the Earth's geospheres as given in Table 1.2.

Thus, also taking into consideration the laws of Earth's matter differentiation, Sorokhtin and Ushakov (2002) computed the rate and total magnitude of radiogenic energy emanated in various geospheres (Figs. 1.15 and 1.16).

At present, the greater part of tidal energy is released in shallow seas. It is considerably higher than that released in the deep oceans and the Earth's asthenosphere. MacDonald (1975) estimated that the contemporary rate of tidal energy release is approximately 0.25×10^{20} erg/s, with about 2/3 of it being dissipated in the shallow seas due to friction of intense deep oceanic currents with the oceanic floor. According to our estimates, at the present time, about 0.287×10^{20} erg/s of tidal energy is dissipated in the Earth's body. In the mantle, the rate of dissipation is 0.018×10^{20} erg/s, whereas the

Table 1.2: Radioactive Isotope Contents in the Earth's Body.

Elements	Continental Crust, 2.25×10^{25} g		The Earth's mantle, 4.07×10^{27} g		The Earth's total, 5.98×10^{27} g	
	Element Contents (g)	Emanated Energy (erg/s)	Element Contents (g)	Emanated Energy (erg/s)	Element Contents (g)	Emanated Energy (erg/s)
^{238}U	3.64×10^{19}	0.341×10^{20}	1.047×10^{19}	0.098×10^{20}	4.69×10^{19}	0.439×10^{20}
^{235}U	0.026×10^{19}	0.015×10^{20}	0.008×10^{19}	0.004×10^{20}	0.034×10^{19}	0.02×10^{20}
^{232}Th	15.18×10^{19}	0.408×10^{20}	2.89×10^{19}	0.078×10^{20}	10.07×10^{19}	0.486×10^{20}
^{40}K	5.24×10^{19}	0.146×10^{20}	5.62×10^{19}	0.157×10^{20}	10.86×10^{19}	0.303×10^{20}
K/U	1.23×10^{4}	N/A	4.60×10^{4}	N/A	1.98×10^{4}	N/A
K/Th	3.00×10^{3}	N/A	1.67×10^{4}	N/A	5.16×10^{3}	N/A
Th/U	4.0	N/A	2.74	N/A	3.83	N/A
Total emanated energy	—	0.91×10^{20}	—	0.337×10^{20}	—	1.248×10^{20}

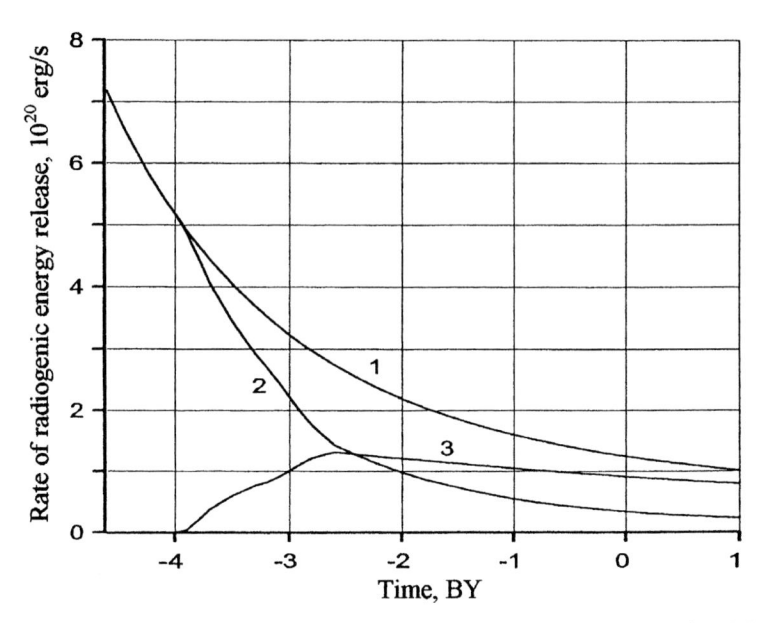

Figure 1.15: The Rate of Radiogenic Energy Generation: In the (1) Earth; (2) Mantle; and (3) Continental Crust.

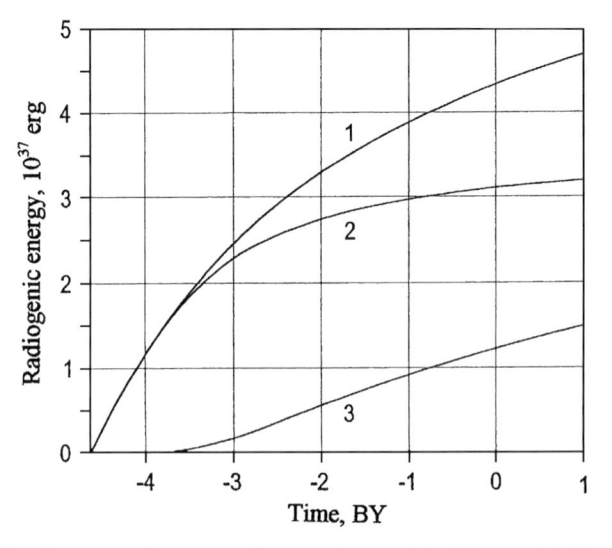

Figure 1.16: Radiogenic Energy Generation with Time: (1) Earth; (2) Mantle; and (3) Continental Crust.

rate of dissipation in hydrosphere is about 0.269×10^{20} erg/s or 94% of the total amount of tidal energy. Inasmuch as the total present-day heat flux through the Earth's surface reaches 4.3×10^{20} erg/s (Gorodnitskiy and Sorokhtin, 1981), one can conclude that the fraction of tidal energy dissipating in "solid" Earth at present time does not exceed 0.5% of the total energy generated inside the Earth. Consequently, one can conclude that the Moon tides play only a minor role in sustaining the Earth's tectonic activity. However, the tidal deformations of lithospheric shell, reaching tens of centimeters in amplitude, in some cases can trigger earthquakes and tsunamis. It is noteworthy that the influence of solar tidal interactions with the Earth does not exceed 20% of the Moon–Earth tidal interactions.

If the amplitude of the solar tides was always insignificant, the influence of Moon tides in the past geologic epochs was quite high. The intensity of such an influence varies inversely with the sixth power of the distance between planets (MacDonald, 1975; Ruscol, 1975). Therefore, in the past, when the Moon was much closer to the Earth, the Moon's tidal influence on the Earth was considerably higher. Moreover, one can hypothesize that at the early stages of Earth development, when the Moon was positioned at the Roche limit, the amplitude of Moon tides could reach 1.5 km, and the tidal energy prevailed over all other sources of endogenic energy. Thus, in the past, the tidal energy played an important role in the tectonic evolution of Earth.

To determine the evolution of past tidal energy generation, one should use the dependence of the distance between the centers of gravity of Earth and Moon (see Fig. 1.5) and apply the basic equations of planetary movement. These equations allow one to determine the relationship between the angular velocity of Earth's rotation and the distance between the centers of gravity of Earth and Moon. The results of calculation show that the rate of tidal energy release in the Earth, $E' = dE/dt$, is proportional to the difference between the angular velocities of the spinning of Earth (Ω) and the Moon (ω), and varies inversely with the Earth's merit factor (Q_μ) and the sixth power of the distance between the planet and the satellite L:

$$E' = \frac{-3/2 \gamma k_2 m^2 R^5 (\Omega - \omega)}{Q_\mu L^6} \qquad (1.16)$$

where γ is the gravitational acceleration, k_2 is the second number of Lyava, and R is the Earth's radius.

In Eq. (1.16), dependence of the tidal energy on time is determined through the distance between Earth and Moon, which is changing in time (see Fig. 1.5). The Earth's merit factor Q_μ was computed using the mass of water and the depth of oceans and epicontinental seas (Sorokhtin and

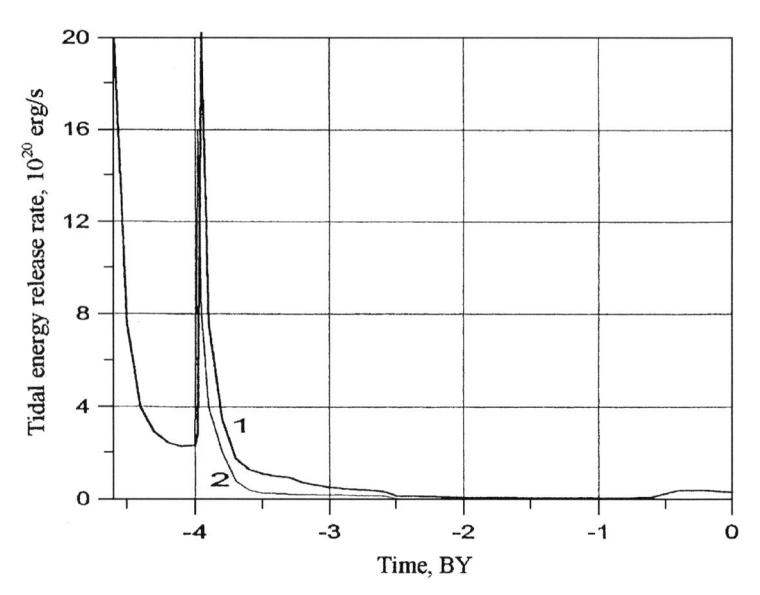

Figure 1.17: The Rate of Tidal Energy Generation in the Earth: (1) the Total Rate of Energy Generation in the Mantle and Hydrosphere, and (2) the Rate of Energy Generation in the Mantle.

Ushakov, 2002). All the other parameters were either constant or determined by the chosen model of the Earth's evolution. Using the Eq. (1.16), we computed the rate of tidal energy generation in the Earth (Fig. 1.17) and its total value (Fig. 1.18).

As shown in these figures, the most intensive generation of tidal energy occurred during the Pre-Archaean time (about 4.6–4.5 billion years ago) and the Early Archaean time (about 4.0–3.7 billion years ago). Generation of the tidal energy in the end of Archaean time and later during the Proterozoic time became more moderate. By that time, the first large oceans formed on the Earth's surface and, because of that, a larger portion of the tidal energy began dissipating in the Earth's hydrosphere. The rate of the tidal energy generation slightly increased again only during the Phanerozoic time, when the increase in tidal interactions with the Moon was linked to the evolution of Earth's hydrosphere, which led to the first wide transgressions of oceans on continents and formation of shallow epicontinental seas. At the present time, a larger portion of the Earth's tidal energy is dissipating in these epicontinental seas.

As shown in Fig. 1.18, during the first 600 million years of the Earth's evolution, the total tidal energy released amounted to about 2.1×10^{37} erg. During the Archaean, Proterozoic, and Phanerozoic times, the amount of generated tidal energy was about 1×10^{37} erg. After formation of the Moon (about 4.6 billion years ago), the total amount of generated Earth's tidal energy

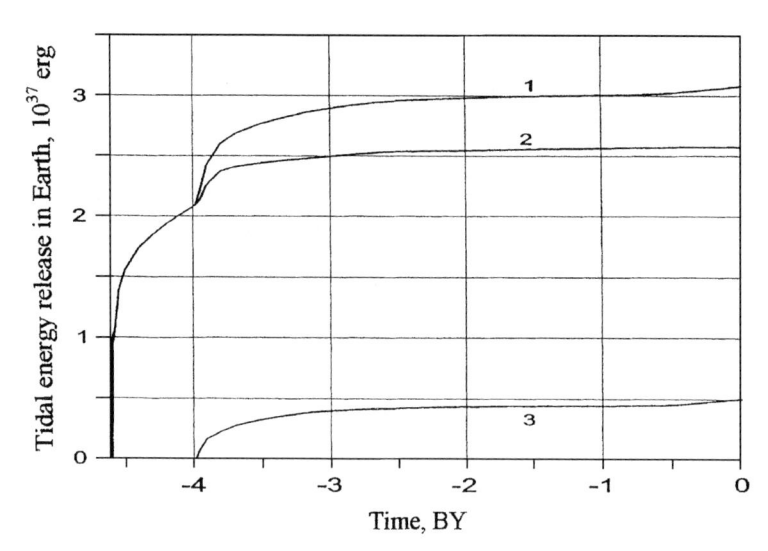

Figure 1.18: Tidal Energy Generation in the: (1) Earth; (2) Mantle; and (3) Hydrosphere.

was about 3.08×10^{37} erg, out of which 2.24×10^{37} erg of energy dissipated in the mantle.

The following amounts of energy were generated in the mantle during the Earth's existence: (1) about 16.85×10^{37} erg in the process of gravitational matter differentiation; (2) about 3.11×10^{37} erg of the radiogenic energy; and (3) about 2.24×10^{37} erg of the tidal energy (not counting its dissipation in the Earth's hydrosphere). Thus, the total amount of heat energy generated in the mantle was about 22.2×10^{37} erg. This amount of energy was sufficient for additional overheating of the Earth up to $4,400\,°C$. Such overheating, however, did not occur, which means that the Earth lost a larger portion of the generated energy as a result of heat radiation into space. During the Earth's existence, the Earth lost about 13.42×10^{37} erg of energy generated in the mantle due to heat radiation into space. This heat radiation characterizes the total energy driving the Earth's tectonomagmatic processes throughout the evolution (Fig. 1.19).

Energy balance of the Earth, driving its tectonic activity is determined by a simple equation:

$$Q' = E' - W' \qquad (1.17)$$

where Q' is the rate of the Earth's inner heat loss, E' is the rate of the total energy (gravitational, radiogenic, and tidal) generation in the mantle, and W' is the rate of change of the Earth's heat resource.

One can determine the present magnitude of the Earth's inner heat flux by subtracting the rate of radiogenic energy generation in the continental crust (about 0.91×10^{20} erg/s) from the total Earth's heat losses (about

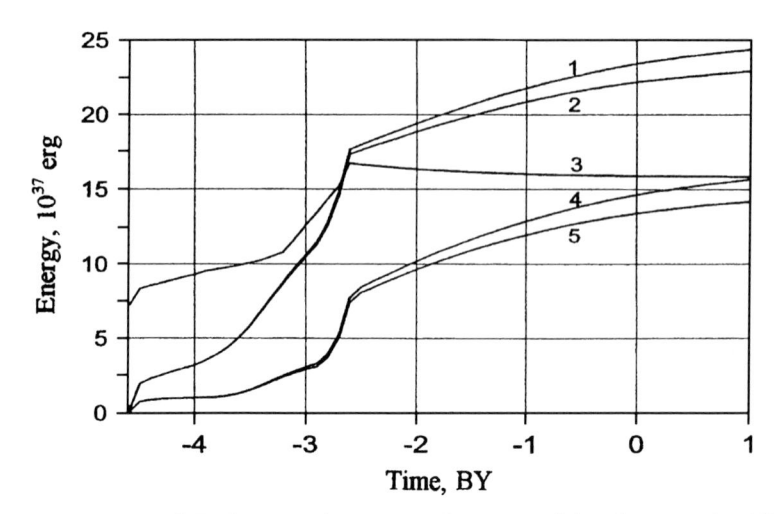

Figure 1.19: Integral Balance of Energy Generated in the Earth. (1) Total Energy Generated in the Earth (without the Tidal Energy Dissipated in the Oceans and Seas); (2) Total Energy Generated in the Mantle; (3) the Earth's Thermal Resource; (4) Total Heat Losses of the Earth; and (5) Total Heat Losses of the Mantle. The Difference between the Curves (1) and (2) and also the Difference between the Curves (4) and (5) Determine the Radiogenic Energy Generated in the Continental Earth's Crust.

4.3×10^{20} erg/s). Consequently, one can estimate the inner heat flux to be equal to 3.39×10^{20} erg/s. Taking into consideration the estimated magnitudes of gravitational, radiogenic, and tidal energy generated in the mantle by the present time ($16.85 \times 10^{37} + 3.11 \times 10^{37} + 2.24 \times 10^{37} = 22.2 \times 10^{37}$ erg) and the total inner heat flux (3.39×10^{20} erg/s), one can estimate that correction for the present-day rate of change of the Earth's heat resource is equal to -0.27×10^{20} erg/s. This indicates that after the Archaean time overheating of the upper mantle, the Earth continues to cool slightly. Summarizing the above findings, one can determine the total present-day energy generation in the bowels of Earth, which reaches $E_0' = 3.12 \times 10^{20}$ erg/s.

1.6. Tectonic Activity of Earth

The Earth's tectonic activity is determined by the transfer of matter and heat in the Earth's body. This activity manifests itself in (1) magmatic intrusions into the Earth's crust (e.g. in oceanic rift zones), (2) the crust deformations (e.g. in mountain belts), (3) the secondary melting of crustal rocks (e.g. in zones of plates movement or during formation of granite plutons), (4) earthquakes, and (5) many other ways of the Earth's matter transformation. All these relocations of the Earth's masses result in the transformation of kinetic energy of the moving masses into heat, which in time dissipates into space. Thus, one can use the heat flux through the Earth's surface as a natural measure of the Earth's tectonic activity (Fig. 1.20). Inasmuch as the transfer of Earth's matter and its partial melting began only in the Archaean time, we need to analyze the evolution of tectonic activity during the past 4 billion years. The Pre-Archaean time (4.6–4.0 billion years ago) should be considered as a cryptotectonic period. Evolution of the Earth's tectonic activity in time is presented in Fig. 1.20.

The curve 1 in Fig. 1.20 characterizes an average tectonic activity for the entire Earth. During the Archaean time, the tectonic activity most probably concentrated inside a longitudinal circular belt of the Earth, with the width gradually increasing with time. At the end of the Archaean time, the tectonic activity expanded throughout the entire Earth. In order to reconstruct a "local" tectonomagmatic activity that characterizes the intensity of tectonic activity in the Archaean longitudinal circular belt of the Earth, which experienced the gravitational matter differentiation, it is necessary to correct total heat losses of the Earth taking into account the width of this belt. On assuming that the width of zonal differentiation belt of the Earth was proportional to the energy generated in it, and that by the end of the Archaean time this belt engulfed the entire Earth, one can calculate such a correction.

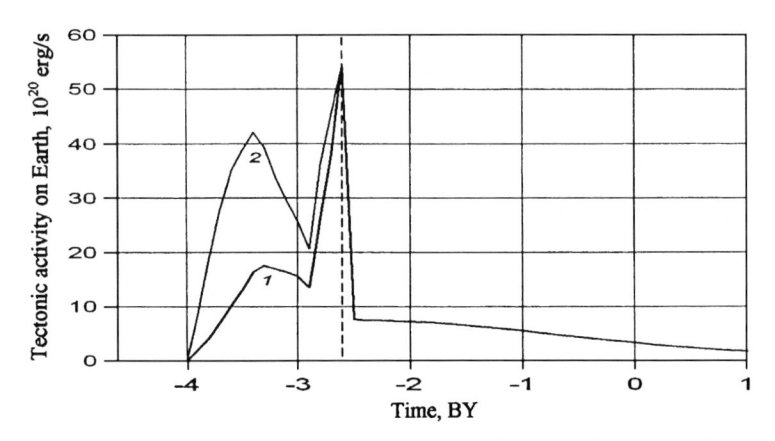

Figure 1.20: Averaged Tectonic Activity of the Earth with Time: (1) the Mantle Heat Flux through the Entire Earth's Surface; (2) the Specific Mantle Heat Flux across Low-Latitude Circular Belt of Tectono-Magmatic Activity during the Archaean Eon (Q'_m).

Thus, one can calculate the "local" tectonic activity of the Earth during the Archaean time (curve 2, Fig. 1.20).

As shown in the figure, the tectonic activity of Earth in the latitudinal belt during the early Archaean time (between 3.7 and 3.5 billion years ago) was quite intensive and exceeded 8–11 times the present-day tectonic activity. In the middle of Archaean time, the tectonic activity decreased considerably and exceeded "only" by about six times the present-day level. During the late Archaean time, however, one could observe a new and the most intensive peak of tectonic activity, amplitude of which exceeded by almost 16 times the present-day level. This intense peak can be explained by initiation of the process of Earth's core formation (see Fig. 1.7).

In Proterozoic and Phanerozoic time, the generation of gravitational energy occurred owing to barodiffusion differentiation of the mantle matter, with considerably lower rate of change in time. That is why the Earth's tectonic activity decreased at that time. However, one should remember that the curves in Fig. 1.20 represent the average values, with the averaging time of about one hundred million years. Real change of tectonic activity in time can be more complex, although the amplitude of superposed oscillations of the tectonic cycles on the averaged curve is probably not significant. Because of the gradual exhaustion of core matter in the mantle, the tectonic activity of Earth during the Proterozoic and Phanerozoic time also gradually decreased. One should expect the same tendency in the future.

The intensity of the mantle convective mixing changed in time similarly to the Earth's tectonic activity (Fig. 1.20, curve 2). In the Archaean time, one can detect two periods of increased convective and tectonic activities. The first one, associated with the zonal differentiation of metallic iron, started at the beginning of Archaean Eon. At that time, the convective mass transfer was driven by the heat exchange and comprised only the upper mantle and its transitional layer (depths of 400–800 km) in relatively narrow equatorial circular belt of the Earth. The first peak of convective and tectonic activity occurred in the Early Archaean time not only because of the high rate of generation of gravitational energy, but also (mostly) because this energy was then dissipated in small volumes of convective mantle. Therefore, in the early Archaean time, the existing convective cells should have been shallow with sizes not exceeding several hundreds or two–three thousands of kilometers. Thus, at that time no less than 80 convective cells existed in the equatorial belt. If one takes into account that the first nuclei of the future continental shields formed over the areas of downward flows in such cells, then one can conclude that no less than 40 (80/2) such continental nuclei formed in the early Archaean time. It is noteworthy that about the same number (33) of primary most ancient nuclei (composed of gneiss, trondemite, and tonalite) of the continental crust is also identified on the basis of geologic data (Glukhovskiy, 1990). When the frontal surface of differentiation moved to the center of Earth, the size of convective cells increased, the separate cells merged with each other, and the number of cells decreased. By the end of early Archaean time, the number of such continental nuclei probably did not exceed 20 (Fig. 1.21).

The second period of pronounced increase in the convective and tectonic activity of the Earth during the late Archaean time was caused by involvement of iron oxides in addition to iron (with formation of the eutectic alloys of Fe · FeO) in the process of zonal differentiation of the Earth's matter. By

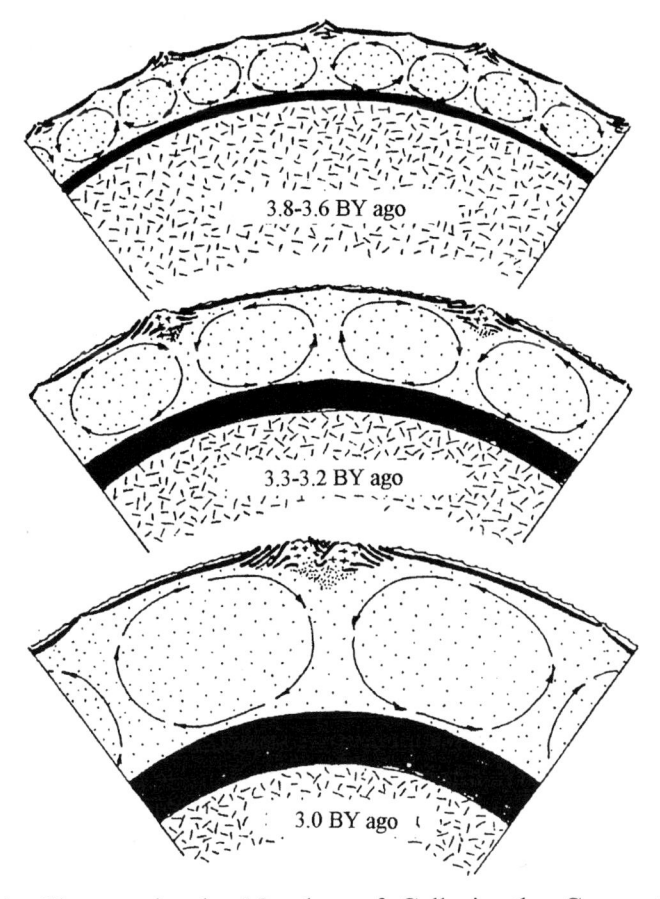

Figure 1.21: Changes in the Number of Cells in the Convective Mantle during the Archaean time and the Formation of Nuclei of the Archaean Continental Shields (Cross-Sections Pertain to the Equatorial Zone and are not Drawn to Scale).

that time, the zonal differentiation belt widened considerably and, as a result, the mass of convective mantle was increased. The number of nuclei of the future Archaean continental shields was probably reduced to 12–14. After this transition, the process of gravitational differentiation of the Earth's matter accelerated significantly. This process was intensified further, especially after the beginning of Earth's core formation in the late Archaean time, at about 2.8 billion years ago (see Fig. 1.8d and e). At the same time, drastic transformation of the convective flows in the mantle occurred and a powerful mono-cell convective structure started to form in the Earth. The mono-cell convection led to the collision of all continental shields formed earlier and their merger into a single unified supercontinent (Fig. 1.8e). Most probably, this event occurred 2.6 billion years ago (Kenoran Orogeny). According to the laws of mechanics on stable rotation of free bodies, this supercontinent must have been positioned above the location of downward flow of mantle matter at the Earth's equator.

After reduction of the tectonic activity of Earth in the Proterozoic Eon, the mantle convection became more ordered. This kind of convection was investigated by A.S. Monin and his associates (Monin et al., 1987a, b) in the Moscow Institute of Oceanology. In the mid-1990s, Yu. O. Sorokhtin carried out a numerical simulation of this convection (Fig. 1.22). As shown in Fig. 1.22, the chemical convection with continuously changing structure is not stationary. During the Earth's evolution, this structure changed from mono-cell convection to two-cell (sometimes even more complex) convection structures, every time returning to the mono-cell structure. In the numerical experiment, the periodicity of entire convective megacycles was about 1 billion years, whereas for the planet Earth this periodicity was close to 0.8 billion years. This periodicity is marked by the times of formation of ancient supercontinents: Monogaea, Megagaea, Mezogaea, and Pangaea (Fig. 1.23), which were also reconstructed based on geologic data (Sorokhtin and Ushakov, 2002).

Numerical modeling demonstrated that the chemical–density mantle convection is not stationary with alternating mono-cell and poly-cell structures. This provides a good explanation for the changing tectonic regimes of the Earth's evolution and their periodicity. The alternating mono- and poly-cell convective structures also provide a persuasive explanation for the temporary mergers of separate continents into a single supercontinent, such as Megagaea by Shtille (about 1.8 billion years ago) or the Pangaea by Wegener (about 300–230 million years ago). The merger of the continents into the Wegener's Pangaea and its subsequent splitting with continuous centrifugal drift of the existing continents is studied thoroughly based on paleomagnetic and geologic data. Drifting of continents continues to the present day.

t = -2.1 BY   t = -1.3 BY   t = -0.9 BY

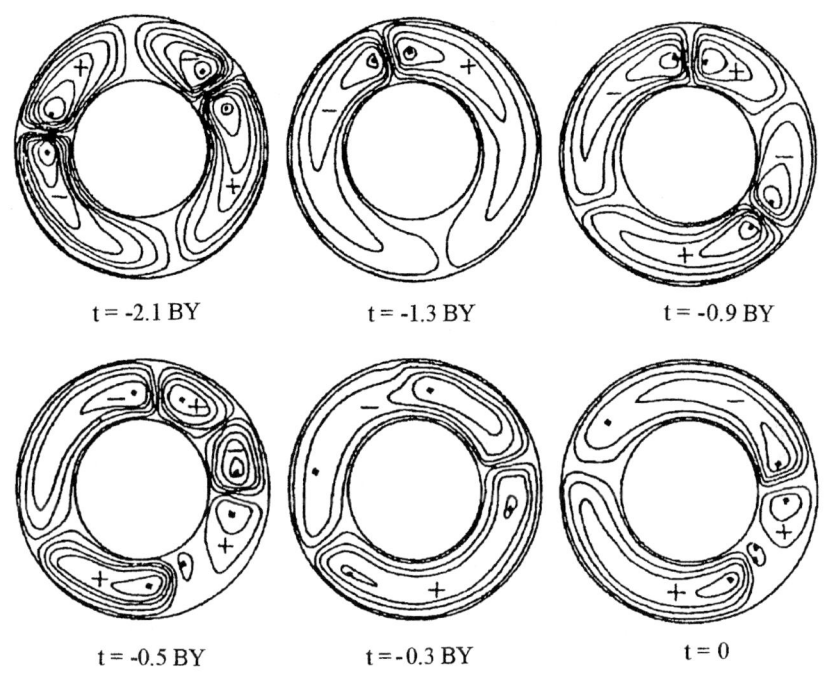

t = -0.5 BY   t = -0.3 BY   t = 0

Figure 1.22: The Results of Mathematical Modeling of the Flux Functions in Cylindrical Coordinates for the Chemical–Density Convection in Mantle (after Yu. O. Sorokhtin).

1.7. Evolution of the Chemical Composition of the Mantle

Identification of the chemical composition of the mantle is very important for understanding the origins of ores, abiogenic methane gas, increasing oxygen content in the atmosphere, and the Earth's climates in general. In studying the chemical evolution of the mantle, first of all one should consider the chemical composition of the convective mantle. After the completion of Earth's core formation, in the Proterozoic and Phanerozoic times, the concept of "convective mantle" becomes equivalent to the concept of "Earth's mantle". The situation was different in the Archaean time. At that time, the convective mantle consisted only of the parts of Earth's shell that underwent zonal differentiation of the Earth's matter and were involved in convective flows. In the early Archaean time, the convective mantle was relatively thin and probably existed as a ringlike geosphere positioned under the equatorial belt of the Earth (Fig. 1.8b). By the end of Archaean time, the convective mantle widened to the size of spherical shell of the Earth and its mass became

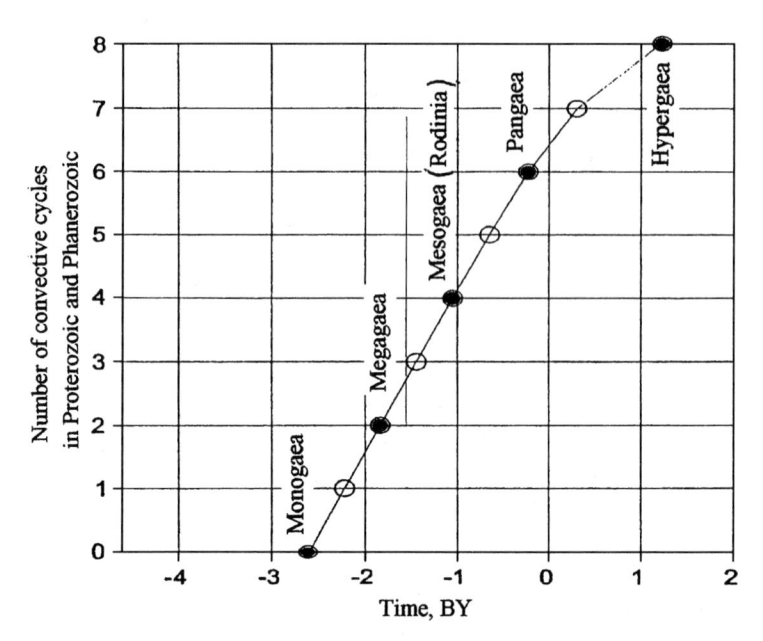

Figure 1.23: Number of Convective (Tectonic) Megacycles in Phanerozoic Eon versus the Time. Solid Circles Indicate the Mono-Cell Structures and Times of Supercontinent Formation; Open Circles Indicate Double-Cell Convective Structures and the Times of Most Pronounced Disintegration of Supercontinents.

maximal. Further, after the Earth's core separation, the mantle mass began decreasing because of the growth of the core (Fig. 1.24).

In the Archaean time, because of progressing zonal matter differentiation, the convective mantle was considerably impoverished in iron, and siderophilic and chalcophilic elements, which were transferred at that time into the molten zone of differentiation (Fig. 1.8). Extraction of these elements from primordeal Earth matter and their transfer into the zone of separation of heavy fractions and transfer of mobile and lithophilic elements into the Earth's crust should have caused considerable changes in the chemical composition of convective mantle. In this process, extraction of the heavy fractions of core matter had to lead to the relative increase in concentrations of remaining elements and compounds (Fig. 1.25).

In the Proterozoic and Phanerozoic times, after initiation of the barodiffusion mechanism of the Earth's matter differentiation, the remaining concentrations of the most abundant elements and compounds of low mobility in the mantle was gradually rising (curve 1, Fig. 1.25) due to continuous transfer of iron and iron oxides into the Earth's core. On the other hand, the total

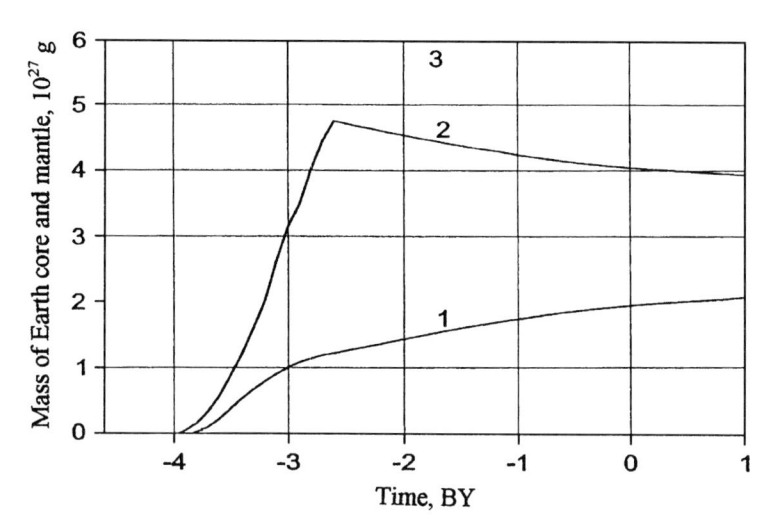

Figure 1.24: Evolution of the Earth's Principal Geospheres. (1) Mass of the Earth's Core (in the Archaean Eon it was the Mass of "Core" Matter Separated from the Mantle); (2) the Mass of Convective Mantle; and (3) the Earth's Mass (5.977 × 10^{27} g).

concentration of siderophilic and chalcophilic elements in the Proterozoic and Phanerozoic mantle was gradually reduced (curve 4, Fig. 1.25). Similar tendency should be assumed for the concentrations of nickel, gold, platinoids, iron sulfides, copper and some other elements, which were also gradually transferred into the Earth's core.

Combining changes with time of the relative concentrations of various elements and compounds (Fig. 1.25) with their initial concentrations in the juvenile Earth (Table 1.1), one can compute the changes in concentrations of these elements and compounds in percents (Fig. 1.26). These changes in concentrations describe the evolution of chemical composition quantitatively.

After the beginning of zonal differentiation of the Earth's matter about 4 billion years ago (Archaean Eon), the concentration of metallic iron in the convective mantle (over the zones of iron separation) had to be reduced rapidly to the equilibrium value of iron in the silicate alloys of overheated upper mantle. Because of low miscibility of such alloys and considerable difference in their densities, one can assume that during a larger portion of the Archaean time the concentration of metallic iron in the convective mantle was rather low. It is noteworthy, however, that the metallic iron is sometimes present in ancient basalts (e.g. it was discovered on the Disco Island in Western Greenland).

At the same time, the total concentration of bivalent iron oxide in the mantle of early Archaean age could have reached 20–23%. Because of the

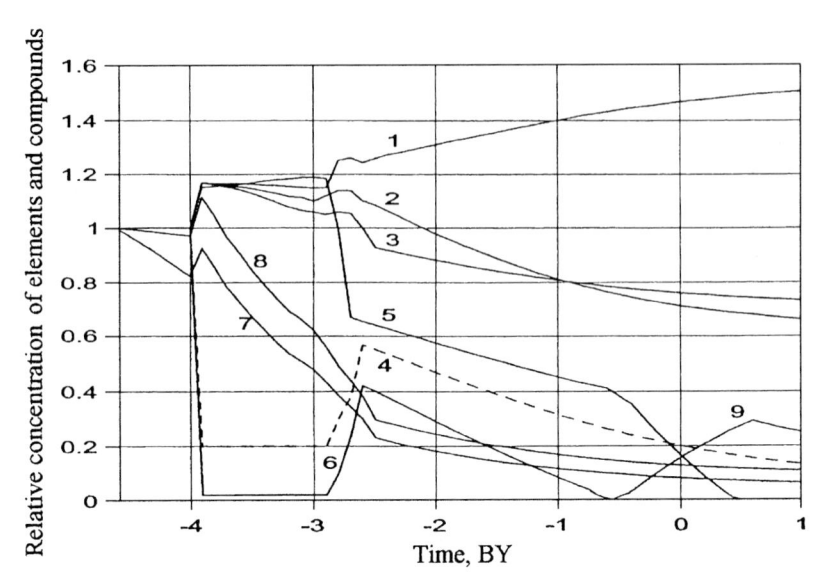

Figure 1.25: Evolution of Chemical Composition of Convective Mantle in Relative Concentrations (the Concentration of an Element in the Primordeal Matter of the Earth is taken as One). (1) SiO_2, TiO_2 MgO, CaO, Al_2O_3, Na_2O; (2) H_2O; (3) K_2O; (4) Ni and the Rest of Siderophilic and Chalcophilic Elements and Compounds: FeS, (Fe, Ni)S, $CuFeS_2$, Co, Cr, Pt, Pd, Os, Ir, Au; (5) FeO; (6) Fe; (7) U; (8) Th; and (9) Fe_3O_4.

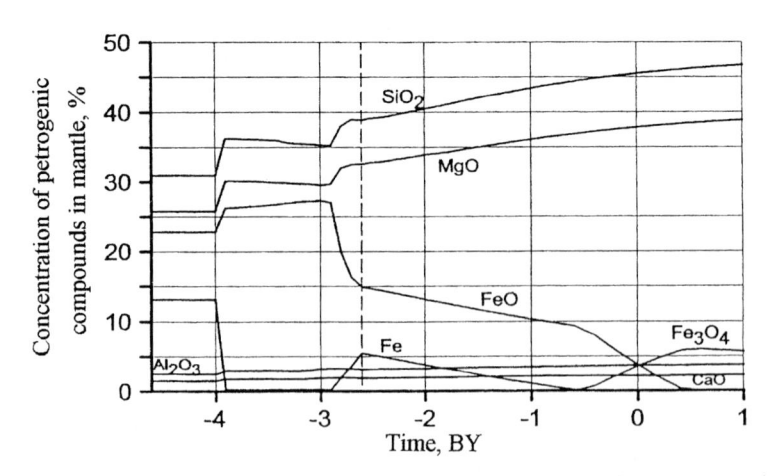

Figure 1.26: Evolution of the Principal Petrogenic Elements and Compounds in the Convective Mantle.

high density of fayalite (member of olivine group; density = $4.3\,\text{g/cm}^3$), which is the major carrier of FeO, and its low melting temperature (1,100–1,200 °C), it is probable that under overheating (over 1,600 °C) of the upper mantle in the Early Archaean time (Fig. 1.7), iron oxides concentrated mostly in the lower convective mantle. In spite of this fact, in the comatites and basalts of the Early Archaean age, the total concentration of the iron oxides usually exceeds 10%, and in some cases reaches 15–25% (and even 27%).

In the late Archaean Eon (about 2.8 billion years ago), the average concentration of metallic iron in the convective mantle began increasing and by the end of that eon reached 5.5%. As mentioned above, this was caused by the entry of a large amount of primordeal Earth matter into the convective mantle, which floated to the upper layers of the mantle from the former "core" of the Earth (Fig. 1.8e). In the Proterozoic time, however, the concentration of the metallic iron in the mantle started to decrease again. Changes in the concentrations of iron and its oxides during the Post-Archaean time can be estimated on the basis of main chemical reactions of the "nucleolus" matter formation. In the Proterozoic time as well as the late Archaean Eon (time interval between 3.2 and 2.6 billion years ago), this matter formed according to the following reaction:

$$\text{FeO} + \text{Fe} \rightarrow \text{Fe} \cdot \text{FeO} \quad (1.18)$$

Quantitative estimates of the changing concentration of metallic iron in the mantle allow one to arrive at an important conclusion: *the metallic iron could have stayed in the mantle no later than 600–500 million years ago*. These estimates were obtained based on assumed composition of the Earth's core and the concentrations of bivalent and trivalent iron oxides (in magnetite) in the contemporary mantle: approximately 4.4% and 4.1%, respectively. The estimated time of disappearance of metallic iron from the mantle is close to the age of main biological evolution milestone, i.e. development of highly organized life on Earth at the boundary between the Proterozoic and Phanerozoic eons (about 545 million years ago). This is a remarkable coincidence. The proximity of the times of occurrence of these two important events corroborates our theory of the Earth's global evolution.

One should note that the elimination of metallic iron from the mantle and spreading zones of the Earth (where iron came in contact with the oceanic water) determined a radical change from reducing to oxidizing conditions in the outer layers of Earth. The latter occurred because the metallic iron (the main chemical reagent) consumed oxygen from the hydrosphere (and atmosphere) during the entire Precambrian time of the Earth's development. Only after almost complete disappearance of the metallic iron from the convective mantle, oxygen started to accumulate in the Earth's atmosphere.

It was produced by plants in quantities sufficient for the creation and normal functioning of the animal species on the Earth's surface (Sorokhtin, 1974).

After disappearance of metallic iron from the mantle, the "nucleus" matter in the lower mantle formed according to a different reaction:

$$2FeO \rightarrow Fe \cdot FeO + O \qquad (1.19)$$

As a result of this reaction, the produced oxygen exceeded the level of existence of stable solutions of iron oxides with silicates (Fig. 1.10). Under high pressure and supply of compaction energy (because of the lower volume of magnetite molecule), oxygen started to react with the iron oxides again, forming the magnetite component of the mantle:

$$3FeO + O + 76.48 \text{ kcal/mole} \rightarrow Fe_3O_4 \qquad (1.20)$$

Consequently, the magnetite concentration in the mantle matter started to increase to about 4% by the present time (Fig. 1.26). Combining Eqs. (1.19) and (1.20), one can deduce that the complete oxidation of bivalent iron in the mantle forming the "nucleus" matter had to occur according to the following reaction: $5FeO \rightarrow Fe_2O + Fe_3O_4$.

Another important biological milestone will be reached when the entire mantle content of bivalent iron oxidizes to magnetite, because under high pressure the trivalent oxide is not stable and transforms into denser version (magnetite): $Fe_2O_3 + 56.2 \text{ kcal/mole} \rightarrow Fe_3O_4 + O$. Such critical situation could occur in the future in about 600 million years. Consequently, at the third stage of functioning of barodiffusion mechanism the "nucleus" matter should form according to the magnetite decomposition reaction generating a large amount of oxygen:

$$2Fe_3O_4 \rightarrow 3Fe \cdot FeO + 5O \qquad (1.21)$$

After complete oxidation of the entire mantle iron to magnetite, oxygen will start entering the hydrosphere and atmosphere. With time, accumulation of a huge volume of oxygen will drastically change the environmental conditions on the Earth. Partial oxygen pressure will be rapidly increasing and the limited solubility of oxygen in oceanic water will only slightly delay this process. Because of continuously increasing influx of oxygen into the atmosphere, the total atmospheric pressure will greatly increase amplifying the greenhouse effect enormously. The latter will create irreversible conditions of the Venus-type hot climate on the Earth, which will result in extinction of organized life species.

The considered mechanism of the core formation allows one to make one more important conclusion. Judging from Eq. (1.19), beginning with the Phanerozoic Eon the oxidizing potential of the mantle matter had been continuously increasing, as indicated by the absence of hydrocarbons because of the following reaction:

$$CH_4 + 4O \rightarrow CO_2 + 2H_2O + 76.17 \text{ kcal/mole} \tag{1.22}$$

More complex hydrocarbons, which are unstable at high pressures, decompose into less complex hydrocarbons right down to methane. This implies that, in the Phanerozoic time, the mantle could not contain any hydrocarbons and, therefore, the hydrocarbons could not enter into sediment layers of the continental crust from the mantle. Thus, there is no degassing of inorganic hydrocarbons along the deep faults reaching the upper layers of mantle at the present time. The generation of abiogenic methane as a result of hydration of the iron-containing silicates of the oceanic crust, however, is a realistic process generating approximately 10×10^6 ton/year (see Chapter 5).

The above discussion leads one to conclude that iron is the main element, which determines not only the structure and global evolution of the Earth but also its general geologic evolution.

1.8. Formation of Oceanic and Continental Lithospheric Plates

The estimates of the mantle heat flux Q' (Fig. 1.20, curve 2) and the area of oceanic crust (S_{oc}) (Fig. 1.36) allow one to compute several other important characteristics of the Earth's tectonic activity. One of these characteristics is the average life span of the oceanic plate τ, which is proportional to the ratio of the square of area of plate to the square of mantle flux:

$$\tau \sim \left(\frac{S_{oc}}{Q'}\right)^2 \tag{1.23}$$

Another important characteristic is the average thickness H_L of the oceanic plates upon reaching the age limit τ:

$$H_L = k\sqrt{\tau} \tag{1.24}$$

where τ is expressed in million of years, H_L is expressed in km, and k is a constant ($k = 7.3$).

Next, let us assume that the present average life span of oceanic plates is about 120 million years, and their thickness is about 80 km. Thus, one can estimate that the thickness of such plates at the time of maximum tectonic activity (in the early Archaean time, about 3.6 million years ago) was about 6.2 km and the life span was about 700 thousand years! In the middle of Archaean time (about 3.3 billion years ago), the average thickness of the oceanic plates increased to about 27 km, whereas the life span increased to almost 14 million years. In the late Archaean Eon (about 2.9 billion years ago), the value of H_L was reduced again to approximately 8 km, whereas the life span of oceanic plates was reduced to about 1.2 million years.

Equation (1.25) can be also used to estimate the averaged relief of the oceanic trenches. The depth Δh_{oc} of such trenches with respect to the level of elevation of the top of mid-oceanic ridges is equal to:

$$\Delta h_{oc} = k_1 \sqrt{\tau} \qquad (1.25)$$

The best fit of the theoretical function (1.25) to the experimental data for a time interval 0–80 million years occurs for the value of $k_1 = 0.35$ (for more ancient regions of the oceanic floor, the relief flattens). In Fig. 1.27, the critical thickness of the oceanic lithospheric plates is also presented, which was computed based on their density. As the thickness of a realistic oceanic plate is greater, it becomes heavier than the underlying hot mantle and can submerge into the mantle. If the thickness of plate is below critical value,

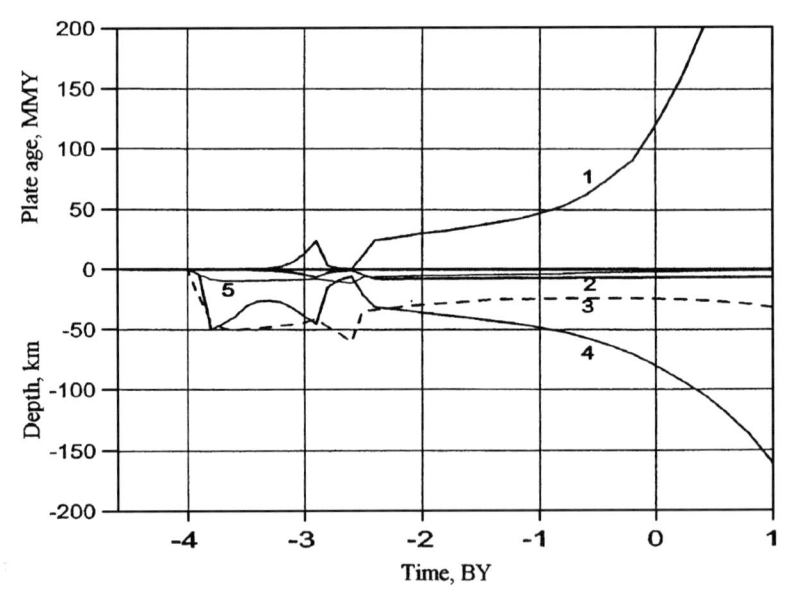

Figure 1.27: Evolution of the Structure of Oceanic Lithospheric Plates and Average Time of Their Existence on the Earth's Surface. (1) Average Life Span of the Plates; (2) Thickness of the Oceanic Crust; (3) the Critical Thickness of Lithospheric Plates that Determines the Possibility Submergence into the Mantle of Thicker Plates; (4) Thickness of the Oceanic Plates at the End of Average Life Span; (5) Thickness of the Basaltic Layer (in the Archaean Eon, it was the Thickness of Basalts and Gabbro).

then the plate cannot submerge into the mantle. Upon collision with another plate, the plate can slide over the other plate or over the edge of a continent. These theoretical considerations lead one to several important geologic conclusions. First, the subduction zones could not have existed in the Archaean Eon. The compression of oceanic plates was compensated by their merger and by sliding over the edge of continental plates (Fig. 1.31). Second, the subduction zones could appear only in the Proterozoic time, i.e. about 2.5 billion years ago. Therefore, the Earth's tectonic evolution in the Archaean time progressed according to the laws of thin lithospheric plates, and the diamond-bearing rocks could not form under the Archaean continental shields. Tectonic activity of lithospheric plates started in the early Proterozoic time (at about 2.5 billion years ago). Only later (about 2 billion years ago), the diamond-bearing melts could form under the Archaean continental shields.

Having calculated the Earth's tectonic activity, which is determined by the mantle heat flux, one can also evaluate the average thickness of lithospheric plates under the ancient continents. Our calculations (Fig. 1.28) indicate that the average height of continental shields over the sea level was close to 6.5 km during almost the entire Archaean time. This circumstance determined the high level of erosion for practically all of the Archaean shields.

These calculations imply an important geological conclusion: a great amount of sediments should have accumulated throughout the Archaean Eon. Very few sediments, however, are preserved. Drilling results in all oceans showed that there are no deposits older than 160–190 million years (Sorokhtin and Ushakov, 2002). Where did the Archaean sediments go?

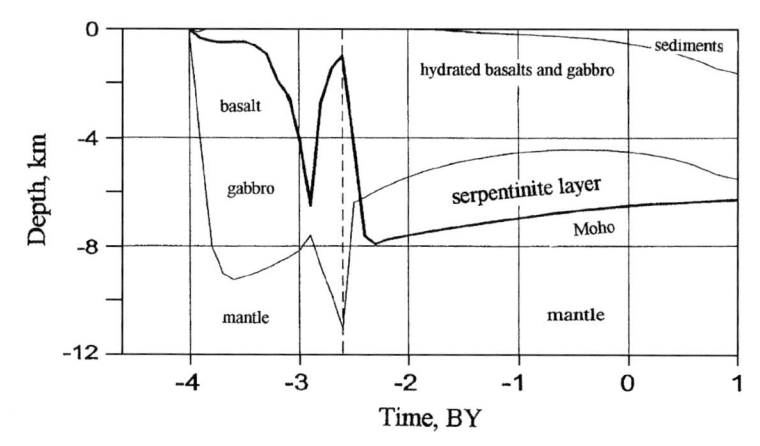

Figure 1.28: Generalized Model of the Evolution of Oceanic Crust Composition.

The main mass of sediments should accumulate on the oceanic plates at the foot of continents (Fig. 1.29). In the Archaean time, however, the speed of movement of the oceanic plates was extremely high, and the oceanic plates collided and intensively subducted under the continental shields. In the process of this movement, sediments, which accumulated at the edge of oceanic plates, were sucked in under the island arcs and active continental margins within the subduction zones. There, the sediments were melted and partially joined the continental crust as granitoid intrusions, whereas the rest were transferred into the convective mantle under the continental plate. This process is called "sediment recycling".

In order to understand the conditions of mass transfer between the lithosphere and mantle, one should consider the physical conditions of existence of oceanic and continental lithospheric plates. A generalized graph of the solidification temperature of the mantle matter is presented in Fig. 1.30 (based on the data of Takahachi (1986) and Ringwood (1981)), together with geotherms of oceanic plates of different age and the geotherm of ancient (Archaean) continental plate calculated by the authors. The temperature distribution in the hot mantle was computed assuming that the temperature of the mantle reduced to the surface was equal to 1,320 °C.

The graphs in Fig. 1.30 indicate that the curve of adiabatic temperature of contemporary mantle intersects the curve of temperature of solidification of

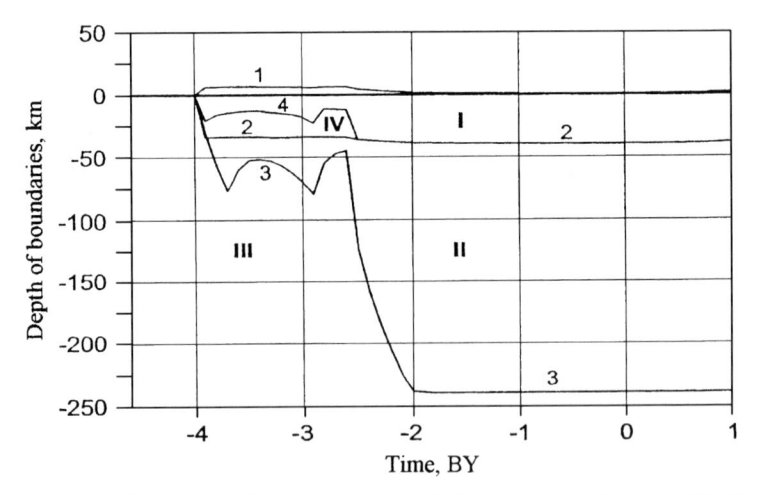

Figure 1.29: Evolution of the Structure of Continental Plates. (I) Continental Crust; (II) Continental Lithosphere; (III) Hot Mantle Underlying the Lithosphere; (IV) Crustal Asthenosphere (Lower Crust); (1) Surface of Continents; (2) Base of Continental Crust (Moho Boundary); (3) Base of Continental Lithosphere; (4) the Top of Crustal Asthenosphere.

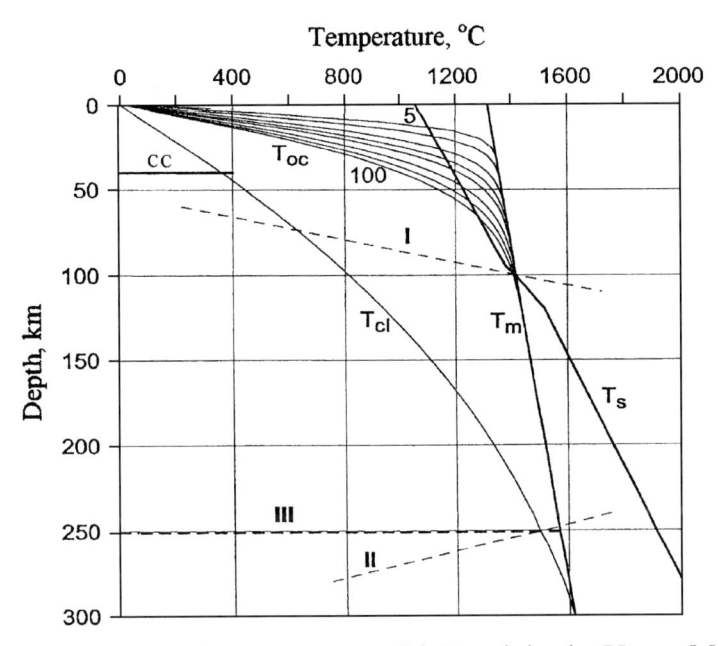

Figure 1.30: Variation of Temperature with Depth in the Upper Mantle and Position of Geothermal Gradients of Lithospheric Plates Depending on their Age; Numbers on the Geotherms show the Age of Lithospheric Plates in Million Years. T_{oc} is the Geotherm of Oceanic Plates; T_m is the Adiabatic Temperature of the Upper Mantle; T_s is the Solidification Temperature of the Mantle Matter; T_d is the Geotherm of Archaean Continental Lithosphere Plates; CC is the Base of Continental Crust. (I) Boundary of the Phase Transition of Basalts into Eclogites; (II) Endothermal Transition of the Firm Lithosphere under the Continents into the Plastic State; and (III) Base of the Archaean Regions of Continental Lithosphere.

the mantle matter at a depth of about 80–100 km. This implies a very important geological conclusion: *the mantle melts (basalt) in the juvenile mantle could not have formed deeper than 80–100 km* (*top of eclogite transition zone*). These depths determine also the position of the bottom of asthenosphere under the oceanic lithospheric plates.

Consequently, a widely used simple physical model of the firm crystalline plate lying on a plastic layer of partially molten mantle matter can be used only for the oceanic lithosphere. In many respects, such a model is analogous to the ice floating on a water layer in a pond starting to freeze, with a single exception: ice is always lighter than water, whereas the cold lithospheric plates can be heavier than the hot partially molten mantle matter. This explains the relatively short life span of the oceanic plates in comparison with a long life span of the continental plates, the average density of which is always lower than the density of convective mantle.

Physical conditions of transfer of the continental lithosphere into the underlying convective mantle are more difficult to define. This is caused by the fact that continental geothermal curve approaches asymptotically the mantle temperature in the regions of the mantle where the partial melting of the mantle matter is impossible. The mantle temperature is 200–300 °C lower than the melting temperature of the mantle matter. Therefore, one can conclude that under thick continental plates (reaching 250 km in thickness), the position of lithospheric bottom is determined not by the commencement of melting of the mantle matter, but by the pressure and temperature of transition of the matter from firm to plastic state. If this is true, then the phase transition at the base of continental plates should have the properties of endothermal boundary, as shown in Fig. 1.30. Consequently, at depths of about 250 km a marked temperature drop should occur in the convective mantle.

If the regular tectonics of the lithospheric plates did not function in the Archaean time, then what was the mechanism of continental crust formation? As a result of compressive forces caused by the convective currents of mantle matter, the thickness of piled lithospheric plates increased. Because of that, the roots of such pile-up structures had to submerge into a hot mantle to depths of 50–80 km. In the Archaean Eon, the upper mantle was considerably overheated, with temperatures exceeding the present level by 400–500 °C (Fig. 1.7). The base of asthenosphere was positioned at a depth of about 400 km. As a result, the roots of piled-up thin oceanic plates were melted again upon submerging into the overheated mantle. In this process, melting of the water-bearing basalts of the former oceanic crust (with subsequent differentiation of the molten matter) led to the formation of lighter plagio-granitic melts. These relatively light melts moved upward in the form of diapirs and domes breaking through the entire thickness of piled-up plates, forming the famous greenstone belts of the Archaean eon – the most ancient parts of the continental crust (Fig. 1.31). Theoretical explanation of the formation of Archaean continents as a result of piling-up and partial melting of relatively thin (thickness of several kilometers) oceanic crust was suggested by Sorokhtin in 1991. In 1992, this model was corroborated by the data of field investigation of the Kaapvaal Archaean Craton in South Africa (Wit et al., 1992).

High heat fluxes during the Archaean time led to the partial melting of the lower part of continental crust. Consequently, one can assume the existence of crustal asthenosphere in this layer (Fig. 1.29). Convective processes should have developed in this anatectic layer, which were accompanied by eutectic melts of granitoids and their transferal into the upper crust together with

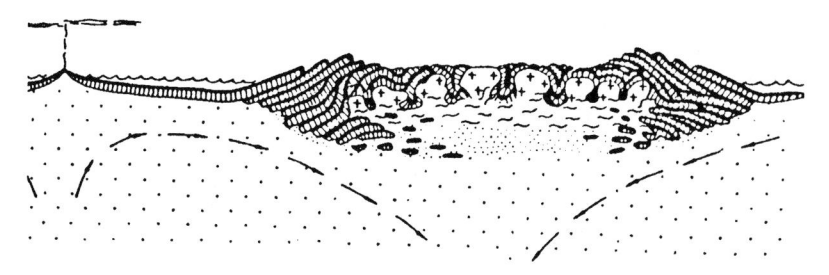

Figure 1.31: Schematic Diagram of the Formation of Continental Crust during the Archaean Eon.

volatile, alkaline, and lithophilic elements. These processes explain (1) the origin and wide occurrence of granitoid intrusions, which invaded the upper crust during the Archaean time; and (2) the origin of granulites of lower crust (Sorokhtin, 1996).

In the Archaean time, because of high heat fluxes, thick and dense lithosphere of the ultrabasic composition could not form under the continental crust. Thus, the relatively light continental crust was floating directly on the surface of hot mantle. Consequently, the top of continents was positioned 4–6 km above the sea level during the Archaean and beginning of Proterozoic time (Sorokhtin, 1997). The latter explains the high level of erosion of all Archaean shields.

Thus, the Archaean continental crust formed as a result of two basic tectonic processes: (1) formation of thin basalt plates of oceanic crust and (2) their piling-up with subsequent melting caused by the convection of mantle matter. These two processes leading to the formation of continental crust in the Archaean Eon correspond to two different stages of petrogenesis (Moralev and Glukhovskiy, 1985), based on empirical data obtained on studying the structure and composition of the Aldan Shield rocks. First stage is the formation of primary basic crust as a result of partial melting and differentiation of mantle matter. Second stage is a partial (15–20%) melting of the lower layer of the basic crust at a pressure of 708 kbar, i.e. at a depth of about 25–30 km, with the formation of silicon dioxide and alkali in quantities sufficient for the formation of first (in the Earth's history) high-temperature enderbites with low potassium content. Closely related are several other types of ancient anorthosites, which were formed during the third stage as a result of differentiation of secondary melts.

Thus, one can describe the Archaean tectonics as "*tectonics of thin basalt plates*". Unlike the tectonics of lithospheric plates, which prevailed on the Earth starting at the early Proterozoic time, the tectonomagmatic processes

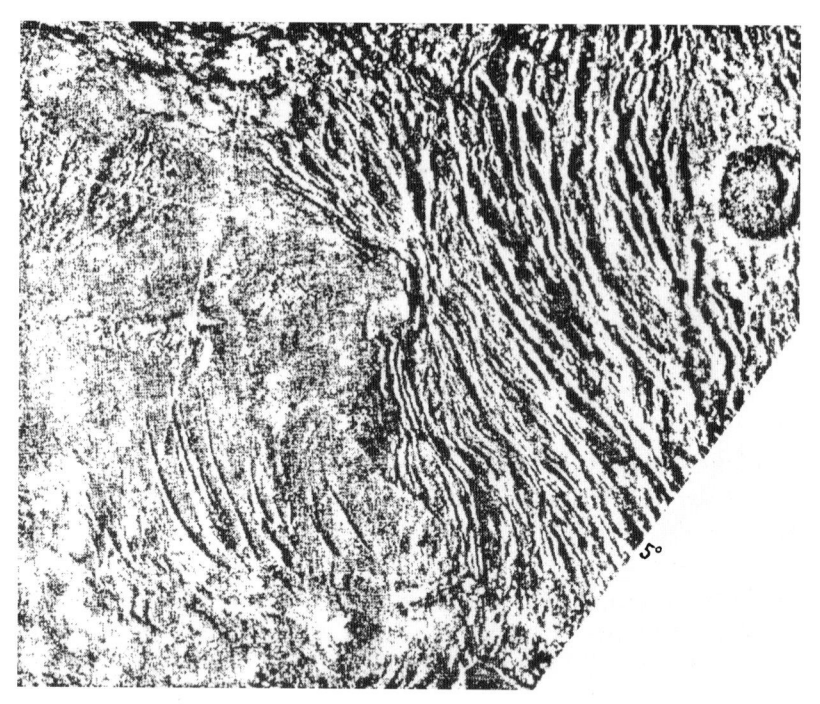

Figure 1.32: Radiolocation Image of the Venus Surface (500 × 550 km) at the Coupling of Maxwell Mountains (Analogous to Zones of Piling-Up of Thin Basaltic Plates) with Lakshmi Plateau (Analogous to Continental Mountain Range). At the Upper Right Corner of the Picture, the Image of Large Meteorite Crater Cleopatra is shown (Planet Venus, 1989).

during the Archaean Eon developed according to different mechanisms probably close to those on the present-day Venus (Fig. 1.32).

The radio-scanning images of the Venus surface indicate the presence of rift zones and structures similar to the mid-oceanic ridges with the absence of subduction zones. Instead, one can observe zones of compression and piling-up of the crustal material with typical structures of small scale (tesserae) or extended ridges as if flowing around the massive mounds. These can be considered as analogs of the Archaean massifs and shields. Region encircling the Lakshmi Plateau with the Mackswell Mountains represents a typical landscape on the Venus surface (Fig. 1.32). The boundary between these two different structures is identified with an abrupt transition from an even surface of the plateau (4–5 km above the average planet level) to the steep slope of Mackswell Mountains, reaching a height of 10–11 km and bordering the plateau from the East and North East.

At the end of Archaean Eon, after completion of the turbulent process of iron oxide core separation in the bowels of the Earth, the tectonic activity decreased considerably. Pronounced reduction of the tectonic activity in the Proterozoic time had led to considerable increase in the life span and thickness of oceanic lithospheric plates. As a result, the density of cooled plates became higher than the density of mantle and, consequently, zones of crushing and piling up of thin basalt plates were replaced by the regular subduction zones of the present type. Besides, about 2.5 billion years ago, the structure and composition of oceanic crust changed significantly. Instead of predominantly basaltic Archaean crust, at the beginning of the Proterozoic Eon, the third crustal layer (serpentine) formed, which became the main reservoir of bound water in the oceanic crust. All these changes led to establishment (starting at the Proterozoic Eon) of a regular tectonic regime, which is described by the "tectonic theory of lithospheric plates".

Because of drastic changes in the tectonic regime of transformation of oceanic crust into continental crust, the process of formation of continental crust in the post-Archaean time underwent radical changes. Local crustal magmatism (mainly of tonelite-trondjemite composition) originating in the zones of piling-up of thin oceanic plates was replaced by the carbonate–alkaline magmatism of linear subduction zones. The processes of granitoid formation had also changed. In the post-Archaean time, granitoids formed mainly (1) as a result of repeated melting of sand–clay sediments (i.e. by being sucked into the plate subduction zones) or (2) due to metamorphic transformation of the sediment layers by overheated fluids rising from the subduction zones as shown in Fig. 1.33.

In the post-Archaean time, the smelting temperature of original crustal rocks was reduced considerably. Indeed, heating of the rocks of oceanic crust in the subduction zones occurs mostly not as a result of inner heat flux but due to internal friction resulting from the movement and deformation of tectonic plates. The process of heat generation, however, is controlled by the temperature of melting of silicates of the crust upon its initiation. Partial melting of silicates significantly lowers the viscosity of rocks involved in the subduction process and reduces the further heat generation. As a result, this self-controlling process automatically stabilizes the temperature of carbonate–alkaline magma generation at a level slightly exceeding the temperature of anatectic state of water-saturated basalts of oceanic crust. This explains the absence of eruptions of overheated magma in the subduction zones: its usual temperature is 1,100–1,200 °C (for the granitoid magma temperature is even lower: 800–900 °C) and never exceeds 1,250–1,300 °C. In the Archaean time, smelting of crustal magma could occur under very high overheating, up to 1,500–1,600 °C.

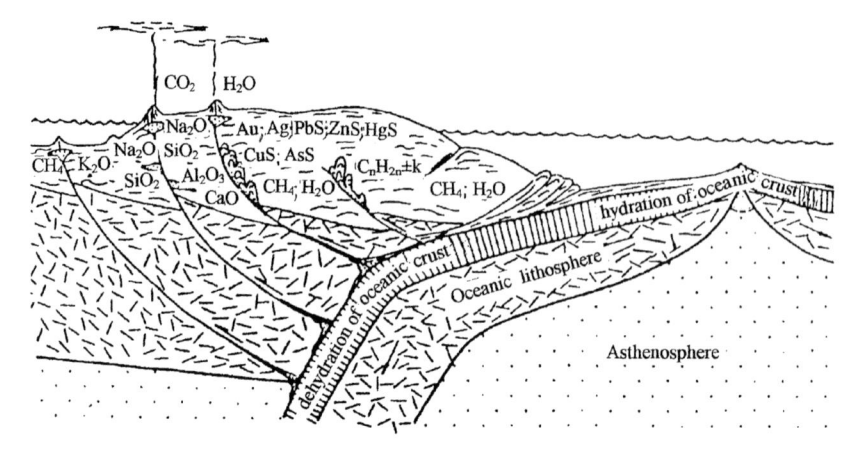

Figure 1.33: Formation of the Continental Crust in the Proterozoic and Phanerozoic Times as a Result of Partial Remelting and Dehydration of Oceanic Crust (with Overlying Pelagic Sediments) in the Subduction Zones of Oceanic Plates under the Island Arcs.

The process of formation of continental crust in the post-Archaean time involved smelting of the crustal magma with metamorphic processes progressing in the presence of excess of water. The latter originated at the subduction zones as a result of dehydration of the serpentine of oceanic crust. Our estimates show that during the Proterozoic and Phanerozoic times about 2.3×10^{25} g of water moved through the subduction zones. This amount of water is about 16 times greater than the water mass in the present World Ocean! The abundance of water is very important in facilitating the mass transfer between geospheres, because it facilitates mineralization and is an active chemical reagent and mineralizer transferring all lithophilic and hydrophilic elements into the continental crust.

Altered conditions of the crust, which formed during the post-Archaean time, are often shown by the geochemical ratios of rocks of the same type of different age. Weiser (1980) pointed out K_2O/Na_2O ratio changes in the crustal eruptions of different age (Fig. 1.34). All the erupted rocks of early Archaean time are characterized by a typical $K_2O/Na_2O \approx 0.5$ ratio of basalts. In the late Archaean Eon, this ratio was gradually rising, because of the beginning of smelting of potassium granitoids. The fastest accumulation of potassium in the crustal rocks occurred in the early Proterozoic time when newly-formed serpentine layer of the oceanic crust began to be saturated with water. By the end of Early Proterozoic time, however, because of reaching the water saturation limit, the K_2O/Na_2O ratio in the oceanic crust reached equilibrium values of 1.2–1.5.

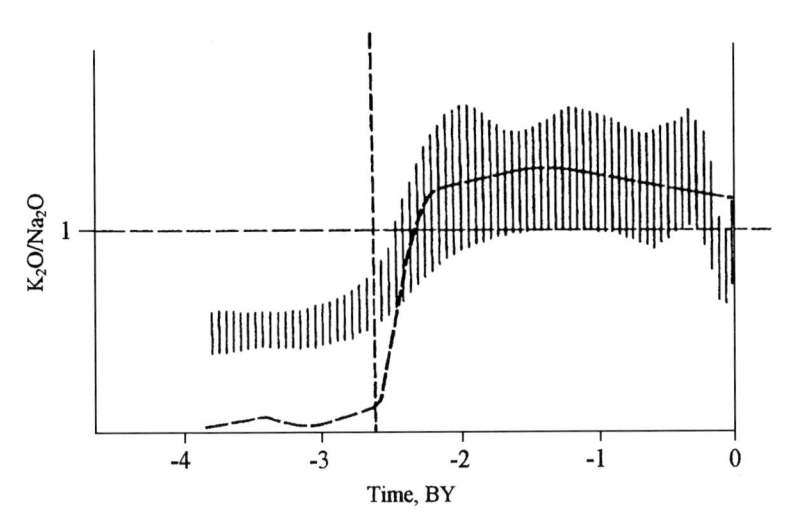

Introduction to the Global Evolution of the Earth   71

Figure 1.34: Evolution of K$_2$O/Na$_2$O Ratio in the Rocks of Continental Crust (Modified after Weiser, 1980) Compared to the Evolution of Concentration of Bound Water in the Oceanic Crust (Dashed Curve).

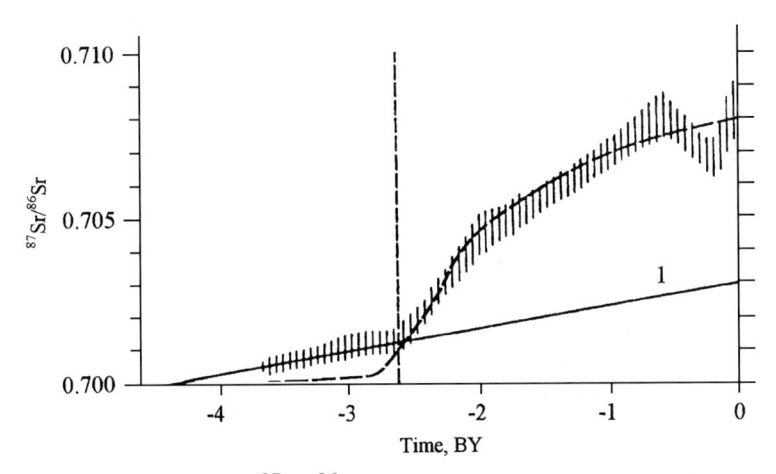

Figure 1.35: Evolution of ^{87}Sr/^{86}Sr Ratio in the Oceanic Sediments (Modified after Weiser, 1980) Compared to the Trend of this Ratio in the Mantle Rocks (Line 1) and the Concentration of Bound Water in the Continental Crust (Dashed Curve).

The ratio of strontium isotopes (^{87}Sr/^{86}Sr) in the limestones of oceanic origin is also quite revealing (Fig. 1.35). This ratio is determined by the process of averaging of the isotope contents of continental rocks in oceanic water as a result of weathering and transferral by rivers to the ocean.

Consequently, the isotopic content of such sediments, which accumulated under conditions of equilibrium with oceanic water, had to reflect the isotopic content of their sources. These sources were the continental crust and oceanic basalts, which were erupting in the rift zones of mid-oceanic ridges. Figure 1.35 shows that during Early Archaean time the initial $^{87}Sr/^{86}Sr$ ratio of the crustal rocks was equal to the $^{87}Sr/^{86}Sr$ ratio of the mantle source. In the late Archaean time, because of smelting of potassium granitoids and beginning of the recycling of crustal material (which led to the prevailing accumulation of potassium and rubidium in the crust) the value of $^{87}Sr/^{86}Sr$ ratio in the crustal rocks increased slightly. However, the contribution of this process to the general composition of late-Archaean crust remained insignificant, and $^{87}Sr/^{86}Sr$ ratio of the crust was approximately equal to its initial level in the mantle. Starting at the early Proterozoic time, drastic changes in the Earth's tectonic regime occurred, which led to the formation of subduction zones. The sediments washed out from continents were sucked-in under the plates and the $^{87}Sr/^{86}Sr$ ratio increased greatly accompanied by rapid accumulation of rubidium (and, consequently, potassium) in the crust.

At the boundary between Archaean and Proterozoic times, the concentration of rare elements changed dramatically (Taylor and McLennon, 1988). For example, the ratio of the sum of concentrations of light rare elements to the sum of concentrations of the heavy elements in fine-grained sediments increased greatly, approximately from 6 to 11. The Th/Sc ratio increased from 0.4 to 1.1; the La/Sc ratio increased from 1 to almost 3; and the thorium concentration rose from 1.5×10^{-6} to 3.5×10^{-6}.

These above-presented examples show that geochemical indicators clearly mark the transition from Archaean Eon to Proterozoic Eon. In conclusion, one can state that the nature of most important geological transition at the boundary of Archaean and Proterozoic eons was determined by the completion of formation of dense Earth's core. By that time, about 63% of contemporary mass of the Earth's core was accumulated. In addition, after the initial core separation, the mechanism of zonal separation of iron and its oxides was replaced with a more moderate mechanism of barodiffusion differentiation. The direct cause of all these important evolutionary changes in geochemical and tectonic conditions of the formation of terrestrial crust was transition from the tectonics of thin basalt plates in the Archaean Eon to the tectonics of lithospheric plates with the subduction zones in the Proterozoic and Phanerozoic times. Formation of the serpentine layer of oceanic crust in the early Proterozoic time also changed considerably the water regime of smelting of the continental crust. After the Archaean Eon, the continental crust began forming in the presence of excess of overheated water solutions rising from the subduction zones. The sediments, sucked into the subduction zones, began playing an important role in smelting-out of the crustal magma (especially of granitoid and alkaline composition).

Based on the hypotheses presented here, systems of differential equations describing the crustal evolution were developed. Integration of these systems allowed us to describe quantitatively the evolution of areas of oceanic and continental crusts (Fig. 1.36). The theoretical curve of growth of mass of continental crust, which is based on our computations, is presented in Fig. 1.37. The most popular model of the continental crust growth by Taylor and McLennon (1988) (considering the geochemical data, the estimates of rates of sediment accumulation in oceans, and the estimates of rates of sediments consumption in the subduction zones) is also presented in Fig. 1.37 for comparison.

Figure 1.37 shows a good fit of the theoretical curve presenting our geodynamic model (curve 1) and the empirical model by Taylor and McLennon (curve 2). The fact that these two models, which were constructed based on different approaches (using independent assumptions and different data), are close to each other is indicative of validity of both models. Both models clearly show that *the continental crust started to form gradually just about 4 billion years ago, i.e. approximately 600 million years after the formation of Earth itself.* That time marks the beginning of *gradual separation of the Earth's core.* This important observation refutes the concepts of early differentiation of the Earth's core (about 4.6–4.5 billion years ago).

Theoretical graph of the rate of continental crust formation as a function of time is presented in Fig. 1.38. This graph was obtained by differentiation

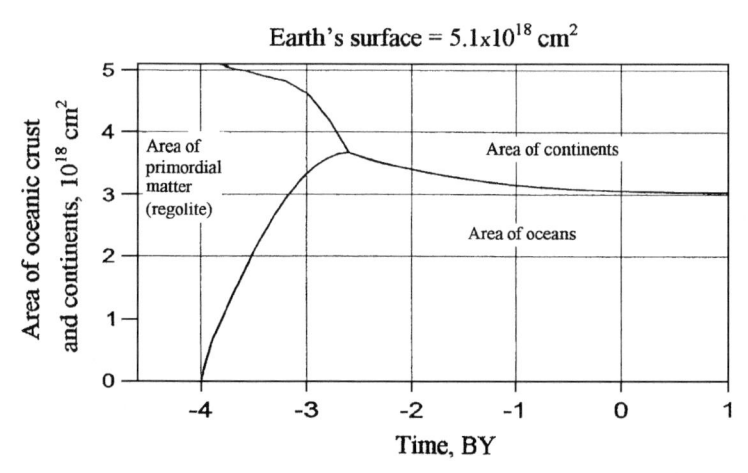

Figure 1.36: Evolution of the Area of Original Earth's Surface and Changes in the Areas of Oceanic and Continental Crusts, in $10^8 \, km^2$.

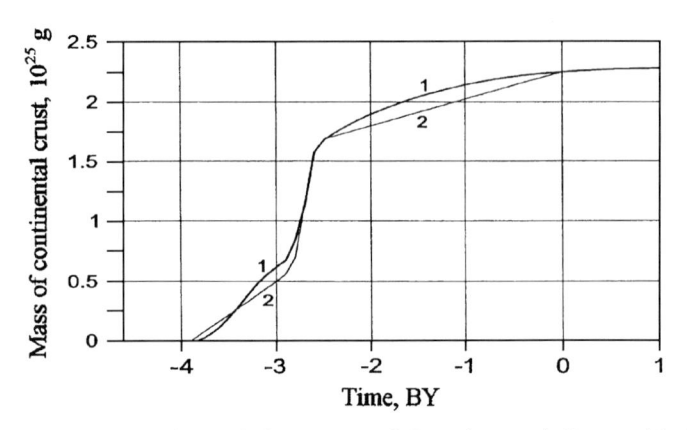

74 *Global Warming and Global Cooling: Evolution of Climate on Earth*

Figure 1.37: Accumulation of the Mass of Continental Crust: (1) According to our Energy Balance Model and (2) According to the Model of Taylor and McLennon (1988).

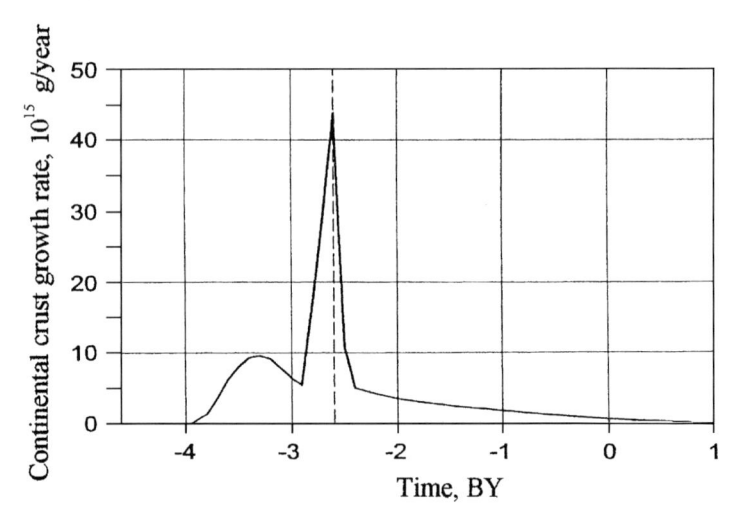

Figure 1.38: Evolution of the Rate of Growth of Continental Crust Mass with Time.

of the theoretical function of crust growth. It shows that in the early Archaean Eon, the rate of growth was comparatively high and reached almost 6×10^{15} g/year or about 2.1 km^3/year. This rate of growth gradually decreased in time because of lowering of the zonal differentiation front into the mantle. In the middle of Archaean Eon (about 3.0–2.9 billion years ago), during the period of general reduction in the Earth's tectonic activity (Fig. 1.20), the rate of the Earth crust formation also decreased. One cannot exclude that in this short time interval (which could be called the litho-plate period of the crust development in the Archaean time), the geodynamic regions typical for the tectonics of lithospheric plates (with normal subduction zones and island-arc type carbonate–alkaline magmatic process), which was close to the present-day analogues, could arise at some places. In addition, during the same period, the formation of Archaean greenstone belts should cease or slow down considerably. Instead, separate ophiolitic layers could form, which were typical for the Phanerozoic Eon.

The tectonic activity intensified considerably during the late Archaean time, about 2.8–2.7 billion years ago. This peak coincided with the beginning of core formation and the transition of differentiation of the Earth's matter from the high-temperature melting regime of metallic iron separation to the more energy-efficient mechanism of eutectic Fe · FeO smelts. Our computations indicate that the maximal rate of crust formation reached 4.5×10^{16} g/year or up to $16 \, km^3$/year (Fig. 1.38). After completion of the formation of juvenile core of the Earth (about 2.6 billion years ago), the tectonic activity of the Earth decreased considerably during the Early Proterozoic eon and, consequently, the rate of formation of continental crust also decreased to approximately 6×10^{15} g/year or $2 \, km^3$/year. In the Proterozoic and Phanerozoic eons, the rate of accumulation of continental crust mass continued to decrease. During the last 800 million years, this rate did not exceed 8×10^{14} g/year. Using analogous approach, one can also determine the rate of formation of oceanic crust (Fig. 1.39).

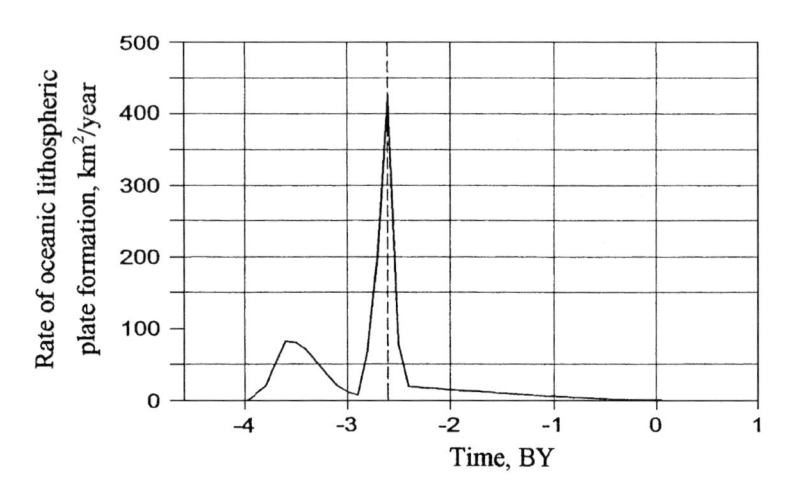

Figure 1.39: The Rate of Formation of Oceanic Crust Area (Contemporary Value $\approx 2.55 \, km^2$/year). Vertical Dashed Line Indicates the Time of Earth's Core Separation.

1.8.1. Breaking-up of Monogaea and Formation of Megagaea at the End of Early Proterozoic Time

After formation of a mono-cell convective structure, all continental shields were drifting to the side of downward flow, forming a single supercontinent of the Pangaea type. After completion of the process of unification, this supercontinent was surrounded from all sides by the subduction zones. In these zones, the oceanic plates of the World Ocean surrounding the supercontinent submerged into the mantle. At a regular average rate of subduction (5–10 cm/year), after several tens of millions of years, these oceanic plates descended to the core level and disintegrated due to barodiffusion mechanism of differentiation of iron oxides and filling-up of the intergranular space in the mantle matter by their smelts.

Inasmuch as the matter of submerging plates is always slightly cooler, it is denser than the surrounding mantle matter. Thus, the mantle matter should penetrate the core under the subduction zones, forming the structure resembling the roots of downward flows (Fig. 1.11). This implies that after differentiation the disintegrated mantle matter of these roots in the form of liquid "magmatic porridge" should "run down" (i.e. rise) on both sides from the locales of penetration of former oceanic lithospheric plates into the core. As a result, large masses of differentiated lighter matter of the mantle start to accumulate gradually under the center of the supercontinent (Fig. 1.40). Thus, after 20–30 million years, vigorous upward-directed convective flow forms under the supercontinent. This convective flow uplifts and breaks open the lithospheric shell at the locale of former downward flow. This leads to the splitting of supercontinent and centrifugal drifting of its fragments in various directions from its former center. The described mechanism probably explains the instability of all previous supercontinents formed in the past geologic epochs and relatively short time of their existence as unified continents (no longer than 100–150 million years). In fact, this is a general rule that can be explained only by the chemical-density convection in the mantle.

An asymmetric distribution of density of the mantle matter, which appeared at the boundary between the Archaean and Proterozoic eons (Fig. 1.8e), should have led to significant asymmetry in the process of barodiffusion differentiation of the mantle matter over the surface of newly formed core. Inasmuch as the original matter was rich in iron (about 13%) and iron oxides (about 24%), the most intensive differentiation should have occurred in the "dense" hemisphere with formation of strong downward-directed convective flows. In contrast to that process, in the "light" hemisphere, equally strong

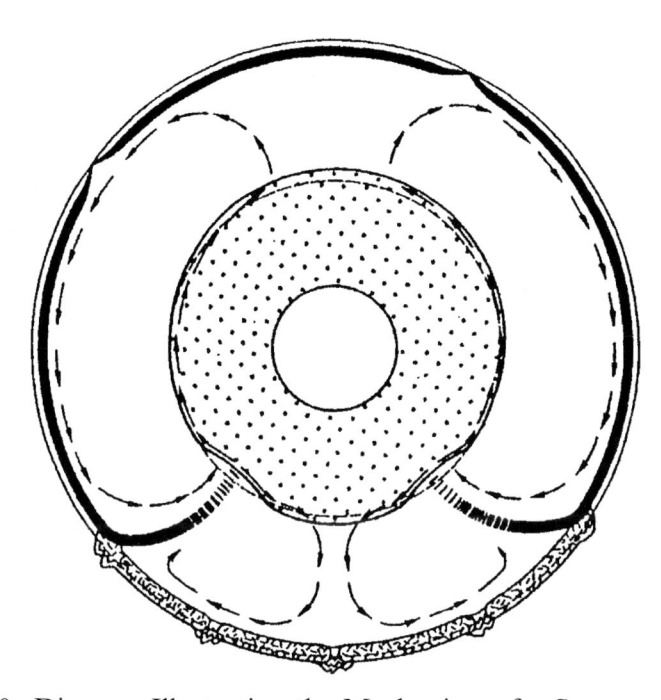

Figure 1.40: Diagram Illustrating the Mechanism of a Supercontinent Disintegration as a Result of Formation of Upward Mantle Matter Flow Instead of Previously Existing Downward Flow.

rising convective flow should have formed under Monogaea, which finally led to the splitting of the supercontinent (Fig. 1.42).

The initial stretching forces developed probably about 2.4 billion years ago, which is supported, in particular, by the age of the Great Dike of Zimbabwe. The main phase of supercontinent splitting occurred somewhat later, about 2.3 billion years ago. After overheating of the mantle in the Archaean Eon, the continental lithospheric plates were comparatively thin (less than 150 km in thickness) and less solid than the plates of contemporary continents, thickness of which under the Archaean shields reaches 250 km. Because of that, one can hypothesize that, under conditions of high tectonic activity of the early Proterozoic Eon, the Monogaea supercontinent split into small blocks – cratons. This splitting occurred mainly along old seams of merging of the Archaean shields (Keran diastrophism) into a unified supercontinent. As a result, in the middle of early Proterozoic Eon, many of the continental shields separated again and began drifting in the centrifugal directions from the center of former Monogaea (Fig. 1.42).

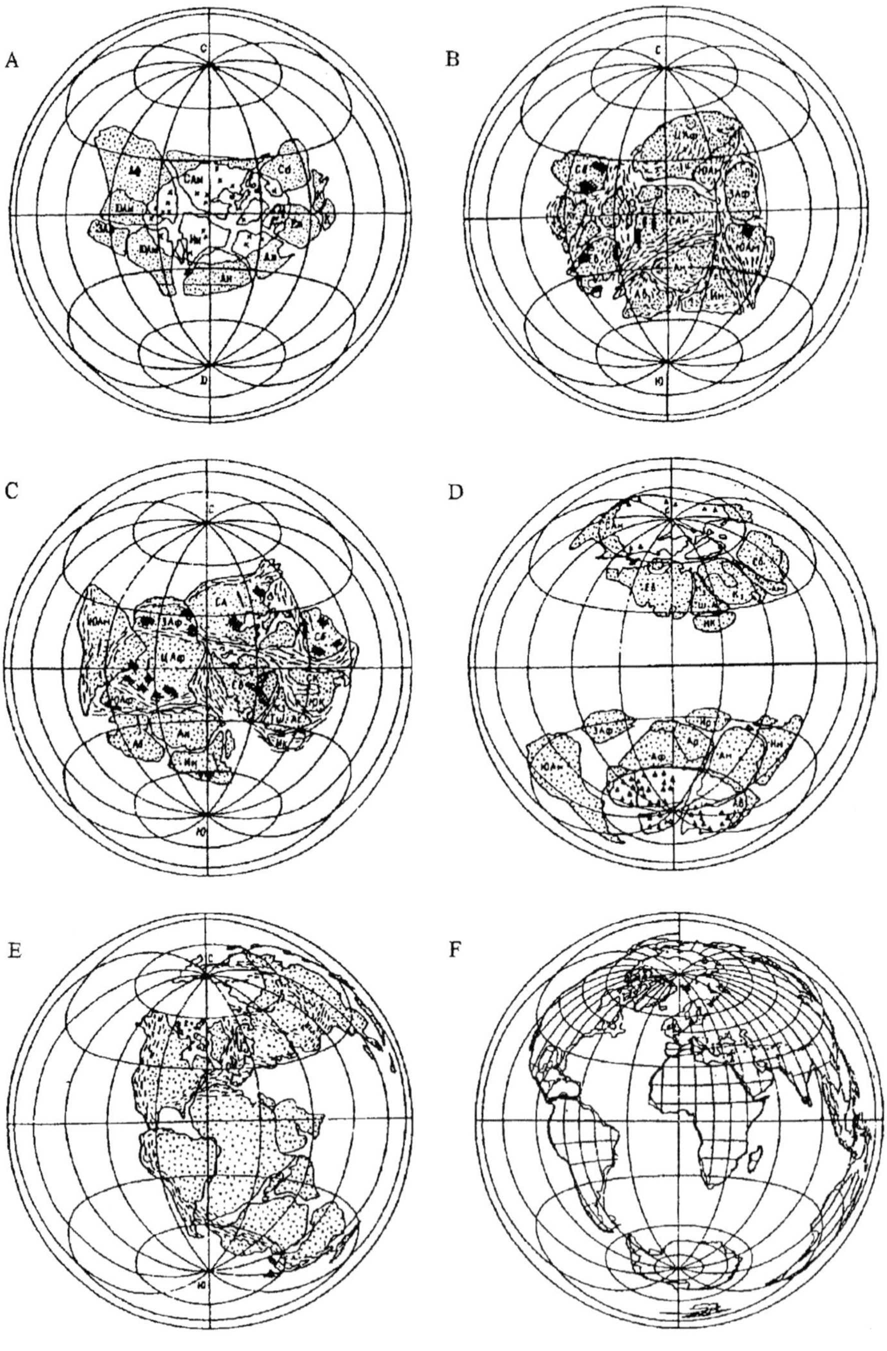
A
B
C
D
E
F

One can reconstruct the cratons (fragments of the former Monogaea supercontinent), separated in the early Proterozoic eon, based on distribution of the Karelian (Svecofennian) orogenic belt and other orogenic belts of the same age, which welded these cratons again into a new supercontinent of Megagaea about 1.9–1.8 billion years ago. For our reconstruction we used the research work on the pre-Cambrian tectonic activity of continents by Khain and Bozhko (1988) and Khain (2001). We also used additional criteria to determine the boundaries of the Archaean shields and platforms in the early Proterozoic Eon. For example, we took into account (1) that the kimberlitic (and similar smelts) formed mostly over the zones of subduction of the plates of Karelian age (Sorokhtin and Mitrofanov, 1996; Sorokhtin, 2004) and (2) the fact that the largest deposits of iron ore of the same age originated over the outskirts of continental blocks in the upwelling zones. In addition, we took into account the probability that kimberlitic rocks formed due to smelting of the tropical and temperate zone sediments, which were rich in carbonates, whereas the lampoites formed as a result of smelting of the sediments of boreal and polar zones poor in carbonate rocks.

Based on geologic data (Khain and Bozhko, 1988; Khain, 2001), one can conclude that, in the North American Platform, the Wyoming, Churchill, Caminac, and Slave cratons of Archaean age on one side, and the Superior and North Atlantic cratons on the other side are separated by the Trans-Hudson and Labrador orogenis with the folding age of 1.9–1.8 billion years. Taking the above observation into consideration, the North American Platform of Proterozoic time was divided into three parts: Eastern, Western, and Greenland (with the Baffin Land and Newfoundland provinces) (Fig. 1.41).

The European platform was divided into three cratons: (1) Kolsko-Karelian, (2) Central Russian (including the South-Western Fennoscandia, Voronezh Massif, Near-Dniper and Near-Azov blocks of the Ukrainian shield), and (3) basements of Belorussia and Baltic Sea area (including

---

Figure 1.41: Paleoreconstruction of the Positions of Continents and Oceans According to Lambert. (A) Monogaea, 2.6 Billion Years Ago (White Areas on the Continents Indicate the Glaciers, whereas Crosses Indicate Tillites and Tilloids – Chumakov, 1978); (B) Megagaea by Shtille, 1.8 Billion Years Ago (Wavy Shading Indicates the Folding Belts; Blackened Areas show Redbed Formations – Anatolieva, 1978); (C) Mezogaea, 1 Billion Years Ago; (D) Disintegration of Mezogaea and Formation of Laurasia and Gondwana, 750 Million Years Ago (White Areas on the Continents Indicate the Glaciers, whereas Small Triangles show the Location of Tillites and Tilloids – Chumakov, 1978); (E) Pangaea of Wegener, 200 Million Years Ago; (F) Present-Day Position of Continents and Oceans.

Kirovograd and Belozersk blocks of Ukrainian Shield). The Siberian platform is divided into four parts: Tunguss, Anabar, and Olenec megablocks, and Aldan Shield with the adjoining territories. The Australian continent is divided into three cratons: the Yilharn and Pillbar blocks with the adjoining territories and also a group of Northern blocks (Pine Creek, Kimberly, etc.). Africa is divided into seven cratons: Southern shield of Kalahari, Congo Protoplatform in Equatorial Africa, Central African Protoplatform (possibly comprising several individual shields) and the Protoplatform of Western Africa. South America is divided into two Protoplatforms: Gwiana Shield with the Amazon Craton and the Eastern Brazilian Craton, including an array of smaller blocks with the Archaean crust basement. The Eastern Brazilian Craton is unified with the Congo Craton of the Eastern Africa, because complete separation of these blocks occurred relatively recently, during the Mesozoic time.

Thus, in the middle of early Proterozoic Eon, the Archaean crust split into an array of separate small plates (Fig. 1.41). According to Khain (2001),

Figure 1.42: Breaking up of Monogaea about 2.2 Billion Years Ago. Numbers on the Diagram Indicate the Archaean Continental Blocks and Shields: (1) Australia; (2) Antarctica; (3) Africa; (4) Baltic Shield; (5) Eastern Siberia; (6) Greenland; (7) West Africa; (8) India; (9) Chinese Shields; (10) Russian Platform; (11) North America; (12) Ukrainian Shield; and (13) South America.

there could be over 30 separated plates (our reconstruction suggests 26 plates). Khain characterized the early Proterozoic time as an era of small plates.

Because of a pronounced nonuniformity in the composition of early Proterozoic mantle, the downward-directed flow of the next mono-cell convective structure, which formed about 1.9 billion years ago, had to be situated opposite to the former supercontinent of Monogaea. Taking the above observation into consideration, the reconstruction of the second supercontinent (Monogaea) was made by translating the continental blocks of Monogaea to the opposite side of the Earth with their consequent centripetal drift until formation (merging) of a unified continental massif. In that process, all the ancient continental massifs surrounded by the Karelian folding belt (or folding of the same age) and the Archaean crust zones of activization and transformation (which occurred about 1.9–1.8 billion year ago) were combined together (Fig. 1.42). The processes of formation and disintegration of supercontinent are of great interest in deciphering the Earth's evolution because they determine the positions of subduction zones of the plates of Svecofennian age. The latter, in turn, indicate deep magmatic processes leading to the formation of mineral deposits (diamonds, for example).

Chapter 2

Degassing of Earth and Formation of Hydrosphere and Atmosphere

2.1. Formation of Earth's Hydrosphere

The processes of development of hydrosphere and atmosphere determine the evolution of the Earth's climates and determine the origins of mineral resources (origin and evolution of coal, oil, and natural gas, for example). In the pre-Archaean time, the juvenile Earth was deprived of hydrosphere. This external and very dynamic geosphere formed on the Earth due to degassing. The latter process could start only after the beginning of mantle matter differentiation in the bowels of the Earth with the first signs of endogenic tectonomagmatic activity about 4 billion years ago. One should expect also that the Earth's degassing, and, more exactly, the mantle degassing, depended considerably not only on the tectonic activity but also on the chemical composition of mantle. Main features of changes in the chemical composition of convective mantle are presented in Figs. 1.25 and 1.26.

By the end of the last century, especially after publication by Rubey (1951) on geologic history of seawater, the world scientific community accepted the idea that the origin of water and causes of water accumulation in the oceans were totally determined by the mantle degassing. Subsequent publications assumed that the early degassing started at the time of Earth's formation.

The concept of Earth's global evolution, together with the tectonic theory of the lithospheric plates, forms a theoretical basis for the quantitative description of the processes of formation of the oceans on the Earth. The first quantitative models of the water mass accumulation in the World Ocean were proposed in the middle of 1970s and early 1980s (Sorokhtin, 1974). These models were developed using two basic assumptions: (1) the Earth's degassing rate is proportional to the rate of convective mass transfer in the mantle and (2) the convection process in the mantle is determined mostly by the energy generated in the process of gravitational Earth's matter differentiation into the dense iron-oxide core and remaining silicate mantle. In these models, the beginning of degassing was set at the time of completion of the Earth's body formation (about 4.6 billion years ago).

In the later publications (Monin and Sorokhtin, 1984; Sorokhtin and Ushakov, 1991, 2002) more sophisticated models of the hydrosphere formation were developed, which were based on the barodiffusion and zonal mechanisms of Earth's matter differentiation (see Chapter 1). In these models, the beginning of degassing process occurred by about 600 million years later than the time of Earth's formation (i.e. by the time of formation of first asthenosphere). This primary asthenosphere could form only after the preliminary heating of originally cold bowels of the Earth to the temperature of melting of silicates.

Probably, the primary mantle degassing was caused by the reduction of solubility of volatile components in the silicate smelts upon decrease in temperature and at relatively low pressures. As a result of tectonic activity, the mantle smelts (mostly basalts), which erupted onto the Earth's surface, started to boil releasing the surplus of volatile components into the atmosphere. Besides, the volatile components could also be released as a result of weathering of the volcanic rocks (i.e. their destruction on the Earth's surface). The main cause of water degassing, however, was the reduction of its solubility upon cooling and crystallization of the basaltic smelts containing water under reduced pressures (Fig. 2.1).

Thus, the degassing rate is proportional to (1) the rate of eruption of mantle mass to the Earth's surface, (2) the content of volatile components in the mantle matter, and (3) mobility of volatile components. As a first

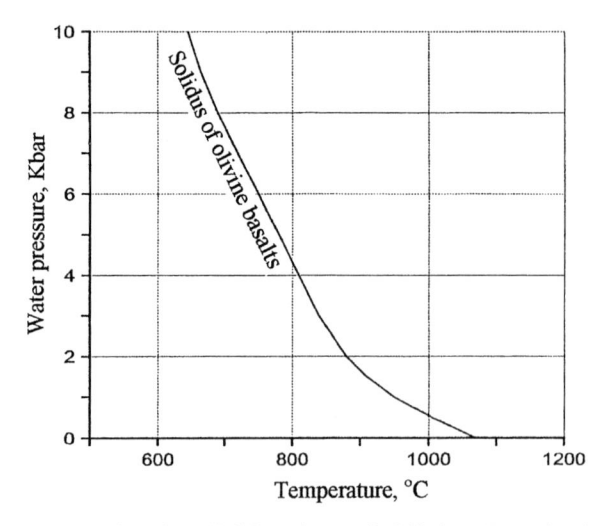

Figure 2.1: The Graph of Solidification of Olivine Basalts in the Temperature–Water Pressure Coordinates (Modified after Yoder, 1979). Upon Crystallization, the Water Content in the Basalt Smelts Decreases.

approximation, one can assume that the rate of mantle mass eruption is proportional to the Earth's tectonic activity, which can be determined by the rate of total Earth's heat release q_m (see curve 1, Fig. 1.20) or by the derivative of the tectonic parameter z with respect to time t. The tectonic parameter z is defined by the equation $z = Q_m/Q_{m0}$, where Q_m and Q_{m0} are the total mantle heat release at the time t and at the present time, respectively (see Fig. 1.19).

Under the above assumptions, the degassing rate is proportional to (1) the mass content of the considered component $(m_i)_m$, (2) the index of mobility χ_i, and (3) the rate of convective mass transfer in the mantle dz/dt:

$$\frac{dm_i}{dt} = -(m_i)_m \chi_i \frac{dz}{dt} \exp\{-\chi_i z\} \tag{2.1}$$

Integrating Eq. (2.1), one can evaluate the mass of a degassed volatile component i and its accumulation in the outer geospheres of the Earth using the following equation:

$$m_i = (m_i)_0 \chi_i [1 - \exp\{-\chi_i z\}] \tag{2.2}$$

where $(m_i)_0$ is the total mass of ith volatile component in the Earth's body, and z is the Earth's tectonic parameter. To determine the mass m_i of a volatile component, water for example, which was degassed out of the mantle, one needs to substitute the initial and boundary conditions of the mass of this component in the outer geospheres of the Earth into Eq. (2.2).

Because the mechanisms of the Earth's matter differentiation in the Archaean and post-Archaean times differed in principle, one should expect that the indices of mobility (χ_i) (or at least some of them) have to differ considerably. In the Archaean time, the silicate matter of the convective mantle, together with all of its volatile components, inevitably passed through a layer of molten iron (Fig. 1.8). In this process, the oxides, the heat of formation of which was less than that for FeO (63.64 kcal/mole), had to dissociate supplying oxygen for oxidation of iron to FeO. The heat of water vapor formation is equal to 57.8 kcal/mole, whereas for the carbon dioxide it is 94.05 kcal/mole. Therefore, the water vapor, passing through the layer of molten iron in the Archaean zones of Earth's matter differentiation had to dissociate (oxygen combining with iron), whereas the carbon dioxide could freely cross the layer of differentiation. This implies that the index of mobility of water vapor in the degassing equations (Eqs. (2.1) and (2.2)) was significantly lower in the Archaean time than that in the post-Archaean time, whereas for the carbon dioxide this index could stay constant during the entire degassing period from the mantle. During the Archaean time, besides water in the iron smelts many other oxides and sulfides (having low heat of formation) were also dissociated and reduced to free elements.

The fact that mobility indices in the Archaean and post-Archaean times can differ considerably implies that one needs to write two equations of degassing (Eq. (2.2)) with different mobility indices combining them with conditions of continuity of degassing process at the boundary between the Archaean and Proterozoic eons. This system of two equations contains three unknown parameters: the two mobility indices and the initial mass of water in the primordeal Earth's matter, m_{w0}. Therefore, for a quantitative solution of the problem, one needs to determine the three boundary conditions and substitute them into the system.

As the first one of these conditions, one can use the total mass of water present in the World Ocean and the continental and oceanic crusts. Our estimates based on the data presented by Ronov and Yaroshenko (1978), and supplemented with our own observations and computations, showed that the World Ocean contains about 1.372×10^{24} g of water, the continental crust (together with the continental water and icebergs) contains about 0.446×10^{24} g, and the oceanic crust contains about 0.446×10^{24} g of bound water. The total mass of water in the Earth's hydrosphere is estimated at 2.176×10^{24} g. This mass of water was degassed out of the bowels of the Earth during its geologic life, i.e. during the last 4 billion years of the planet's history. Strictly speaking, however, this statement is not absolutely correct. The water released onto the Earth's surface is partially dissociated in the process of hydration of oceanic rocks and in the upper layers of atmosphere (under the influence of solar radiation). In the Proterozoic and Phanerozoic times, considerable part of the degassed water returned back into the mantle as a result of subduction. If one, however, takes into consideration not the absolute but the effective mass of degassed water (difference between the degassed and subducted water), then all our calculations are still valid, and the indices of mobility (χ_i) are slightly lower than their real values.

To determine the second boundary condition (the total mass of water in the Earth's body, m_{w0}), one needs to determine its mass in the contemporary mantle. The problem of evaluation of water concentration in the mantle matter, which is fundamental for global petrology, is not solved yet mainly because of the fact that practically all of the mantle rocks reaching the Earth's surface are contaminated with the surface water in the process of uplift and outcropping at the Earth's surface. Newly formed basalts of oceanic islands contain only traces of water (less than 0.3%: Yoder and Tilly, 1965), in spite of the possibility of absorbtion of ocean waters, which are filtering through the bodies of strato-volcanoes. Taking this circumstance and other theoretical facts into consideration, the majority of present-day petrologists, studying rocks of mantle origin, have concluded that the mantle contains only traces of water. Thus, Ringwood (1981) estimated that the water content in the mantle is about 0.1%, whereas Pugin and Khitarov (1978) suggested that the water content is 0.025–0.1%. Our estimates (Sorokhtin and Ushakov, 1991, 2002) show that the mantle matter is extremely dry (the water content does not exceed 0.05–0.06%. Thus, the total mass of water in the contemporary mantle reaches 2.007×10^{24} g and the total water mass in the Earth's body is approximately equal to 4.193×10^{24} g.

To determine the third boundary condition, one can use the total water mass in the hydrosphere at some intermediate moment of time. One can estimate this mass using certain additional geologic data: One needs to determine at first a certain pivotal moment in the process of hydrosphere development. Because the volume of the World Ocean gradually increased, there was the time during its development when the oceanic water covered the tops of mid-oceanic ridges where the Earth's rift zones were located. Following this event, the hydration of oceanic crust rock should rapidly increase, with release of ore elements out of the rift zones into ocean. Therefore, after the occurrence of this event, the geochemistry of sediments should change drastically with appearance of ore elements delivered from the mantle. The most representative element is iron, which is an explicit indicator of the occurrence of the described event. In the Precambrian mantle, the metallic (free) iron content was notable (Fig. 1.26). Rising with the hot mantle matter into the rift zones, metallic iron reacted with seawater, forming with carbon dioxide (in the absence of oxygen) a highly soluble iron bicarbonate:

$$FeO + H_2O + 2CO_2 \rightarrow Fe(HCO_3)_2 \tag{2.3}$$

After covering the tops of mid-oceanic ridges, the soluble iron bicarbonate started to spread around the ocean. Reaching the shallow areas of the ocean, in the presence of micro-algae and cyanobacteria, the bivalent iron was oxidized to the trivalent form, giving rise to thick strata of the iron ore deposits.

In the process of losing iron out at the oceanic rift zones, one can identify at least two characteristic stages of the mass accumulation of iron-ores: (1) at the end of Archaean Eon (iron ores of the Kivatine type, about 3 billion years ago) and (2) at the end of early Proterozoic time (ores of the Krivorozhskiy type, about 2.2 billion years ago). The second stage of iron ore accumulation is revealed most explicitly in the Earth's geologic history.

In the Early Proterozoic Eon (2.2 billion years ago), the total water content in the hydrosphere was about 0.819×10^{24} g (Sorokhtin and Ushakov, 2002). Substituting this estimate, together with the estimates of water contents in the contemporary Earth and its hydrosphere and attaching the conditions of continuity of degassing process, we composed a system of equations describing evolution of the water mass degassing from the mantle. In this system the water mobility index is equal to 0.123 for the Archaean Eon and almost 12 times higher (1.45) for the Phanerozoic Eon.

The above system allows one to compute the water mass accumulation in the Earth's hydrosphere. The graph of the water accumulation is given in Fig. 2.2, which shows that the mode of water accumulation changed considerably at the boundary between the Archaean and Proterozoic eons. Most significantly, this change, manifested by the water mass accumulation in the oceanic crust, was caused by the formation of serpentine layer of oceanic crust at the beginning of Proterozoic Eon. This serpentine layer is the largest reservoir of bound water in the Earth.

After estimating the parameters and boundary conditions for Eq. (2.1), one is able to compute the rate of water accumulation in the Earth's hydrosphere. The rate of degassing of water from the mantle is presented in Fig. 2.3, which shows that maximum degassing rate occurred 2.5 billion years ago, i.e. at the beginning of Proterozoic Eon. On the other hand, the maximum tectonic activity occurred in the Archaean time. This seeming discrepancy can be explained by the fact that during the Archaean time the larger part of degassed water was dissociated inside the mantle: the water reacted with the smelts of iron in the process of Earth's matter differentiation according to the reaction $H_2O + Fe \rightarrow FeO + H_2$. During Proterozoic time,

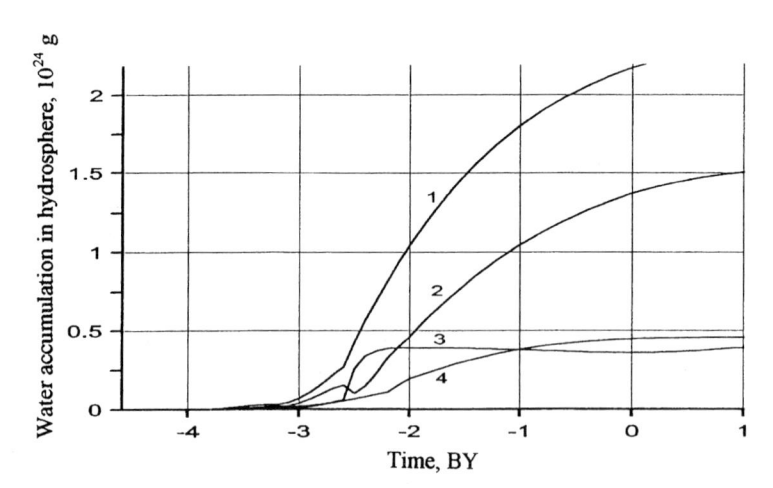

Figure 2.2: The Water Accumulation in the Earth's Hydrosphere: (1) Total Mass of Water Degassed from the Mantle; (2) Water Mass in the World Ocean; (3) Water Mass Bound in the Oceanic Crust; and (4) Water Mass Bound in the Continental Crust.

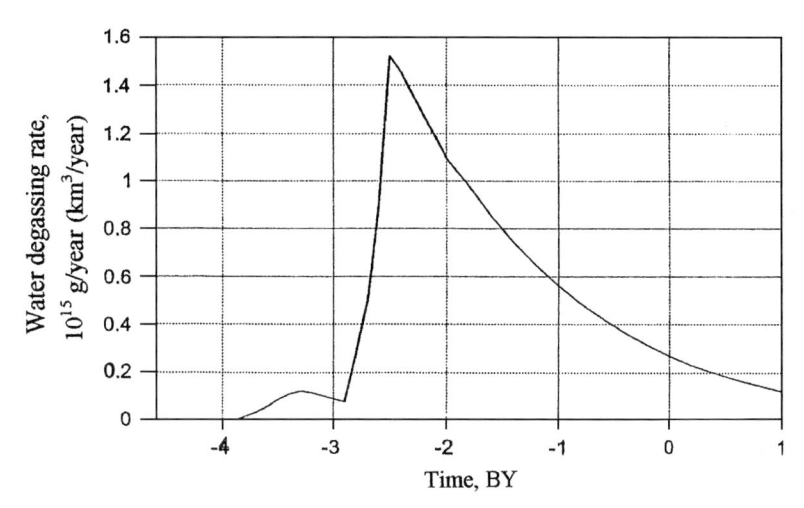

Figure 2.3: The Rate of Degassing of Water from the Mantle into the Earth's Hydrosphere.

and more so in the Phanerozoic time, water did not dissociate in the mantle. Thus, the water was released into the Earth's hydrosphere without losses.

In the early Archaean Eon, the oceans were located in low latitudes. Thus, on determining the average depth of oceanic basins and their area, one can use the estimated oceanic water mass to determine the level of ocean surface with respect to the average level of peaks of mid-oceanic ridges. The results of these calculations are presented in Fig. 2.4.

Figure 2.4 shows that in the early Archaean time the difference between levels of oceanic basin bottoms and the peaks of mid-oceanic ridges was not too large: 80–200 m. At that time, the water content in the oceans was not significant. During the Early and Middle Archaean time the real oceans did not exist. There was a multitude of shallow isolated small sea basins. At that time, the tops of mid-oceanic ridges and, in particular, the crushing zones of lithospheric plates rose above the sea level. In the centers of rift zones, the nuclei of future continental massifs formed with up to 6 km in elevation.

It is noteworthy that such a high elevation of the continental massifs in the Archaean Eon was caused by the overheating of upper mantle (Fig. 1.7) and the high inner heat fluxes. As a result, thick and heavy lithospheric plates could not form under the Archaean continental shields. The crust itself (like a light cork) rose high above the average mantle surface. In contrast to the Archaean one, the contemporary continental crust is underlain by (and welded to) a thick (up to 200 km in thickness) and dense (about 3.3 g/cm^3) lithosphere of ultrabasic composition, which sinks down the continents into the upper mantle.

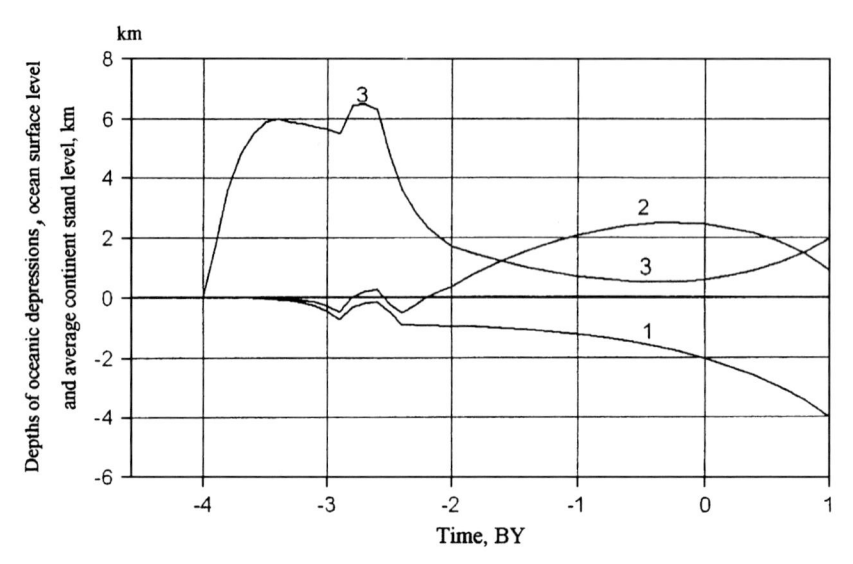

Figure 2.4: The Evolution of Level of Ocean Surface (Curve 2), Depth of Oceanic Trenches (Curve 1), and Elevation of Continental Massifs (Curve 3) with Respect to the Average Level of Peaks of Mid-Oceanic Ridges.

Elevation of the continental shield during the Archaean time led to its intense erosion. One can observe the evidence of this erosion in the many contemporary Archaean shields, surfaces of which exhibit the amphybolite and granular facies of deep-seated (5–10 km) metamorphism. The volume of sediments formed as a result of erosion during the Archaean time had to be huge. The following question arises: where are these enormous volumes of sediments, which should have accumulated during the Archaean time?

It may appear that there were insignificant volumes of Archaean sediments. The sediments (mostly conglomerates–breccias and arcosic–greywacke sandstones), which were deposited mostly at the foot of Archaean continental blocks on the oceanic basaltic plates, could accumulate there only during the life span of the plates. Inasmuch the Archaean tectonic activity was intense, their life span lasted only 0.1–2.0 million years. After that, the oceanic basaltic plates and their sediments were overthrusted over the edge zones of continental massifs (Fig. 1.31). During such a short time interval, only 20–50 m of sediments have accumulated on the ocean floor. Later on, they were drawn into the zones of oceanic plate pileup and then into the hot mantle under the weight of subsequently overthrusted plates. Together with hydrated basalts, they were then remelted giving rise to granitoid intrusions that are now exposed in the granite–greenstone belts of Archaean shields. As a result of this intense recycling, huge volumes of Archaean sedimentary rocks rejoined the continental shields, but now as a part of granitoid intrusions, mass of which is also tremendous.

In the middle of Early Archaean time (3.5–3.4 billion years ago), the ocean water briefly covered the crests of mid-oceanic ridges. That ocean, however, was shallow: no deeper than 150 m. By the end of Early Archaean time (3.3–3.2 billion years ago), the ocean depressions were deepened again and the crests of ridges appeared above the water. Komatiite pillow lavas, which erupted under water in the Barberton greenstone belt, can serve as evidence that the first oceans were rather shallow. The Early Archaean basalts often bear traits of subaerial sheets. Such a geodynamic environment during the Early Archaean Eon suggests the lower primary hydration of basalts (and greenstones) at the beginning and in the middle of Archaean Eon.

The regimes of ocean development described above are reflected in the $^{87}Sr/^{86}Sr$ ratios of oceanic sediments and K_2O/Na_2O ratios of the continental rocks (see Chapter 1, Figs. 1.34 and 1.35). The $^{87}Sr/^{86}Sr$ ratio reflects mixing of the matter received from the mantle of oceanic rift zones and the matter washed off from the continents into the ocean, whereas the K_2O/Na_2O ratio depends on the regimes of smelting of continental crust. The ^{87}Sr isotope forms as a result of β-decay of the alkaline metal rubidium – ^{87}Rb isotope. The latter metal is usually concentrated in the continental alkaline rocks. Consequently, the mature continental crust rocks have higher $^{87}Sr/^{86}Sr$ and K_2O/Na_2O ratios.

In the Archaean time, the water volume in the ocean was low, and the oceanic crust consisted of basalt. Consequently, the crustal rocks were hydrated only slightly and the continental crust was melting out under dehydrated conditions (with participation of only juvenile water). As a result, differentiation of Rb and Sr, as well as K and Na, during the formation of both oceanic and continental crusts, occurred in the environment close to the one of smelting of slightly hydrated basalt from the mantle. That is why the $^{87}Sr/^{86}Sr$ and K_2O/Na_2O ratios of the Archaean time in the oceanic and continental crust rocks remained close to those in the mantle. The same isotope ratios have been inherited by the Archaean marine sediments.

In the Proterozoic time, after the emergence of serpentine oceanic crust layer and its saturation by water, the water supply into the newly-formed subduction zones began increasing significantly. Growing supply of the water led to the accelerated transport of lithophilic elements (especially alkalis, including rubidium and potassium) from the oceanic crust to the continental crust. Consequently, the continental crust in the early Proterozoic time began to be enriched significantly in ^{87}Rb and K and, therefore, in ^{87}Sr (which is a product of ^{87}Rb β-decay). Thus, $^{87}Sr/^{86}Sr$ and K_2O/Na_2O ratios in the rocks of continental crust began to increase rapidly in the Early Proterozoic time (Figs. 1.34 and 1.35).

After the continental crust had been completely saturated with oceanic water at the end of Early Proterozoic time, the melting regime of the continental crust stabilized. At that time (about 2.2 billion years ago), the K_2O/Na_2O ratios in the crustal rocks also stabilized (Fig. 1.35). Thus, as a first approximation, one can conclude that this ratio is proportional to the water content in oceanic crust. At the same time, the radiogenic strontium accumulated in the oceanic crust at a rate proportional not only to the rate of ^{87}Rb β-decay but also to its mass. The accumulation of rubidium in the continental crust occurred along with the water accumulation. Therefore, one can assume that the $^{87}Sr/^{86}Sr$ isotope ratio in the continental crust is proportional to its bound water content. When strontium isotopes supplied from the mantle are mixed with those brought-in from the continents, the resulting $^{87}Sr/^{86}Sr$ ratios are close to those presented in Fig. 1.34.

By the end of Proterozoic Eon, the surface of the growing ocean rose to the average level of continental plains. After that (already in the Phanerozoic Eon), the first global marine transgressions occurred. They shrank considerably the river runoff and the continental sediment provenance areas. Because of that, a local $^{87}Sr/^{86}Sr$ ratio minimum is observed in the Phanerozoic oceanic deposits at the time of maximum transgression (Fig. 1.34). Somewhat later (the "life" span of oceanic lithosphere, Late Cretaceous), the continental clastic sediments (with characteristic ratios of $K_2O/Na_2O \approx 1.2$) were swallowed at the subduction zone. Due to the marine transgressions, their volume in the Mesozoic time turned out to be smaller with respect to the basalts of oceanic crust. Inasmuch as the K_2O/Na_2O ratio of basalts is about 0.5, this ratio was considerably lower in the young continental crust resulting from smelting of basalts and clastics. This is illustrated by the local minimum in the Cenozoic time as presented in Fig. 1.35.

2.2. Formation of Iron-bearing Deposits

As shown in Chapter 1, iron was the main chemical element that determined the process of geologic development of the Earth. Iron movement to the surface in the rift zones also determined the atmospheric compposition and climates of the Earth. First, let us consider the conditions of iron rock deposits formation in the Precambrian time.

Geologists are well aware about the unique periods of iron ore accumulation at the end of Archaean Eon (about 3–2.6 billion years ago; Kivatine ore type) and in the early Proterozoic time (about 2.2–1.8 billion years ago; Krivorozhskiy ore type) (Baily and James, 1975; Goldich, 1975a, b;

Voitkevich and Lebed'ko, 1975; Starostin et al., 2000). In both cases, the density of iron rock deposits (≈ 3.5–$4\,\text{g/cm}^3$) exceeded the density of continental lithospheric plates.

In analyzing the origins of unique Precambrian sedimentary iron ores, one needs to answer two main questions. First, what were the sources of iron supply into the sediments? Second, what were necessary conditions and mechanisms of the iron transfer to the areas of its accumulation? In addition, one needs to explain the nonuniformity of the process of massive formation of iron ore deposits in time, because the predominant masses of these deposits accumulated in relatively short time intervals in the Precambrian time only.

The following model provides simple logical explanation of the origin of unique iron ore deposits at the end of Archaean and beginning of Proterozoic eons. Throughout the larger part of Archaean time, the iron concentration in the convective mantle was relatively low (Figs. 1.25 and 1.26), because at that time iron was almost entirely concentrated in the differentiation zones of Earth matter underlying the convective mantle. By the end of Archaean Eon, however, a juvenile matter with high concentrations of iron and its oxides began entering into the convective mantle from the central area of the Earth (Fig. 1.8d and e). According to our estimates, at the end of Archaean and beginning of Early Proterozoic eons, the average metallic iron concentration in the mantle could have reached 5.5% and that of bivalent iron reached 15%. In the oceanic rift zones, metallic iron, which was uplifted to the Earth's surface, reacted with oceanic water. Hot iron was oxidized at the expense of water dissociation, generating abiogenic methane and then combining with carbon dioxide to form iron bicarbonate (highly soluble in water) and methane:

$$4Fe + 2H_2O + CO_2 \rightarrow 4FeO + CH_4 + 41.8 \text{ kcal/mole} \tag{2.4}$$

$$4Fe + 2H_2O + 9CO_2 \rightarrow 4Fe(HCO_3)_2 + CH_4 \tag{2.5}$$

$$FeO + 2CO_2 + 2H_2O \rightarrow Fe(HCO_3)_2 \tag{2.6}$$

Apparently, the bicarbonate of iron was spread around the ocean. In the near-surface sediment layer, due to cyanobacteria and algae, the bivalent iron was oxidized to the trivalent form and was transferred into sediments:

$$2Fe(HCO_3)_2 + O \rightarrow Fe_2O_3 + 2H_2O + 4CO_2 \tag{2.7}$$

As a result of metabolism of iron-reducing bacteria, the trivalent iron could be reduced to stoichiometry of magnetite (Slobodkin et al., 1995). Silicon dioxide was carried out of the rift zones together with iron. The silicon dioxide was formed as a result of hydration of pyroxenes, for example, according to the following reaction:

$$6\underset{\text{enstatite}}{MgSiO_3} + 4H_2O \rightarrow \underset{\text{serpentine}}{Mg_6[Si_4O_{10}](OH)_8} + \underset{\text{silicon dioxide}}{2SiO_2} + 27.38 \text{ kcal/mole} \tag{2.8}$$

This explains the paragenesis of iron oxides and silicon dioxide in jespilites of the Precambrian iron ores (Fig. 2.5).

Relationship between the accumulation of iron ores and development of biosphere and generation of abiogenic hydrocarbons (methane in particular) was discussed by Galimov (1988). In addition, Galimov concluded that during the Archaean Eon the ocean water was significantly impoverished in ^{18}O isotope.

The massive transfer of iron and other metals from the mantle into the hydrosphere could have occurred only (1) if the mantle matter contained a significant amount of these metals and (2) when the ocean surface approached the average level of rift zones at the mid-oceanic ridges or even overtopped them. It is important to emphasize that only the joint combination of these two conditions could ensure the transfer of iron from the mantle into the hydrosphere and further into the Earth's sediments. Changes in the iron

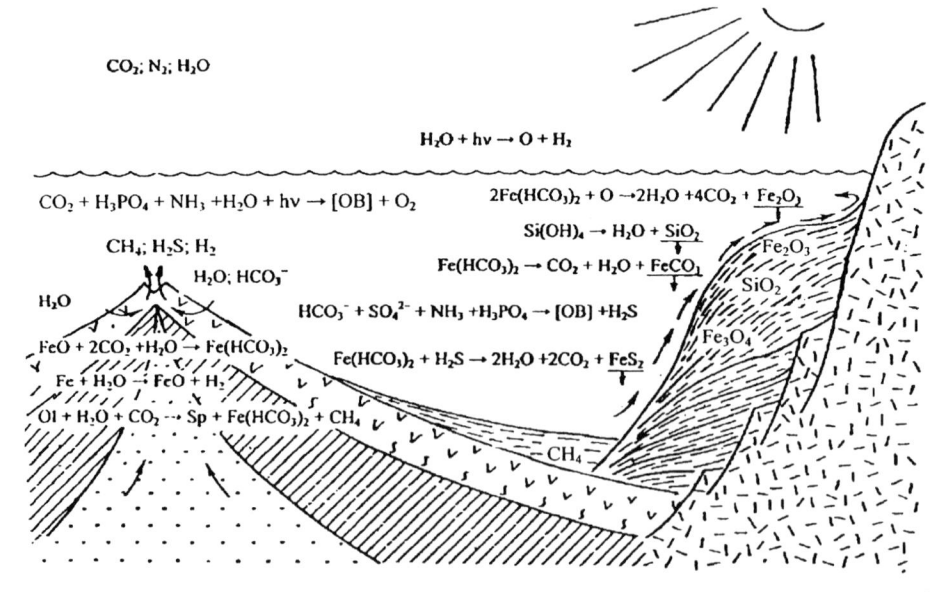

Figure 2.5: The Processes of Iron Transfer from the Mantle into the Rift Zones and Oceans and Formation of Iron Ores at the Continental Margins during Early Proterozoic Time.

concentration in the convective mantle are presented in Fig. 1.26, whereas changes in the position of ocean surface with respect to the peaks of mid-oceanic ridges are shown in Fig. 2.4. Besides the two mentioned conditions, the metals transfer was influenced considerably by the composition and content of oceanic crust (Fig. 1.28). In the Precambrian time, the iron content of the basaltic crust was considerably lower (about 10 times lower) with respect to serpentinites, formed as a result of hydration of the restive areas of the mantle matter. Taking all of the above factors into consideration, the authors developed a model of the relative rate of iron ore accumulation in the Precambrian time provided that these formations contained 50% iron and only 50% of iron was extracted from the oceanic crust.

In this model, the authors assumed that the transfer of iron from the oceanic crust into the ocean water was proportional to the water contents in the serpentinites (m_{wsp}) and basalts (m_{wbs}), as well as the average rate of area increase for lithospheric plates ($u_{oc} = kQ_m^2/S_0$, where Q_m is the heat flow indicating the tectonic activity of Earth (Fig. 1.20, curve 2) and S_0 is the area of oceanic crust). Using this model, the authors estimated the rates of iron transfer from the mantle into sediments throughout the Precambrian time (Fig. 2.6).

Figure 2.6 shows that during the Precambrian time there were at least five periods of intense accumulation of iron ores. The earliest iron ores formed about 3.8–3.5 billion years ago (Isua Formation in Western Greenland). The

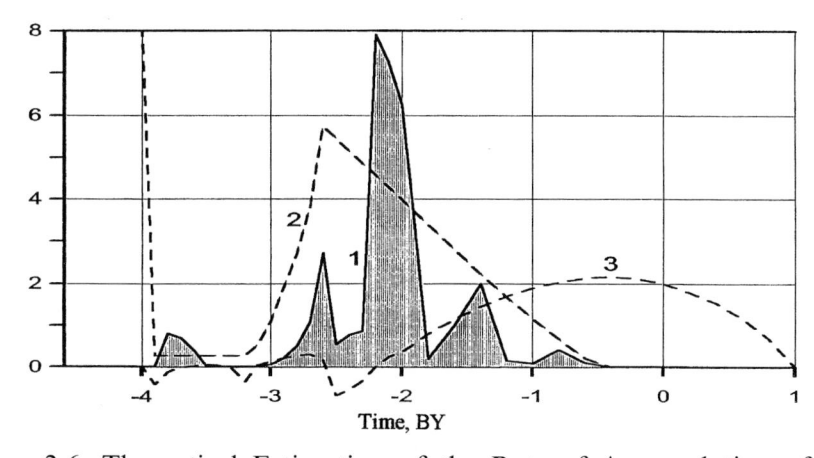

Figure 2.6: Theoretical Estimation of the Rate of Accumulation of Iron Ores in the Precambrian Time: (1) Total Rate of Accumulation of Iron Ores, 10^9 tons/year; (2) Concentration of Metallic Iron in the Convective Mantle, %; (3) Elevation of Ocean Surface with Respect to the Average Level of Peaks of Mid-Oceanic Ridges, km.

second stage was the late Archaean Eon (about 3.0–2.6 billion years ago) when sedimentary volcanogenic iron ore massifs of the Kivatine type were formed. One can also refer to the ore of Kostamuksha and other regions of Karelia and Kolskiy Peninsula, the Taratash iron ores in the Ural Mountains, and Starooskolskiy ores of the Voronezh crystalline massif (in Russia).

The most intensive period of iron ore formation, however, was at the end of early Proterozoic eon (2.2–1.8 billion years ago). The iron ore formations at the end of early Proterozoic Eon were found on practically all continents; many of them were formed at the same time. The unique deposits of jespilites in the Krivoy Rog region of Ukraine, the Kursk Magnetic Anomaly in Russia, the Karsakpay deposits in Kazakhstan, Hamersley deposits in Western Australia, deposits of the Great Lakes region in the USA and Canada and many others are the formations of that age. During this period, which lasted only about 5–7% of the total time of the Earth's geologic development, 70–75% of the total world deposits of iron ores were formed. According to our estimates, during formation of the early Proterozoic iron ores, the rate of iron accumulation reached 3.3 billion tons per year, which is close to the previous estimates of $(1-3) \times 10^9$ ton/year (Holland, 1989). The total amount of iron formations, which was accumulated during the Precambrian time, should reach 3.3×10^{18} tons, which is six orders of magnitude greater than the amount of discovered iron ore deposits (about 3×10^{12} tons according to Bykhover, 1984). This amount exceeds (by over 30 times) the iron oxides content in the sedimentary ores of continents $(0.1 \times 10^{18}$ tons according to Ronov and Yaroshevskiy, 1978). Probably, somewhat more iron is present in the metasedimentary rocks and granite layer of the continental crust. Therefore, one can conclude that the greater part of sedimentary iron sank during the Precambrian time into the mantle through the subduction zones. This implies that during the Early Proterozoic time large volumes of heavy iron ore sediments had to sink into the mantle through the subduction zones. Smelting of these sediments at great depths under the ancient continents led to the formation of diamond-bearing kimberlites and similar rocks.

An important feature of this unique time of iron ore accumulation is the fact that this process started almost simultaneously on all continents (about 2.2 billion years ago). This occurred due to the following reason: at that time the oceanic crust had become saturated with water, after which the mid-oceanic ridges disappeared under the ocean surface (curve 3 in Fig. 2.2). Then, the water-soluble iron oxides from the rift zones began entering the ocean as shown in Fig. 2.5.

By the end of early Proterozoic time, after the formation of second supercontinent Mezogaea (about 1.8 billion years ago), massive accumulation of the sedimentary iron ores stopped almost as abruptly as it started. The causes of this event are not clear yet. The concentration of metallic iron in the mantle of that time reached 3.5%, which was sufficient for carrying out huge masses of iron hydroxide into the oceanic water. The large and medium size deposits of iron ores are not known in the mid-Proterozoic and early and mid-Riphaean age, but there is information in the scientific literature about small and medium size deposits (Goldich, 1975a, b; Pelymskiy and Shishova, 1985). This can be probably explained by the rise of ocean surface to about 400 m above the mid-oceanic ridges (i.e. to the level surpassing the thickness of active oceanic layer) about 1.8 billion years ago. Probably, during the Proterozoic and Riphaean times, the ocean waters were characterized by the stable stratification and stagnant deep waters. As a particular evidence, one can consider large-scale development of black oil shale deposits during that time. As a result, the hydroxides of iron, entering from the rift zones, saturated the deep waters only and could not be oxidized to the insoluble state. In turn, the saturation of deep oceanic waters with iron prevented its further transfer into the ocean from the rift zones.

One more slight increase in the iron ore accumulation occurred about 1.6–1.4 billion years ago at the time of splitting of the Megagaea supercontinent (Shtille). This occurred because of formation of many riftogenic structures, which later transformed into the relatively shallow intercontinental oceans (Fig. 2.7). The juvenile oceans were shallow because splitting of the Megagaea supercontinent occurred above a strong rising convective flow of mantle matter (Fig. 1.40). This convective flow is similar to the currently observed one in the North Atlantic region (Iceland).

Inasmuch as the early Riphaean intercontinental oceans were shallow, the metallic iron (by that time the mantle contained about 2–2.5% of iron, Fig. 1.26), reaching the oceanic water through the rift zones, entered the active layer of the ocean and was further spread throughout the basin. After oxidation in the shallow regions of the continental periphery having high oxygen content, the iron oxides were deposited in sediments, gradually forming the iron ores. This process, however, lasted only from 1.6 to 1.4 billion years ago, until a new downward-convective flow arose at the place of splitting of Megagaea, which led (about 1.0 billion years ago) to the formation of a new supercontinent (Mezogaea).

It is noteworthy that at about the same time (1.5–1.4 billion years ago) huge anorogenic volcano-plutonic belts of enormous size formed, similar to those along the eastern border of North American platform or the Western periphery of Russian platform. During the Riphaean time, thousands of large (up to 100 km in size) differentiated plutons of gabbro, anorthosites, rapakivy granites, and potassium granites and syanites formed in these broad

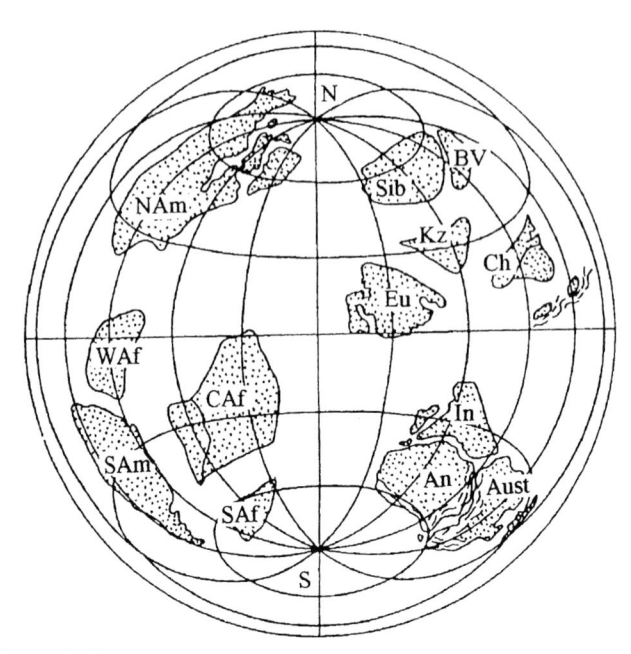

Figure 2.7: Splitting of Megagaea at the Time about 1.4 BY Ago.

belts, extending up to several thousand kilometers. This unique phenomenon of the early Riphaean time is not explained yet. One, however, can assume that these plutons were formed as a result of secondary remelting of sedimentary rocks, which accumulated during 200–300 million years on the passive periphery of continents (fragments of Megagaea broken apart about 1.7–1.6 billion years ago). It is possible that remelting of gigantic masses of shelf sediments (thickness up to 12–15 km) was facilitated by the high content of iron oxides (at the beginning of early Riphaean time there was an intense formation of iron sedimentary ores on the continental shelves). If the density of such formations ($\approx 3.2\,g/cm^3$) exceeded the average density of oceanic lithospheric plates, then as a result of their pressure on the oceanic plates in the zones of their linking with the continental crust, these deposits split these plates and collapsed into the hot mantle. It is noteworthy that the thickness of mid-Proterozoic oceanic plates did not exceed 40 km, which is about twice lower than the thickness of contemporary plates (Fig. 1.27). After remelting of sediments, iron sank into the mantle. The lighter silicate magma floated to the surface and crystallized at shallow depth in the form of granitoid or alkaline plutons (Sorokhtin and Ushakov, 2002).

Stagnant stratification of the deep water of World Ocean probably lasted only until the formation of ice shields on the continents of Laurasia and

Gondwana at the end of Riphaean time. During the ice spells, the entire oceanic water experienced intensive mixing. This was caused by the sinking of cooled water from the circumpolar basins, which engulfed frozen continents, into deep oceanic trenches. Therefore, during these periods, iron could be transferred again into an active layer of oceans, undergo oxidation and precipitate.

Distribution of tillites (Chumakov, 1987) indicates that ice shields covered equatorial and South Western Africa about 740–780 million years ago. The tillites of Australia and North America are of the late Riphaean age. The Laplandia glacier of Europe and the glacier of the shores of North Atlantic Ocean are younger (650–670 million years). The West African horizon of tillites is even younger (615–620 million years). These ages are well correlated with the ages of large iron ore formations: 750 million years in the geosynclinal zone of Adelaide and about 600–800 million years in the North Baikal areas (Russia) and the provinces of Mackenzie, Yukon, and North West Territories of Canada. The fifth (last) period of intensive sedimentary iron ore accumulation in the Precambrian time was more modest than the previous ones (Fig. 2.6), because by that time less than 1% of metallic iron remained in the mantle matter (Fig. 1.26). The primary sedimentary iron ore deposits of this type probably did not form in the Phanerozoic time.

2.3. Nature of Global Transgressions

Averaged characteristics of the Earth's tectonic activity discussed in Chapter 1 did not include the periodic fluctuations. During the geologic history, the fluctuations in the intensity of the mantle convection and the rate of movement of lithospheric plates had to occur. For example, this could have been due to (1) nonstationary nature of the chemical–density convection and changing structure of convection or (2) continental collisions and destruction of lithospheric plates that influenced considerably the rate of mutual displacement of plates. The data on the age identification of oceanic floor magnetic stripe anomalies show that the average rate of movements of Pacific plate in the Late Cretaceous time, for example, was 1.5 times the current rate, whereas it was close to the current one during the Late Jurassic and Early Cretaceous time.

Vestiges of marine transgressions on and regressions from continents are helpful in the reconstruction of Earth's tectonic activity. It was shown (Menard, 1966; Pitman and Heyes, 1973; Sorokhtin, 1976; Turcotte and Burke, 1978) that global marine transgressions on continents and the subsequent regressions might have been caused by the pulsations in the Earth's tectonic activity. In addition to the relatively slow World Ocean level changes lasting hundreds of millions of years, there were shorter eustatic level changes due to water accumulation and melting of sheet glaciers, which formed during the ice periods.

Prior to development of the plate tectonics theory, it was customary to ascribe the global marine transgression and regressions to the vertical motions of continental platforms, which supposedly experienced periodic uplifts and subsidence. That was an old traditional explanation going back to Strabo and Aristotle. The concept appeared to be self-evident and considered to be axiomatic. It was not even discussed as controversial in the handbooks on geology and tectonics. Major vertical fluctuations of the platforms that resulted in global marine transgressions and regressions were usually associated with periodic heating and cooling of the upper mantle matter under the continents (Joley, 1929; Belousov, 1959). The nature of these cyclic heating and cooling, however, remained unclear and was not amenable to quantitative calculations.

The theory of plate tectonics provided a completely different explanation for the origins of global transgressions and regressions. This approach was pioneered by the marine geologists. Menard (1964) suggested that substantial fluctuations in the ocean level may have occurred due to periodic changes in the volume of mid-oceanic ridges. In particular, he showed that the formation of present-day ridges could have resulted in the rise of ocean level by over 300 m.

The plate tectonics theory implies that the thickness of lithosphere is determined by the depth of cooling down and crystallization of the mantle matter and, consequently, depends on the time of exposure of hot mantle matter at the Earth's surface (Eq. (1.24)). As the lithospheric plates spread at the rift zones, their edges build up continuously due to cooling down and crystallization of the rising asthenospheric matter. Inasmuch as the density of silicates increases upon crystallization, with increasing thickness of oceanic lithosphere its surface subsides as a function of the square root of age of lithosphere (Eq. (1.25)). Therefore, the faster the ocean floor spreading in the rift zones (i.e. the higher the Earth's tectonic activity), the flatter are the mid-oceanic ridges. Correspondingly, the volumes of oceanic depressions decline and the greater amount of water is pushed out of the oceans onto the continents. Thus, the eustatic changes in the ocean level associated with the tectonic events are determined by the average rate of oceanic lithospheric plate movements, i.e. by the tectonic activity of Earth.

An inverse approach is also valid: if the oceanic eustatic fluctuations are independently determined, then they may be used to describe the Earth's tectonic activity. The eustatic fluctuations of oceanic level may be reconstructed practically for the entire Phanerozoic time based on the areal extent of marine sediments on the continents and analyzing the seismo-stratigraphic cross sections of sedimentary deposits at the continental margins. In Fig. 2.8, the graph of eustatic fluctuations is plotted using this approach.

Interpreting the graph of eustatic fluctuations, one should keep in mind that there are four major causes of fluctuations. First, the water mass in the oceans gradually increases due to the mantle outgassing. Over 4 BY of geologic history, on the average the ocean level rose by 4.5 km. Second, fluctuations occur owing to evolutionary changes in the tectonic activity. These are slow changes, with the characteristic period of fluctuations of ocean level of about 1 BY. The only exception was the Archaean time when the period of fluctuations was on the order of 100 MMY, but with a significant amplitude of 1–2 km. Third, the fluctuations occur due to periodic restructuring of the chemical–density convection in the mantle. The respective ocean level fluctuations have a characteristic period of 100 MMY and the deviation (δh) of $\approx \pm 200$–400 m. The fourth major cause is the

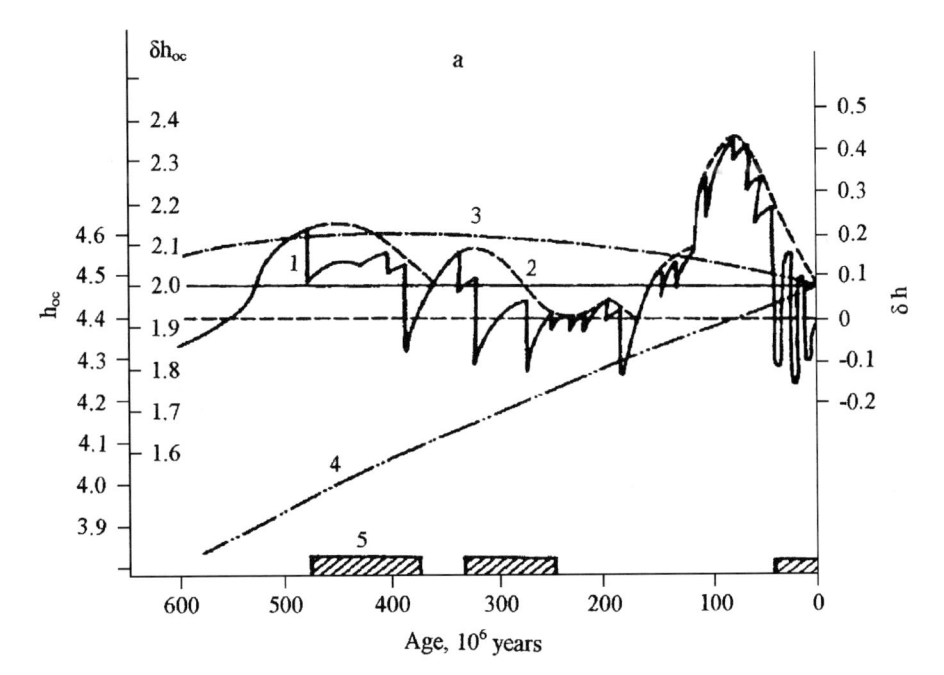

Figure 2.8: Eustatic Changes of Ocean Level in the Phanerozoic Time: (1) According to Vail et al. (1976); (2) Averaged Curve; (3) the Curve of Evolutionary Changes in Ocean Surface Level, δh (km); (4) Curve of Evolutionary Deepening of Ocean with Respect to the Average Level of Elevation of Mid-Oceanic Ridges, h_{oc}; (5) Periods of Glaciation.

emergence and melting of glaciers on the continents in the near-polar regions. This causes the fastest (duration of several thousand years) ocean level changes. Usually, the period of glacio-eustatic fluctuations is about 100 MY and their amplitude reaches $\approx \pm 100$–150 m.

Curve 1 in Fig. 2.8 (eustatic changes in ocean level) was obtained by averaging abrupt changes, which are due to either (1) rapid glacio-eustatic changes or (2) skipping or erosion of regressive sedimentary sequence. Figure 2.9(a) shows the results of converting the averaged curve of eustatic fluctuations of World Ocean level (Fig. 2.8, curve 2) to average rate of movement of oceanic lithospheric plates. Inasmuch as the heat flow through the ocean floor varies directly with the square root of average oceanic plate spreading, it was also possible to estimate at the same time the

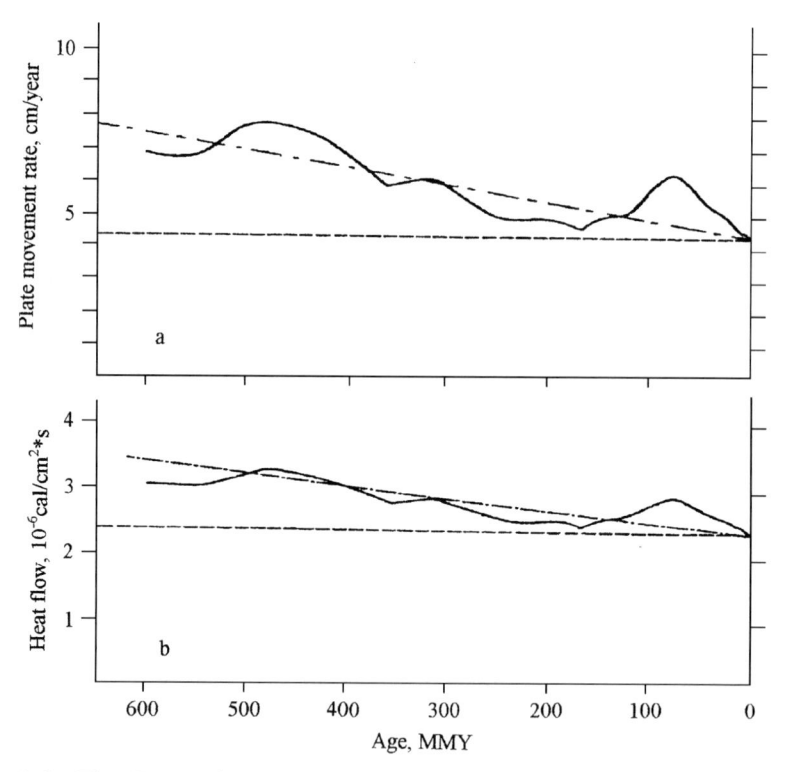

Figure 2.9: The Tectonic Activity during the Phanerozoic Time (Modified after Sorokhtin and Ushakov, 1991): (a) Expressed by the Average Rate of Movement of Lithospheric Plates; (b) Expressed by the Average Heat Fluxes through the Ocean Bottom; Dash-Dotted Lines Indicate the Evolutionary Changes in the Rates of Movement of Oceanic Plates and the Heat Fluxes through the Ocean Bottom.

average heat flux through the oceanic plates in the Phanerozoic time (Fig. 2.9(b)).

As shown in these graphs, some deviations (pulsations) in the tectonic activity may have reached 15–20%. There were supposedly three deviation maxima in the Phanerozoic time. The largest one occurred in the Ordovician and Silurian times. It happened during the Caledonian Orogeny about 500–400 MMY ago when the formation of Wegener's Pangaea started. The second, smaller maximum occurred in the Carboniferous time during Hercynian Orogeny and continuing formation of Pangaea. These active orogenic periods gave way to a brief quiescent period of the Pangaea supercontinent in the Triassic and Jurassic times. The last, Late Cretaceous burst of the tectonic activity was associated with the Pangaea disintegration and disappearance of the paleo-ocean Thetys, instead of which a grandiose Alpine-Himalayan Mountain Belt emerged.

Global marine transgressions and regressions may have resulted in a substantial restructuring of marine biotic communities. As an example of such restructuring is a massive and rapid extinction of many coral species on oceanic islands in the Early Cretaceous time. This was the period of last global transgression caused by an increase in the average rate of oceanic lithosphere growth. This caused transferal of calcium and magnesium carbonates from the ocean water into the shallow-water environment of vast continental seas and deposition of carbonate sediments on their floor. As a result, the open ocean water was impoverished in carbonates, which led to the extinction of corals and many mollusks (rudistids, for example) in the middle of Cretaceous period.

2.4. Formation of the Earth's Atmosphere

Of all the planets of Solar System, only Earth has a unique atmosphere favorable for the emergence and flourishing of higher life forms. Such a favorable composition of atmosphere gradually evolved through a lengthy interaction between the Earth's outgassing and chemical and biologic alterations of volatile elements and compounds. Discussing the origin and evolution of the Earth's atmosphere, one has to take into account the fact that primordial Earth's matter was considerably impoverished in volatile and mobile elements and compounds compared to that of the Sun's. Otherwise, the present-day atmosphere and hydrosphere would have been much thicker. Mason (1971) estimated that the relative hydrogen content in the Earth (relative to silicon) is $10^{6.6}$ times lower than in the outer space, nitrogen content $-10^{5.9}$ times lower, carbon content -10^4 times lower, noble gases -10^6 to 10^{14} times lower. That is why despite prevalence of hydrogen, helium, nitrogen, water, carbon dioxide, methane, ammonium, etc. in the outer space their concentrations and concentrations of their compounds in the Earth's matter are exceptionally low. Apparently, significant primary differentiation of the Earth's matter occurred as early as the pre-planetary stage of Solar System evolution. This has happened when the Sun was at the T-Tauras stage of evolution due to intense sweep of volatile and mobile components from the internal parts of protoplanetary gas–dust cloud to its periphery into the region of formation of giant planets.

Terrestrial volatile compounds and elements (H_2O, CO_2, N_2, HCL, HF, HI, etc.) could get into the Earth only in combined state. Water could be fixed in hydrosilicates, nitrogen in nitrites and nitrates, carbon dioxide in carbonates, halogens in halides, etc. As a result, such compounds were buried in subsurface layers. The rest of reactive volatile compounds released on impact and explosion of planetesimals must have been actively absorbed by the ultrabasic regolith on the surface of growing planet and also must have been buried under new layers of meteorite matter.

In view of the aforementioned, one may conclude that the Earth's primordial atmosphere was composed of inert and noble gases. Thus, due to short escape time of helium from the Earth's atmosphere (about 10^6 years), its partial pressure in the primordeal atmosphere over the time of formation of the primordeal atmosphere (about 10^8 years) had enough time to reach equilibrium and consequently did no exceed the current one. Isotope ^{40}Ar (product of radioactive decay of ^{40}K) also must not have been noticeably represented in the atmosphere. In conclusion, apparently the partial pressure of noble gas in the primordeal atmosphere did not exceed 2×10^{-5} atm.

It is more difficult to estimate the partial pressures for the remaining components of the primordeal atmosphere (N_2, H_2O, CO_2, and CO). To make these estimates, one needs to know the sorption and reactive capacity of these gases with the ultrabasic regolith, which, in addition, contained free metals (Fe, Ni, Co, Cr, etc). One can expect, however, that their pressure, except for the inert nitrogen, did not exceed 10^{-4} atm. According to our estimates, at the boundary between pre-Archaean and Archaean times the average temperature in the equatorial zone at the sea level was positive. At that time, the atmospheric pressure of nitrogen atmosphere should have exceeded 0.9 atm.

The bulk of carbon dioxide, which is currently almost entirely fixed in rocks and organic matter, was undoubtedly degassed from the mantle. The irreversible Earth's matter degassing, however, could have begun only after its internal temperature rose to the level of partial melting of silicates and emergence of the first convective flows in the mantle. This could have happened only after the formation of asthenosphere and the ascent of mantle melts through the formed cracks to the Earth's surface. Judging from the ages of the oldest Earth crustal rocks and the beginning of the basalt volcanism on the Moon, this event could have occurred about 4.0 BY ago.

2.5. Evolution of Partial Pressure of Nitrogen in the Atmosphere

Nitrogen is a moderately active element, reacting weakly with natural inorganic compounds. After the emergence of life on the Earth (about 3.8–4.0 BY ago), however, this gas was continuously fixed in organic matter and buried in oceanic sediments. After the live species moved to the land (about 400 million years ago), nitrogen began to be bound also in the continental deposits. As a result, life activities of organisms over their long evolutionary period could have significantly reduced the partial pressure of nitrogen in the atmosphere and, therefore, changed the Earth's climates. In order to estimate the volume of bound nitrogen, one has to take into account the fact that organic nitrogen (N_{org}) of the oceanic sediments had been continuously removed from the oceans through the zones of Archaean oceanic crust pileup or the subduction zones during the Proterozoic and Phanerozoic times. After that, nitrogen was partially included into the granite–metamorphic rocks of continental crust or was transferred into the mantle. In the process, nitrogen had been partially degassed anew and entered the atmosphere.

Our estimate of the nitrogen mass, buried in the oceanic sediments, is based on its direct correlation with the mass of buried organic carbon (C_{org}). For this purpose, one needs to determine the C_{org}/N_{org} ratio. In the open ocean floor deposits, $C_{org}/N_{org} \approx 25/1$ (Lisitsyn and Vinogradov, 1972). According to Romankevitch (1977, 1984) this ratio is close to 10/1, whereas Fore (1989) arrived at a ratio of 20/1. Based on these estimates, we used the ratio of 100/7 in our calculations.

According to Ronov and Yaroshevsky (1978) and Ronov (1993), about $(2.7–2.86) \times 10^{21}$ g of C_{org} are buried in the oceanic deposits (pelagic plus shelf) and about $(8.09–9.2) \times 10^{21}$ g of C_{org} are buried in the continental deposits. Taking into consideration the effect of avalanche-like deposition of solid terrigenous sediments in the nearshore regions of oceans (Lisitsyn, 1984, 1991) with high organic carbon concentration, in our calculations we used a slightly higher value of the present-day mass of oceanic sediments and, correspondingly, of the mass of organic carbon: $C_{org} \approx 3.36 \times 10^{21}$ g. We estimated that the mass of nitrogen buried in the (1) sediments of oceanic floor and in the shelf regions and (2) continental sediments can be estimated as 2.35×10^{20} g and 5.0×10^{20} g, respectively.

The organic nitrogen in the present-day oceanic sediments accumulated during the time of existence of contemporary oceanic crust, i.e. during the last 150 million years. In the continental sediments, the organic nitrogen accumulated during the last 400 million years. To determine the rate of accumulation of organic nitrogen in the past geologic epochs, one should consider the fact that the factor limiting the development of life in oceans is the content of phosphorus, which has a limited solubility in the ocean water (Schopf, 1982). Consequently, the biomass of the World Ocean was always approximately proportional to the mass of dissolved phosphorus (i.e. the mass of ocean) despite the fact that the development of oceanic life occurred mostly in the photic layer of ocean. This allows one to estimate the entire biomass of ocean (Fig. 2.10). Based on the above assumptions and estimates, it is possible to estimate the mass of N_{org} contained in the oceanic deposits that were extracted through the subduction zones of lithospheric plates throughout the Earth's geologic evolution:

$$(N_{org})^{\Sigma} \approx \frac{N_{org}}{m_0^w} / \tau \times \int_{-4 \times 10^9}^{0} m^w dt \qquad (2.9)$$

where $(N_{org})^{\Sigma}$ is the total mass of organic nitrogen removed from the oceanic reservoir from $t = -4 \times 10^9$ years to $t = 0$ (present time); τ is the average age of ocean floor ($\tau \approx 120$ million years); and m^w and m_0^w are the current value and present-day mass of water in the World Ocean ($= 1.37 \times 10^{24}$ g), respectively (Fig. 2.2).

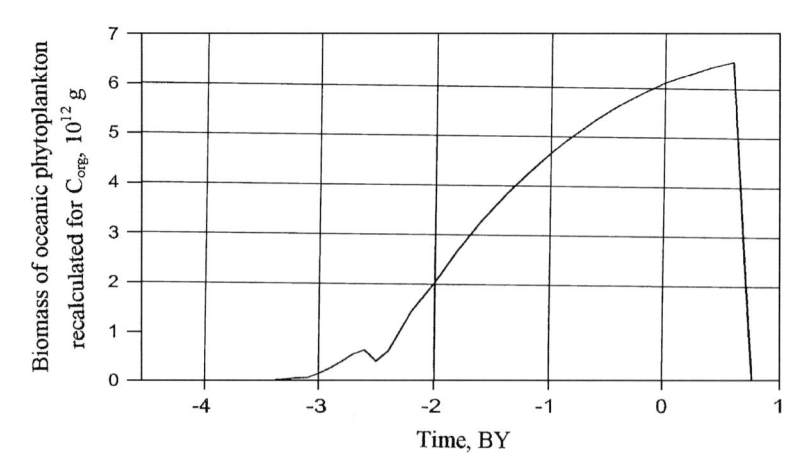

Figure 2.10: The Evolution of Biomass in the World Ocean throughout the Geologic History.

Numerical integration of Eq. (2.9) shows that 3.3×10^{21} g of nitrogen was removed from the Earth's atmosphere due to the life activities of sea biota during the geologic development of Earth (i.e. during the last 4 billion years). To this amount, one should add approximately 5.0×10^{20} g of nitrogen present in the deposits of continents, which accumulated there for about 400 million years. Thus, during the Earth's geologic history about 3.8×10^{21} g of nitrogen was removed from the Earth's atmosphere, which is equivalent to the reduction of atmospheric pressure by 745 mbar, i.e. about the level of present-day partial pressure of nitrogen (765 mbar).

The nitrogen of contemporary atmosphere could have two origins: (1) primordeal atmosphere and (2) degassing from the mantle. Because of that, let us consider two extreme cases. First, let us assume that there was no nitrogen degassing from the mantle. Then one can determine that the initial effective atmospheric pressure in the pre-Archaean time (4.6–4.0 BY ago) was approximately 1.51 bar (1.49 atm). Second, let us assume that the entire atmospheric nitrogen had been degassed from the mantle over the last 4 BY (Sorokhtin and Ushakov, 1992). Based on the nitrogen mass removed from the atmosphere and using Eq. (2.2), one can determine the evolution of partial pressure of nitrogen in the Earth's atmosphere as illustrated by the two graphs in Fig. 2.11. The actual partial pressure evolution should be represented by some intermediate curve, the shape of which may be specified only by using additional information about the Earth's climates of past geologic epochs.

First, one should take into account necessary conditions for the emergence of life at the beginning of Early Archaean time. The temperature conditions, existing at that time, should have ensured the existence of liquid water at least in the near-equatorial regions, where the Earth's tectonic activity had initially started (Sorokhtin and Ushakov, 2002; Sorokhtin, 2004). In this case, taking into account the lower level of solar radiation at the boundary between pre-Archaean and Archaean times ($S \approx 10^6$ erg/cm^2 s), one can determine that the pressure of nitrogen (primary) atmosphere should have been at least 0.9 atm.

Secondly, one has to take into consideration the fact that in the mid-Proterozoic time, when the Earth's atmosphere consisted almost completely of nitrogen with a little admixture of argon, the climate was considerably warmer than now with no continental ice shields (Chumakov, 2004). Therefore, at that time (2–1 BY ago) at the average continental heights the average yearly temperature did not drop below 0 °C (even at the poles). According to the adiabatic theory of greenhouse effect (Sorokhtin, 2001a, b), in order to have positive temperatures at the surface of continents with the elevations of 1–1.5 km (which at that time were located at the poles of the Earth), it was necessary for atmospheric pressure to be above 1.4 atm. The atmosphere in the Proterozoic time consisted almost entirely of nitrogen. Under these

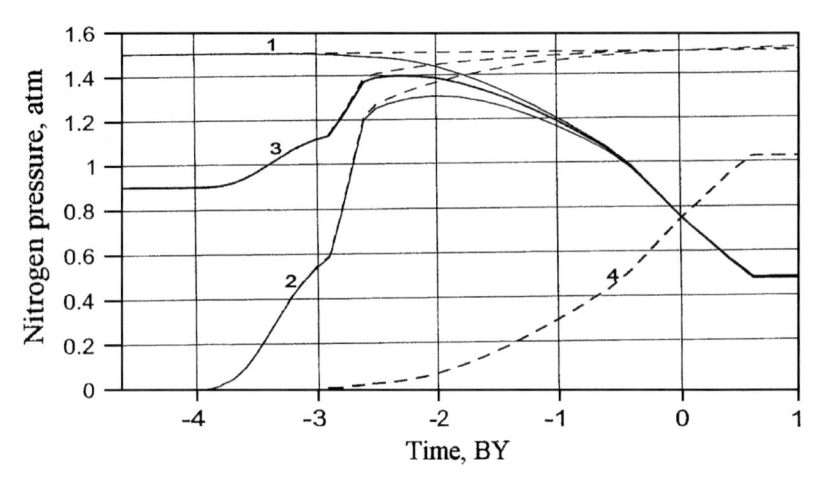

Figure 2.11: The Evolution of Partial Pressure of Nitrogen in the Earth's Atmosphere, Counting on Bacterial Consumption with Subsequent Burial of Nitrogen in the Mantle through the Subduction Zones: (1) Pressure of Nitrogen Assuming the Primordeal Origin of its Entire Mass; (2) Pressure of Nitrogen Assuming that its Entire Mass was Degassed from the Mantle; (3) Evolution of Partial Pressure of Nitrogen Assuming that it Consists of 60% of Gas of Primordeal Origin and 40% of the Nitrogen Degassed from the Mantle; (4) Mass of Nitrogen Removed from the Atmosphere.

circumstances, how could one explain changes in the partial nitrogen pressure from 0.9 atm in the pre-Archaean time to 1.4 atm (and above) in the Proterozoic time and its reduction to 0.755 atm at the present time? There is only one plausible explanation: besides primordeal nitrogen, which remained in the Earth's atmosphere from the time of Earth's formation, this gas was degassed from the mantle and that there was an efficient mechanism of nitrogen removal from the atmosphere (Sorokhtin and Ushakov, 1998a). The best explanation of the above-described changes is given by the hypothesis that the contemporary atmosphere contains about 60% of relic nitrogen and 40% of nitrogen degassed from the mantle. In this case, the evolution of partial nitrogen pressure in the Earth's atmosphere may be illustrated by the curve 3 in Fig. 2.11.

2.6. Evolution of Carbon Dioxide Partial Pressure

Carbon dioxide (and the products of its transformation, i.e. carbonates, kimberlites, hydrocarbons, and organic matter) was the only source of carbon supply to rocks. Studying the process of degassing of carbon dioxide from the mantle, one needs to take into account that in contrast to water the carbon dioxide heat of formation (94.05 kcal/mole) is significantly higher than that of iron oxide (63.05 kcal/mole). Consequently, the carbon dioxide mobility factor (as in the case of nitrogen) should not depend on the Earth's matter differentiation mechanism (zonal iron separation or Earth's core matter: barodiffusion separation mechanism). This important circumstance makes it much easier to determine the pattern of carbon dioxide degassing from the mantle, because it predetermines the constancy of carbon dioxide mobility factor and direct variation between the rate of carbon dioxide degassing and the Earth's tectonic activity.

According to Ronov and Yaroshevskiy (1978), the Earth's crustal carbonates contain about 3.91×10^{23} g of CO_2. In addition, the crust contains 1.95×10^{22} g of organic carbon (C_{org}). Before its reduction by biological processes, this organic carbon fixed 5.2×10^{22} g of O_2. Therefore, the total mass of CO_2 degassed from the mantle [$m(CO_2)$] is approximately equal to $(3.9 + 0.72) \times 10^{23} = 4.63 \times 10^{23}$ g.

It is much more difficult to determine the carbon (or CO_2) content in the mantle. Based on fractional distillation (in oxygen atmosphere), the high-temperature volatile fractions from toleite basalt glass from oceanic rifts contain between 20 and 170 g of carbon (mantle origin with the isotope ratio of about −5‰) per ton of basalt (Watanabe et al., 1983; Sakai et al., 1984; Exley et al., 1986). According to the data by Galimov (1988), the dispersed carbon in volcanic rocks has negligible concentration of 10–100 g/ton and is impoverished in the heavy isotope of carbon ($\delta^{13}C \approx -22...-27$‰). In contrast, the carbon contained in the Earth's crust is isotopically heavier with $\delta^{13}C \approx -3...-8$‰ with an average value of −5‰. One has to keep in mind that the mantle carbon is present partially in an atomic state (dispersed in the crystalline grid of silicates, Watanabe et al., 1983) and, thus, it is not a volatile component of basalts. Consequently, the carbon content in the gaseous phase (CO_2) in basalts may be considerably lower than the one based on the above-presented total carbon concentration. Summarizing, one can assume that the movable carbon content in the mantle is about 30 g/ton or, in terms of CO_2, 110 g/ton. In such a case, the mantle contains about 4.48×10^{23} g of CO_2.

Now, one can estimate that the total CO_2 mass in the Earth (converting C_{org} to CO_2) is about 9.11×10^{23} g. In Eq. (2.2), the index of mobility $\chi(CO_2) \approx 0.71$. Then, using Eqs. (2.1) and (2.2), one can plot (1) the carbon dioxide degassing rate from the mantle versus time (Fig. 2.12) and (2) the mass of degassed carbon dioxide from the mantle into the outer geospheres (atmosphere, hydrosphere, and Earth's crust) versus time (Fig. 2.13).

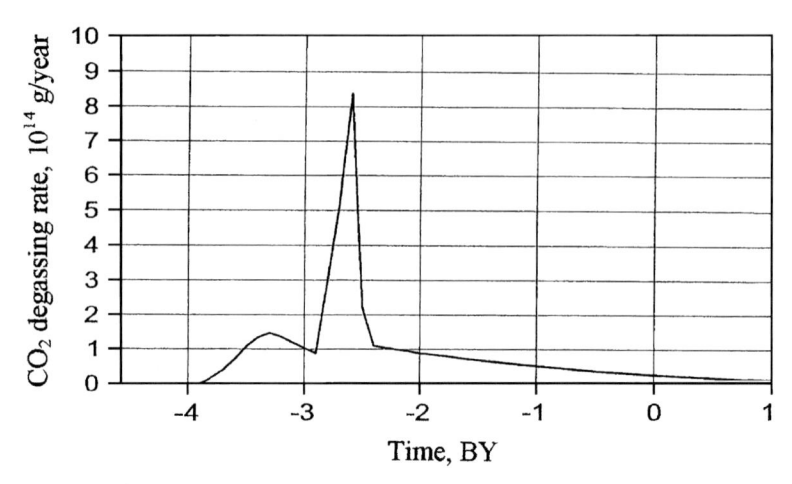

110   Global Warming and Global Cooling: Evolution of Climate on Earth

Figure 2.12: The Rate of Degassing of Carbon Dioxide from the Mantle.

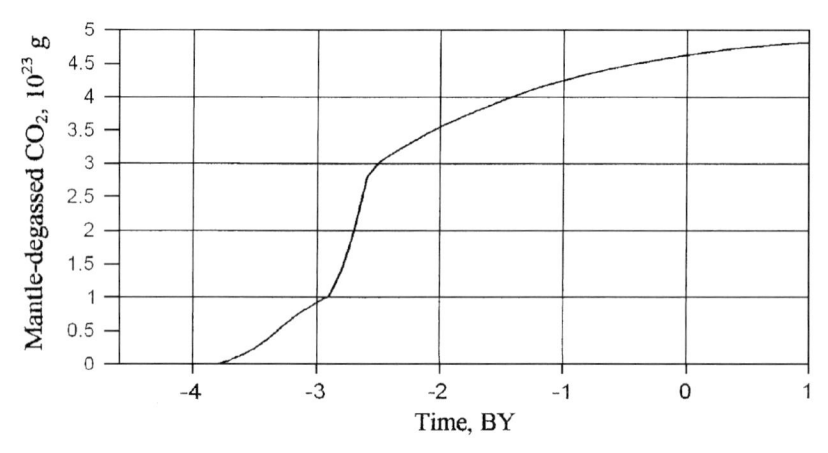

Figure 2.13: The Evolution of Mass of Carbon Dioxide Degassed from the Mantle.

There is a drastic difference between the carbon dioxide and water degassing processes (Fig. 2.3). As mentioned above, this difference is caused by the lower heat of formation of water and its dissociation on the metallic iron in the course of Earth's matter zonal differentiation during the Archaean time. The peak of CO_2 degassing rate coincides with the maximum tectonic activity about 2.7 billion years ago, whereas the water degassing rate reached its maximum about 2.5 billion years ago, i.e. only after formation of a high-density Earth's core. At that time, the mechanism of Earth's matter differentiation had been changed from zonal melting to the more quiescent barodiffusion mechanism of mantle matter differentiation.

If the entire amount of degassed carbon dioxide remained in the atmosphere, then its partial pressure at the present time would have reached about 90 atm, i.e. it would have been the same as that on Venus. Fortunately, for the life on Earth, simultaneously with transferal into the atmosphere, CO_2 was being fixed in carbonate rocks. These reactions could proceed only in the presence of water in liquid phase, because in this case the hydration of silicates is accompanied by CO_2 consumption and formation of carbonates:

$$2\underset{\text{anorthite}}{Ca \cdot Al_2Si_2O_8} + 4H_2O + 2CO_2 \rightarrow \underset{\text{kaoline}}{Al_4[Si_4O_{10}](OH)_8}$$
$$+ 2\underset{\text{calcite}}{CaCO_3} + 110.54 \text{kcal/mole} \quad (2.10)$$

$$4\underset{\text{forsterite}}{Mg_2SiO_4} + 4H_2O + CO_2 \rightarrow \underset{\text{serpentine}}{Mg_6[Si_4O_{10}](OH)_8}$$
$$+ 2\underset{\text{magnesite}}{MgCO_3} + 72.34 \text{ kcal/mole} \quad (2.11)$$

Therefore, the evolution of partial pressure of carbon dioxide in the Earth's atmosphere was considerably affected by the evolution of Earth's oceans (see Figs. 2.2 and 2.4).

Typical reactions (2.10) and (2.11) show that for every two CO_2 molecules fixed in carbonates, four water molecules are used for hydration of the rock-forming minerals in the oceanic or continental crust. Also

$$\underset{\text{fayalite}}{Fe_2SiO_4} + 3\underset{\text{forsterite}}{Mg_2SiO_4} + 5H_2O \rightarrow Mg_6[Si_4O_{10}](OH)_8$$
$$+ Fe_2O_3 + H_2 + 21.1 \text{ kcal/mole} \quad (2.12)$$

$$4Fe_2SiO_4 + 12Mg_2SiO_4 + 18H_2O + CO_2 \rightarrow \underset{\text{serpentine}}{Mg_6[Si_4O_{10}](OH)_8}$$
$$+ 4\underset{\text{hematite}}{Fe_2O_3} + CH_4 + 144.6 \text{ kcal/mole} \quad (2.13)$$

Therefore, in the presence of CO_2 surplus in the atmosphere and hydrosphere, practically all of the rock hydration reactions would be accompanied by the CO_2 fixation in carbonates (approximately 1.22 mass units of CO_2 per unit mass of water bound in rocks). Under the CO_2 deficit, the reaction of hydration will proceed without the CO_2 fixation, e.g.:

$$4\underset{\text{forsterite}}{Mg_2SiO_4} + 6H_2O \rightarrow \underset{\text{serpentine}}{Mg_6[Si_4O_{10}](OH)_8} + 2\underset{\text{brucite}}{Mg(OH)_2}$$
$$+ 33 \text{ kcal/mole} \quad (2.14)$$

In the late Archaean time about 2.8 billion years ago, the surface of the World Ocean already covered the mid-oceanic ridges (Fig. 2.4). As a result, the hydration of basaltic oceanic crust increased considerably and the rate of growth of partial CO_2 pressure in the late Archaean atmosphere was slightly reduced. The most considerable drop in partial CO_2 pressure occurred, however, at the boundary between the Archaean and Proterozoic times, after the separation of Earth's core and the consequent pronounced decrease in the tectonic activity (Fig. 1.20). As a result, the smelting out of oceanic basalts decreased considerably. The basalt layer of oceanic crust became markedly thinner than it was in the Archaean time, with formation of serpentine layer underneath for the first time. The spentine, which is a major continuously renewed reservoir of bound water on the Earth contains up to 12% of constitutional water (Fig. 1.28).

The above-described events caused a pronounced decrease (by about 10,000 times) in the partial carbon dioxide pressure in the Early Proterozoic atmosphere, about 2.4 billion years ago (i.e. by the beginning of Huronian Ice Age). The partial CO_2 pressure dropped to the level of equilibrium of about 0.5 mbar and the total atmospheric pressure decreased from 5–6 bar at the end of Archaean eon to 1.12 bar in the early Proterozoic Eon. This process of CO_2 removal from the atmosphere lasted no longer than 100–150 million years. As a result, the Early Proterozoic atmosphere contained mostly nitrogen (with insignificant admixture of argon with partial pressure of about 9.6 mbar). Reduction in the atmospheric pressure caused considerable cooling of the atmosphere and, in addition, because of high elevation of continents in the Early Proterozoic time, the widest in the Earth's history Huronian Ice Age had occurred. A gigantic glacier covered practically all the continents unified at that time in a single supercontinent of Monogaea (Chumakov, 1978) despite of its location in the equatorial belt (Fig. 1.41). Later, the partial carbon dioxide pressure was controlled by the average temperature of oceanic water according to the Henry's Law. During the ice ages, the partial pressure decreased to 0.3–0.4 mbar; whereas during interglacial periods, and especially in the warm Mesozoic time, it increased to 0.7–1.0 mbar.

The total mass of water bound in the crustal rocks is the sum of curves 3 and 4 in Fig. 2.2. These graphs indicate that the highest rate of hydration of crustal rocks had been taking place at the boundary between the Archaean and Proterozoic times and, especially, at the beginning of early Proterozoic time. This was also the time of massive fixation of carbon dioxide in the carbonate rocks (especially in dolomites). Based on these considerations, one can estimate the mass of CO_2 that may have been fixed in carbonates. The evolution of life in the ocean is limited by the phosphorus content in water, and the solubility of phosphorus is relatively low (Schopf, 1982). This allows one to assume that the absolute mass of organic matter in the ocean water varies directly with the mass of dissolved phosphorus. Thus, the organic carbon mass (recalculated to CO_2) in the Archaean and Proterozoic times may be estimated assuming that (as a first approximation) it is proportional to the water mass in the World Ocean (curve 2 in Fig. 2.2).

Figure 2.14 allows one to compare the mass of CO_2 degassed from the mantle (curve 1) with the mass of CO_2 present in carbonates (curve 2) and the total CO_2 mass in the biosphere (curve 3). Evolution of the CO_2 mass in atmosphere is presented by curve 4. As shown in Fig. 2.14, almost all of the CO_2 degassed from the mantle during the Proterozoic and Phanerozoic times was bound in the carbonate (C_{car}) or biogenic (C_{org}) reservoirs. The scenario during the Archaean time was quite different, because the Archaean oceans contained only small amounts of water. Thus, the mass of bound CO_2 was much lower than that of the CO_2 degassed from the mantle. Consequently, a substantial part of CO_2 in the Archaean time must have been in the atmosphere (Fig. 2.15) and dissolved in the oceanic water.

Emergence of the serpentine layer in the oceanic crust at the boundary between the Archaean and Proterozoic eons had led to a pronounced

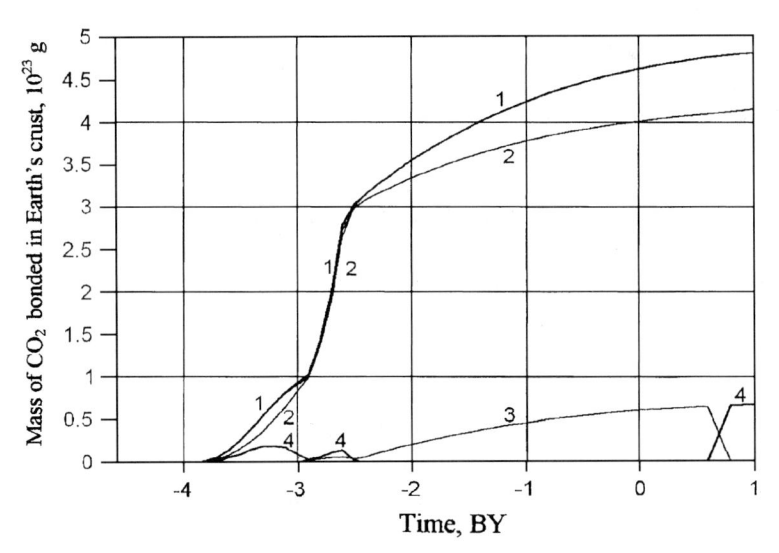

Figure 2.14: The Evolution of Mass of Carbon Dioxide Bound in the Earth's Crust: (1) The Mass of Carbon Dioxide Degassed from the Mantle; (2) The Mass of Carbon Dioxide in the Carbonate Reservoir of Earth's Crust; (3) The Mass of Organic Carbon Recalculated as Carbon Dioxide; (4) The Mass of Carbon Dioxide in the Atmosphere.

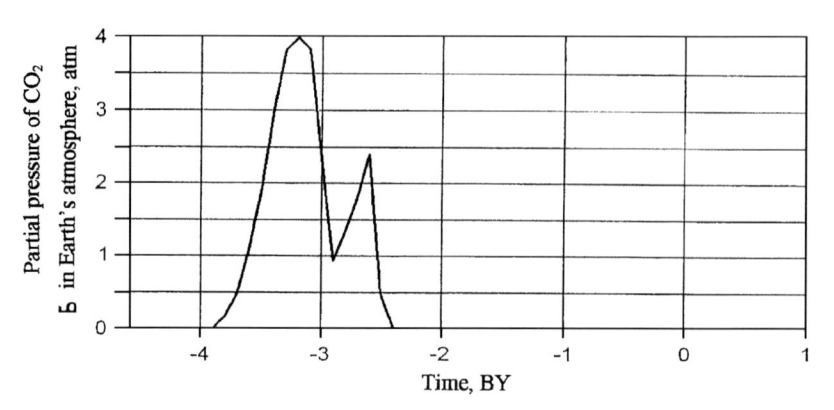

Figure 2.15: The Partial Pressure of Carbon Dioxide in the Earth's Atmosphere (the Expected Increase in its Partial Pressure in the Remote Future is not shown in this Graph).

increase in the rate of CO_2 fixation in the carbonates according to the reactions of Eqs. (2.12) and (2.13), and their deposition in the oceanic sediments (Fig. 2.16). As a result, in the early Proterozoic time gigantic masses of carbonates should have been deposited. Because of high sensitivity of the carbonate rocks to weathering (Garrels and McKenzie, 1974), however, greater part of carbonates was later redeposited in the younger formations. Simultaneously with the described events, the partial pressure of CO_2 decreased considerably, leading to a corresponding reduction of total atmospheric pressure. As a result, the hot Archaean climate was replaced by a moderately warm Proterozoic climate.

Besides the CO_2 bound in carbonates, part of it (in the form of HCO_3^- anion) is dissolved in the ocean water. To estimate the mass of CO_2 dissolved in the ocean water, one should consider the fact that solubility of CO_2 in water is proportional to its atmospheric partial pressure:

$$C(CO_2)_{oc} = K(CO_2)p(CO_2) \tag{2.15}$$

where $C(CO_2)_{oc}$ is the CO_2 concentration in ocean water, $K(CO_2)$ is the Henry's constant, and $p(CO_2)$ is the partial pressure of CO_2 in atmosphere.

Gas solubility in water is an exponential function of temperature T:

$$K(CO_2) = K_0 \exp\left\{\frac{[-V(CO_2)p(CO_2)]}{RT}\right\} \tag{2.16}$$

where $V(CO_2)$ is the partial molar volume of CO_2 dissolved in water (indefinitely diluted solution), $R = 1.987\,\text{cal/mole} \cdot {}^\circ C$, and K_0 is a normalization coefficient.

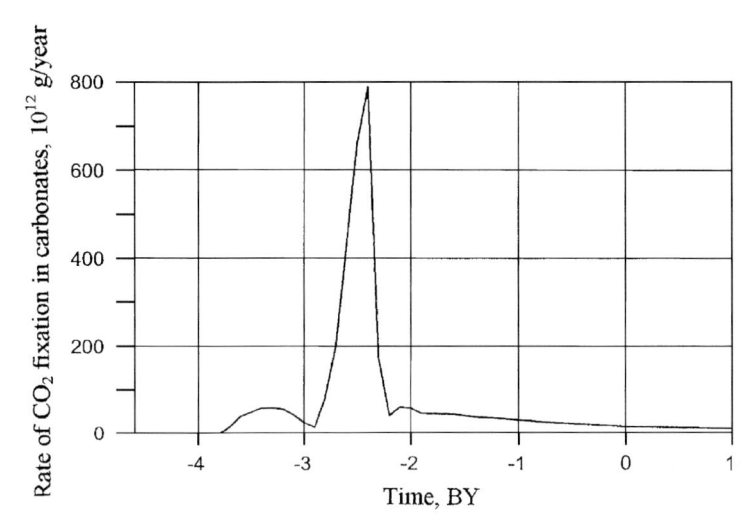

Figure 2.16: The Rate of Fixation of Carbon Dioxide in Carbonates.

To determine the normalization coefficient K_0, one needs to find the effective value of the Henry coefficient for the present-day ocean. Inasmuch as approximately 1.4×10^{20} g of CO_2 is dissolved in the ocean today and its partial atmospheric pressure is 0.46 mbar, using Eq. (2.15) the effective value of Henry coefficient for the ocean $K_{oc} \approx 0.213$. For the current average temperature of the Earth's atmosphere at sea level of 288 K, the normalization coefficient K_0 in Eq. (2.16) is equal to 0.803. After that, one can determine the mass of CO_2, which is dissolved in the oceanic water:

$$\delta m(CO_2) = m(H_2O) \cdot K(CO_2) \cdot p(CO_2) \tag{2.17}$$

During the late Archaean time (about 2.7 billion years ago), the mass of CO_2 dissolved in the oceanic water is estimated at about 10^{23} g. This means that at an average temperature of ocean water of 55 °C in the late Archaean time *the water was hot and acidic and, consequently, it was a very potent solvent for many substances* (*including the ore elements*).

The conditions for emergence of a dense carbon dioxide atmosphere in the past existed only in the Archaean time. During the Proterozoic and Phanerozoic (Paleozoic, Mesozoic, and Cenozoic) times (after the Earth's core separation and pronounced reduction in the Earth's tectonic activity), the serpentine layer was formed in the oceanic crust (Fig. 1.28) and CO_2 was mostly fixed in carbonates. As a result, *in the Early Proterozoic time* (*in about 100 million years*) *practically the entire amount of carbon dioxide was removed from the Earth's atmosphere. Thus, the atmosphere became almost entirely composed of nitrogen* with the total pressure of about 1 atm. The Archaean carbon dioxide was almost completely transferred from the atmosphere and hydrosphere into carbonates and only partially into the organic matter. In the Phanerozoic time, the atmospheric pressure reached its maximum in the middle of Mesozoic time when the oxygen generation rate became maximal because of the wide proliferation of blooming plants (phanerogamia).

2.7. Oxygen Accumulation in the Earth's Atmosphere

In the Precambrian time, oxygen was generated owing to the life activity of micro-seaweed and some bacteria (cyanobacteria, for example). In the Phanerozoic time, oxygen was also generated as a result of life activity of macro-seaweed and land plants. In the process of photosynthesis, oxygen is freed as a result of dissociation of carbon dioxide according to the following reaction:

$$CO_2 + H_2O \rightarrow CH_2O + O_2 \qquad (2.18)$$

where CH_2O is a generalized formula for the organic matter (Schopf, 1982).

Besides the biogenic origin of main mass of oxygen, one should also consider the abiogenic generation of oxygen as a result of photodissociation of water vapor under the solar hard radiation:

$$H_2O + h\nu \rightarrow O + H_2 \qquad (2.19)$$

The latter source of oxygen in the atmosphere seems to be ineffective (Walker, 1977). For example, at the present time, only one-millionth part of the atmospheric oxygen is generated by this process (Walker, 1974). In the Early Precambrian time, however (at low partial pressure of oxygen and the lack of protective ozone layer), the share of abiogenic oxygen generated by the photodissociation process has been significantly higher.

When the organisms generating oxygen die, upon burial in sediments, usually the reverse oxidation reaction of organic remnants occurs. In this reaction, oxygen is consumed forming water and carbon dioxide.

$$CH_2O + O_2 \rightarrow CO_2 + H_2O \qquad (2.20)$$

If, however, the organic matter is accumulated in the environment devoid of oxygen and is buried in the sediments, then its dissociation progresses according to the following reaction:

$$CH_2O \rightarrow C_{org} + H_2O \qquad (2.21)$$

or with formation of hydrocarbons

$$2CH_2O \rightarrow C_{org}H_4 + CO_2 \qquad (2.22)$$

Examination of Eqs. (2.18) and (2.21) shows that for one buried carbon atom there are two atoms of liberated oxygen of biogenic origin. Thus, only the burial of organic carbon leads to the generation of biogenic oxygen in the atmosphere (Walker, 1977).

The atmospheric oxygen also participates in the oxidation of rocks in the process of weathering and in the oxidation of ore elements in water solutions. Consequently, the biogenic oxygen also accumulates in the sediments. Only oxidized sediments and organic carbon on the continents and bottom of present-day oceans are accessible for observation. The amounts of organic carbon and oxides, which sank into the mantle at the subduction zones, cannot be determined directly. It is possible, however, to use indirect estimates based, for example, on the theoretical models of accumulation of the iron ores, because iron is the most active and major consumer of atmospheric oxygen.

In the Precambrian time, oxygen was generated only by the marine microorganisms. Consequently, the oxygen mass generated in such a way has to be proportional to the ocean biomass (Fig. 2.8), or to the mass of ocean water (curve 2, Fig. 2.2). In the Phanerozoic time, after the vegetation migrated to the land (about 400 million years ago), additional oxygen was generated by the land plants. In general, the volume of oxygen entering the atmosphere from ocean must vary directly with the ocean water mass and inversely with the mass of iron ores. In addition, during the Phanerozoic time, one needs to take into account the input of land vegetation:

$$p(O_2) = \frac{aM(wo)}{bR(Fe) + c} + p(O_2)_{lv} \qquad (2.23)$$

where $M(wo)$ is the ocean water mass (Fig. 2.2), $R(Fe)$ is the rate of accumulation of iron ores (Fig. 2.6), and $p(O_2)_{lv}$ is the contribution of land vegetation to the partial pressure of oxygen ($p(O_2)_{lv} \approx 0.1$); a, b, and c are normalizing coefficients that can be estimated based on historical data (so that the partial pressure of oxygen in the present-day atmosphere would be equal to 0.231 atm and 600 million years ago would have been lower than 0.0231 atm. The results of calculations using Eq. (2.23) are presented in Fig. 2.17 (in logarithmic scale).

These theoretical results together with geological data indicate that the partial pressure of oxygen in the Archaean and early Proterozoic atmosphere was very low. Because of the lower content of free iron in the convective mantle above the areas of zonal differentiation during the Archaean time, the partial pressure could have been slightly higher than that in the early Proterozoic time (at the time of massive formation of the iron ores).

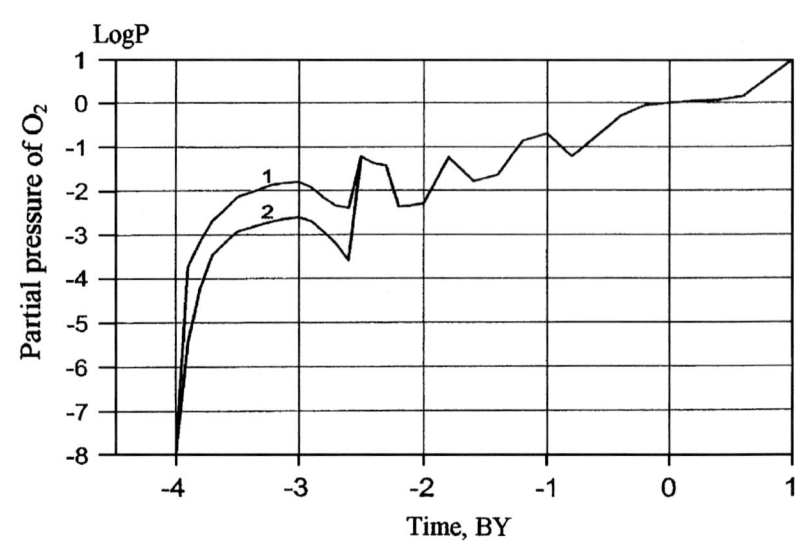

Figure 2.17: The Evolution of Partial Pressure of Oxygen in the Atmosphere (on Logarithmic Scale). In the Precambrian Time, Oxygen was Generated by the Oceanic Biota only. In the Phanerozoic Time, Oxygen was also Generated by the Land Plants. (1) The Pressure of Biogenic Oxygen Assuming that during the Archaean Time the Intensity of its Generation by the Prokaryotic Biota was One Order of Magnitude Lower than its Generation by the Eukaryotic Microalgae; (2) the Pressure of Biogenic Oxygen Assuming that the Intensity of Oxygen Generation by the Prokaryotic Biota was Two Orders of Magnitude Lower than its Generation by the Eukaryotic Microalgae.

Besides, in the early Precambrian, the processes of abiogenic oxygen generation (a result of water dissociation under the hard solar radiation) could have contributed considerably to the content of oxygen in the atmosphere and could have noticeably increased the partial pressure of oxygen. Our estimates show that the partial pressure attributable to photodissociation did not exceed 0.1 mbar.

Geochemical parameters of ancient sediments seem to indicate existence of a reducing environment and almost complete absence of oxygen in the Earth's atmosphere up to 2 BY ago. These geochemical parameters include (1) the presence of uranite (UO_2) and pyrite (FeS_2) in the early Precambrian sediments; (2) high FeO/Fe_2O_3 ratio in the sediments; (3) high content of bivalent iron in authigenic minerals (e.g. siderite); (4) high Mn/Fe ratio in iron ores; and (5) low oxidation rate of europium and cerium in the evaporite deposits of that age (Sorokhtin and Ushakov, 2002). Late Archaean and early Proterozoic conglomerates of the type of Witwatersrand in South Africa or Elliot Lake in Canada formed 3.0–2.1 BY ago (very common in many ancient shields) contain significant amounts of clastic uranite and pyrite. Formation of these minerals cannot be explained by stagnant water in the oceans. During the extensive Huronian glaciation (2.5–2.3 BY ago), for example, stagnant water could not have existed. Therefore, at the time of their formation the uranite and pyrite of the early Proterozoic conglomerates were in geochemical equilibrium with aerated oceanic water, i.e. with the atmosphere. Reducing nature of the ancient atmosphere is also supported by the high ferrous/ferric iron ratios in the Precambrian deposits and weathered rocks, as well as massive chemogenic siderite deposits.

It should be emphasized that the simultaneous deposition of iron oxide ores during the early Precambrian time does not contradict the above-stated arguments on a very low partial pressure of oxygen in the atmosphere of that time. Moreover, the existence of sedimentary Precambrian ore formations supports these conclusions. Indeed, in the oxidizing environment of the atmosphere and ocean the iron oxide solubility drastically decreases with consequent decrease in the iron migration mobility. In such a case, the entire iron content would have been oxidized to the trivalent insoluble form directly in the rift zones of the Earth and deposited there forming metal-bearing ophiolite deposits. However, upon contact with water, the highly soluble bicarbonate of iron was formed (Eq. (2.5)). It was dispersed then all over the ocean, including the passive continental margins, where with the participation of microorganisms the iron was oxidized to an insoluble trivalent state forming the major iron oxide deposits. Therefore, at the time of intense accumulation of iron ores, the partial pressure of oxygen in the Earth's atmosphere should have been lowered even further because almost the entire oxygen content was consumed by the oxidizing iron. In contrast, during the periods of slowing down of the processes of iron ore accumulation, the oxygen content of Precambrian atmosphere could slightly increase. It is difficult to estimate the scale of fluctuations of oxygen partial pressure, which possibly could have reached 2–3 orders of magnitude.

After the Archaean time, significant increase in oxygen content in the Earth's atmosphere occurred about 2.5 BY ago (curve 2 in Fig. 2.15). It was probably the time of emergence of the first eucaryotic unicellular microorganisms. Inasmuch as the metabolism of the eucaryotic microorganisms required a small amount of oxygen, they could proliferate only after an increase of partial pressure of oxygen to the level of about 10^{-3} of its current level (according to Urey). Simultaneously, the efficiency of oxygen generation by biota increased considerably, which contributed to the increase in partial pressure of oxygen. As a result, at the end of early Proterozoic time (about 1.9 BY ago) the first (although primitive) metazoans (multi-cell organisms) emerged (Fedonkin, 2003).

The partial pressure of oxygen continued to increase slightly during the mid and late Precambrian time. As a result, after 2 BY ago, red-bed weathering crusts and red-bed clastic sediments appeared on the continental margins (Salop, 1973; Anatol'eva, 1978). The emergence of free oxygen in the atmosphere (and hydrosphere) is evidenced by these red-bed deposits. As the migration capacity of iron drastically decreased, after oxidation to the trivalent state in the process of weathering of carbonates or silicates, iron was preserved in situ imparting a reddish-brown coloration to the deposits. Another evidence of changing redox potential in the ancient atmosphere is the transition of valence of europium in sediments from 2 to 3 (Eu^{2+} to Eu^{3+}). This transition occurred between 1.9 and 0.8 BY ago (Fryer, 1977). Increase in the partial pressure of oxygen after this (≈ 2 BY ago) was reflected in the Earth's biological evolution: many species of single-cell bacteria and algae rapidly developed.

After complete disappearance of metallic iron in the mantle at the boundary between Proterozoic and Phanerozoic eons, the partial pressure of oxygen in the atmosphere began to increase rapidly, which led to a dramatic reorganization of the Earth's biota. At that time, the highly organized new life forms appeared: multi-cell algae and animals, whose metabolism was based on the consumption of oxygen from the environment. Besides, at the beginning of Cambrian time, the skeletal organisms had appeared together with practically all types of other organisms known today.

Thus, the third pronounced geologic–biologic transition, which occurred during the transition period from the Proterozoic to Phanerozoic time, was clearly reflected in the biological history of Earth. This transformation drastically changed the ecological scenario on the Earth's surface: the Earth's atmosphere became oxidizing instead of basic or neutral. The life forms, whose metabolism was based on oxidation of organic matter (synthesized by plants), proved to be more efficient and started to proliferate rapidly. Thus, the writers envision the evolution of partial pressure of oxygen in the Phanerozoic atmosphere as shown in Fig. 2.18.

2.8. Generation of Abiogenic Methane

Among the minor additives to the atmospheric content, special attention is given to methane because this gas influenced considerably the conditions of emergence of Earth's life species during the early stages of Earth's

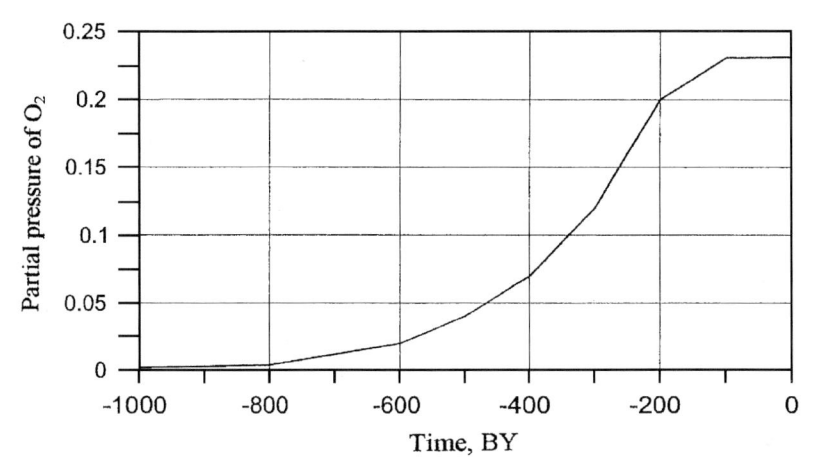

Figure 2.18: Most Probable Scenario of Evolution of Partial Pressure of Oxygen (atm) during the Last Billion Years (the Present-Day Pressure of Oxygen in the Atmosphere is 0.2315 atm).

development. Besides, the abiogenic methane could have fed the ancient bacterial flora and apparently played a significant role in forming the gas-hydrate deposits in the oceanic sediments.

In the Precambrian time, the iron ore deposits formed as a result of iron oxidation at the expense of thermal dissociation of CO_2-saturated oceanic waters and hydration of the iron-bearing rocks of oceanic crust with the same waters. At the same time, methane was generated according to the reactions (2.4) and (2.5). Inasmuch as the contemporary mantle does not contain the free metallic iron, the methane generation occurs mostly as a result of hydration of oceanic crust rocks according to the reaction (2.13).

Due to fractionation reactions of carbon isotopes between carbon dioxide and methane, methane had to be enriched in light isotope ^{12}C, whereas the heavy isotope ^{13}C was mainly transferred into carbon dioxide. Such shifts in carbon isotope are observed, for example, in the hydrothermal wells emerging in serpentines: as a rule, the $\delta^{13}C$ of methane gas of such wells is $\approx -13\ldots-14‰$, whereas HCO_3^- and CO_2 dissolved in the ocean water are characterized by $\delta^{13}C$ close to $\approx -5.5‰$ (Sorokhtin et al., 2001). Therefore, the reaction of isotope exchange between the carbon dioxide and methane

$$^{12}CO_2 + ^{13}CH_4 \rightarrow ^{13}CO_2 + ^{12}CH_4 \tag{2.24}$$

most probably progresses from left to right.

During the Precambrian time, iron ore formations were widespread. The transfer of iron from the mantle into the hydrosphere (Fig. 2.5; Eq. (2.6)) and the hydration of iron-containing silicates are accompanied by generation of abiogenic methane. It is interesting to compare the results of our estimates with the biological evaluations of the above-described events of iron accumulation in the Precambrian time. The periods of maximum rate of iron transfer into the ocean coincided with the times of maximum rate of methane generation, which led to increase in the mass of methane-consuming bacteria. In addition, as the Eq. (2.24) implies, fractionation of carbon isotopes leads to the lightening of isotope content of methane and, consequently, to the lightening of isotopes of organic matter generated by the bacteria-consuming methane. This probably explains why the organic matter of methane-consuming bacteria is usually characterized by extremely low values of $\delta^{13}C_{org}$ isotope ratio lower than $-50‰$. This can also explain the presence of two local minima in the distribution of $\delta^{13}C_{org}$ at the time of maximal formation of iron ores at the end of Archaean time and in the early Proterozoic time (Fig. 2.19).

In the early Archaean time, iron ore formation must have been accompanied by the methane generation according to reaction (2.4). This process was most intense at the earliest stages of Earth's degassing at the beginning of Archaean time after the first precipitation (rains). At that time, most of the Earth's surface was composed of the primordial regolith (see Fig. 1.8b and 1.36) that contained up to 13% metallic iron.

After wetting the regolith by rainwater, methane, which was generated according to the reactions (2.4) and (2.13), entered the atmosphere. Thus, at that time methane could have been more abundant than carbon dioxide in the atmosphere. At the same time, highly soluble iron bicarbonate, which formed according to Eq. (2.5), was washed out by the rainwater into the shallow seas formed at that time. There, iron was oxidized because of water dissociation under solar radiation and, together with silicate gels, was deposited in sediments.

The atmosphere at the beginning of early Archaean time was mainly reducing and contained mostly nitrogen, carbon dioxide, and methane. About 100 MMY later, after complete hydration of primordeal regolith, the reducing potential of Archaean atmosphere must have decreased due to photodissociation of methane by solar radiation with the formation of formaldehyde:

$$CH_4 + CO_2 + hv \rightarrow 2HCOH \qquad (2.25)$$

As a result, the Archaean atmosphere became composed of nitrogen and carbon dioxide with small admixture of methane and water vapor. At the zones of generation (outside of sediments), methane could not accumulate and entered the atmosphere directly. Thus, the composition of organic matter of that time (about 3.8 BY ago) was determined solely by reactions of

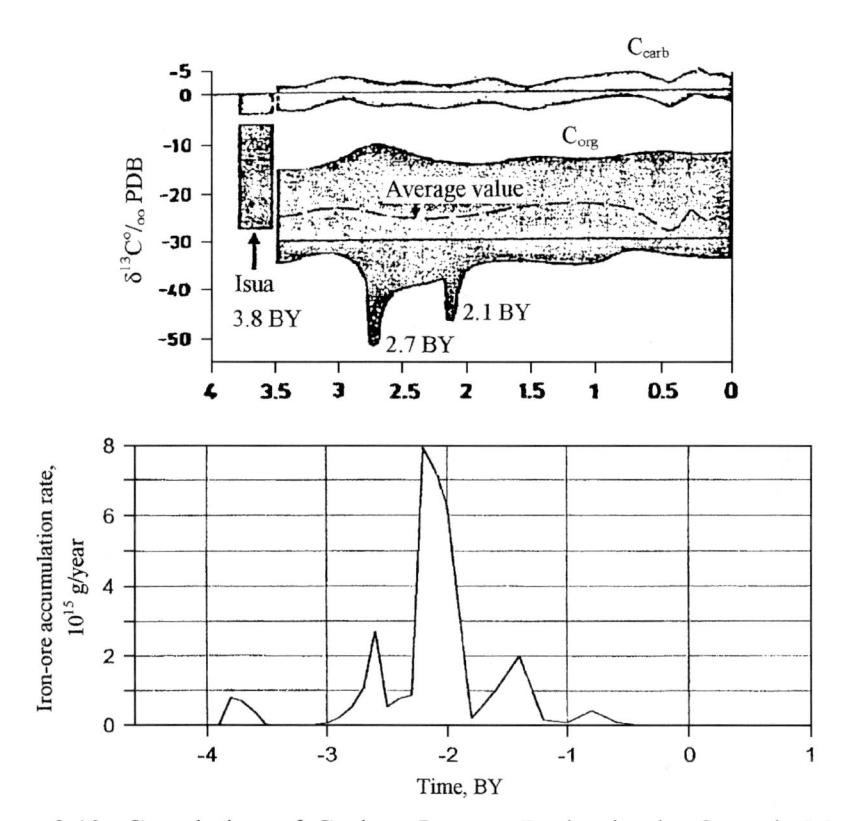

Figure 2.19: Correlation of Carbon Isotope Ratios in the Organic Matter with the Periods of Iron Ore Accumulation during the Precambrian Time. The Upper Graph Presents the Distributions of $\delta^{13}C_{org}$ and $\delta^{13}C_{carb}$ Throughout the Geologic History (Modified after Schidlowskiy, 1987); the Lower Graph Presents the Rate of Iron Ore Formation in the Precambrian Time, shown in Fig. 2.6.

pre-bacterial life forms, without influence of the isotopic composition of methane (Eq. (2.24)). The periods of iron ore formation in the Late Archaean and Early Proterozoic time also coincided with the two broad periods of development of stromatolites (Fig. 2.20). These stromatolites were possibly fed by the abiogenic methane, which was generated in the process of hydration of iron-containing silicates.

In estimating the mass of generated methane, one should note that the rate of its generation R_{me} is usually determined by the rate of oceanic crust formation (Dmitriev et al., 2000; Sorokhtin et al., 2001), which is equal to $U(H_b + H_{sp})\rho$, where U is the rate of increase in area of oceanic crust in the rift zones (Fig. 1.39); H_b and H_{sp} are the thicknesses of basalt–gabbro and

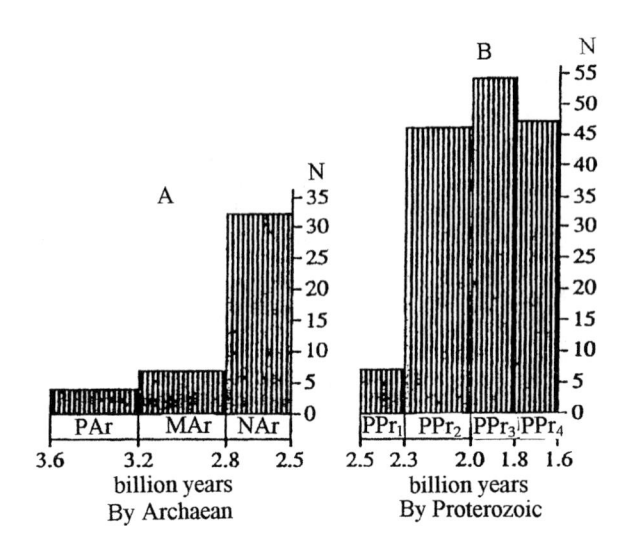

124   Global Warming and Global Cooling: Evolution of Climate on Earth

Figure 2.20: The Histograms of Changes in the Numbers of Stromatolite Formations in the Archaean Time (A) and in the Proterozoic Time (B) (Modified after Semikhatov et al., 1999); N is the Number of Formations with Stromatolites. Comparing the Presented Histograms, One should take into Account that the Mass of Archaean Stromatolites is Significantly Lower than the Mass of Early Proterozoic One.

serpentine layers, respectively; ρ is the average density of oceanic crust ($\approx 2.9\,\text{g/cm}^3$). In addition, at the times of massive accumulation of iron ores, methane was generated at the rate proportional to the rate of iron accumulation R_F in such formations.

In the presence of abundant carbon dioxide, methane is generated in the process of iron oxidation according to reactions (2.4), (2.5), and (2.9). Reaction (2.9) occurs in the oceanic crust when its rocks are hydrated with ocean waters. In the Precambrian rift zones, methane was generated with participation of free metallic iron. Thus, reaction (2.4) occurred at the hot regions of Precambrian rift zones, whereas reaction (2.5) occurred at the areas of outflow of solutions in these zones. It is also probable that during the Archaean time these reactions occurred due to saturation of primordeal regolith with acidic rainwater, because at that time regolith covered a significant part of the Earth's surface (Figs. 1.8 and 1.36).

To conduct calculations at the same scale, all the reactions of methane generation should be reduced to the same unit mass. Thus, according to the reaction of iron oxidation (Eq. (2.5)) in the rift zones and in the primordeal regolith, one mole of methane (16 g) is formed at the expense of recombination of 368 g of initial reagents. Therefore, the scaling coefficient is equal to $k = 16/368 = 0.04348$. By analogy, the reaction of hydration of iron ore containing rocks of the oceanic crust and regolith has the scaling coefficient $n = 16/2,876 = 0.00556$. Taking these considerations into account, one can present the rate of abiogenic methane generation (in the presence of abundant carbon dioxide) in the following form:

$$R_{me} = 0.8 \times 10^{15} U \rho n C(\text{FeO})(H_{sp} + 0.12 H_b) S_0/S_g + 0.1 k R_{Fe} \qquad (2.26)$$

where R_{Fe} is equal to $d(\text{Fe})/dt$; S_0 is the area of oceanic crust (Fig. 1.36); S_g ($= 5.1 \times 10^{18}$ cm^2) is the area of the Earth's surface; $C(\text{FeO})$ is the concentration of iron oxides in the mantle and oceanic crust (Fig. 1.26); the coefficients 0.1 and 0.8 characterize the degree of extraction of the given component (Fe or FeO) from the oceanic crust layers. According to Eq. (2.26), the rate of methane generation in the present-day oceanic crust is equal to 1.86×10^{12} g/year, which is close to the estimates of Dmitriev et al. (2000) of about 2×10^{12} g/year. Correlation between the isotope ratios of carbon in organic matter and the rate of generation of abiogenic methane in the oceanic crust is presented in Fig. 2.21.

This graph shows that generation of abiogenic methane in the oceanic crust started about 4 BY ago and at that time reached about 10 million ton/year. Then, this rate decreased rather rapidly to several million ton/year in the mid-Archaean time. Afterwards, because of the beginning of core separation (Fig. 1.8d, e) and a considerable increase in the tectonic activity of Earth (Fig. 1.20), the methane generation increased again and reached its absolute maximum of about 120 million ton/year at the end of Archaean time. At the boundary between the Archaean and Proterozoic times due to new pronounced decrease in the Earth's tectonic activity, the rate of methane generation decreased again to the level of 40 million ton/year. The massive formation of iron ores, which started in the Early Proterozoic time (Fig. 2.6) about 2.6 BY ago, had led again to the new increase in the rate of methane generation approximately to 90 million ton/year. In about 500 million years from now, when the bivalent iron disappears from the mantle in the process of core formation (Fig. 1.26), the process of abiogenic methane generation will terminate completely.

There is a strong correlation between the isotopic ratio of organic carbon and the rate of methane generation and, consequently, with the mass of iron ores accumulated during the Early Precambrian time. This can be explained by the fact that the iron oxidation through reactions (2.4) and (2.5) and its further accumulation in the iron ore formations is accompanied by the generation of abiogenic methane, which, in turn, is a food base for methane-consuming bacteria and cyanobacteria. Therefore, this relationship should be markedly recorded in the geologic history of Earth. Owing to isotopic

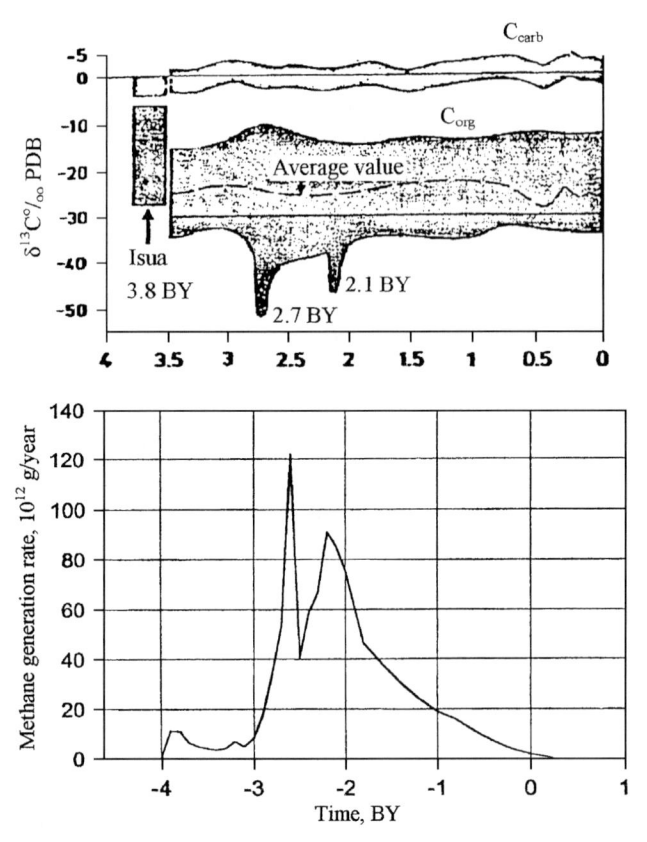

Figure 2.21: The Evolution of Rate of Generation of Abiogenic Methane in the Oceanic Crust, Evaluated According to Eq. 2.26 (Lower Graph), Presented Together with the Evolution of Carbon Isotope Ratios in the Organic Matter (Upper Graph) (Modified after Schidlowskiy, 1987).

carbon fractionation in exchange reactions of carbon dioxide with methane according to Eq. (2.24), methane is always enriched in the light carbon isotope, whereas the remaining carbon dioxide, which later is fixed in the carbonates, becomes isotopically heavier. Therefore, the organic matter, generated by the methane-consuming bacteria, should always be characterized by relatively lighter isotope carbon, in some cases reducing to $\delta^{13}C_{org} \approx -50‰$ and even lower. This phenomenon also allows one to explain the occurrence of local minima in the $\delta^{13}C_{org}$ distribution exactly at the time of deposition of iron ores with maximal rates of accumulation at the end of Archaean and Early Proterozoic time.

Intensified generation of methane at the times of massive accumulation of iron ores had to be reflected in the intensity of development of stromatolite deposits that represent the remnants of bacterial mats (films). According to Semikhatov et al. (1999), the pronounced maxima of stromatolite accumulations occurred during the periods of massive development of iron ores and intense methane generation (Fig. 2.20). It is also possible that part of the abiogenic methane was fixed in the form of gas-hydrate deposits in the oceanic sediments.

In addition to generation of methane in the oceanic rift zones, at the very beginning of Archaean time an isolated pronounced release of abiogenic methane into the atmosphere occurred due to hydration of ultrabasic regolith, which covered a large part of the Earth's surface (Figs. 1.8 and 1.36). The magnitude of this release can be estimated assuming, for example, that during the first 100 million years (from 4.0 to 3.9 billion years ago) the entire layer of primordeal regolith, with initial concentrations of 13% of iron and about 23% of iron oxides (Fig. 1.26), was completely saturated with carbonic acid rainwater. In this case, in the presence of abundant CO_2, the initial partial pressure of methane can be determined using the following expression:

$$p(CH_4) = \rho_r H_r (0.13k + 0.23n) S_r / S_g \times 10^3 \text{ bar} \qquad (2.27)$$

where ρ_r ($\approx 2\,\text{g/cm}^3$) is the density of regolith (by analogy with the Moon regolith); H_r is the unknown thickness of regolith layer saturated with rainwater; S_r is the area of the Earth's surface covered by regolith (Fig. 1.36); S_g ($= 5.1 \times 10^{18}\,\text{cm}^2$) is the area of Earth's surface.

In the reactions of methane generation (Eqs. (2.4) and (2.5)), however, the limiting factor is the mass of degassed carbon dioxide. Our estimates show that by the time of 3.9 BY ago, about 6.4×10^{20} g of CO_2 were degassed from the mantle. If one assumes that the entire mass of carbon dioxide was used for the methane generation, then the generated methane mass should be equal to $\approx 2.33 \times 10^{20}$ g $[m(CH_4) = m(CO_2) \times 16/44]$ and the partial pressure of methane $p(CH_4)$ would not exceed ≈ 45.6 mbar with the nitrogen partial pressure of about 900–1,000 mbar. To generate this partial methane pressure, the thickness of regolith layer saturated with rainwater had to be about 37.5 m (Eq. (2.27)). This thickness should be considered as an upper estimate, whereas the partial pressure of methane at the very beginning of Archaean time was somewhat lower (but not less than 20 mbar).

In the regions of its generation (in primordeal rigolith and outside of the sedimentary deposits), methane could not accumulate in the waters of sea basins of that time and entered directly into the atmosphere, creating a reducing environment. In the wet and hot atmosphere of the Early Archaean time, under the solar ultraviolet radiation methane oxidized according to the reaction $CH_4 + H_2O + hv \rightarrow CO + 3H_2$, with liberation of hydrogen. As a result, the primitive life species (probably viruses) could not consume the abiogenic methane directly.

Galimov (2001) showed that some simple original organic compounds had to be accumulated before the emergence of primitive life. One of such compounds might have been formaldehyde generated like methane in the process of iron oxidation:

$$2Fe + H_2O + CO_2 \rightarrow HCOH + 2FeO + 3.05 \text{ kcal/mole} \qquad (2.28)$$

In this process, formaldehyde remained dissolved in the water, which saturated regolith, and later was washed out of the regolith by rainwater into the newly-formed albeit shallow sea basins. Besides formaldehyde, hydrogen cyanide (another basic compound of primitive life) should have been generated in the basic atmosphere of Early Archaean time with abundant nitrogen and methane:

$$N_2 + 2CH_4 + Q \rightarrow 2HCN + 3H_2 \qquad (2.29)$$

where Q ($= 26.6$ kcal/mole) is the energy of atmospheric electric discharges necessary for reaction (2.29) to occur. The isotopic ratios of organic matter of that time were determined solely by the attributes of reactions (2.4), (2.5), and (2.13) without contribution by the bacterial metabolism. One cannot exclude a possibility that the most ancient Earth's rocks were populated with bacteria somewhat later and without abiogenic methane participation. This supposition allows one to explain why the isotopic carbon ratios at the time (3.9–3.8 BY ago) in the Shydlovskiy diagram are slightly higher than in the subsequent epochs (Figs. 2.19 and 2.21).

Equations (2.26) and (2.27) allowed us to determine the evolution of partial pressure of abiogenic methane in the Earth's atmosphere (Fig. 2.22). Because the partial pressure of methane varied considerably throughout the geologic history (greater than 4 orders of magnitude) the evolution of partial methane pressure is presented on the logarithmic scale. To complete evaluation of the rate of methane gas generation, one needs to consider the bacterial generation of methane in the marshes of Eurasia and Canada and in the soils of tropical zones. In addition, the methane gas is seeping from the hydrocarbon deposits on continents. Assuming that these processes started about 1.8 BY ago and taking into consideration the fact that present-day partial pressure of methane is equal to 1.216×10^{-6} bar, it is possible to estimate the total methane pressure in the atmosphere (Fig. 2.22). Our estimates show that the present-day methane generation on the continents is about 3.3 times higher than in the oceanic crust.

Methane is an unstable gas and under normal conditions, it disintegrates rapidly. Thus, upon degassing from the oceanic water, methane enters the

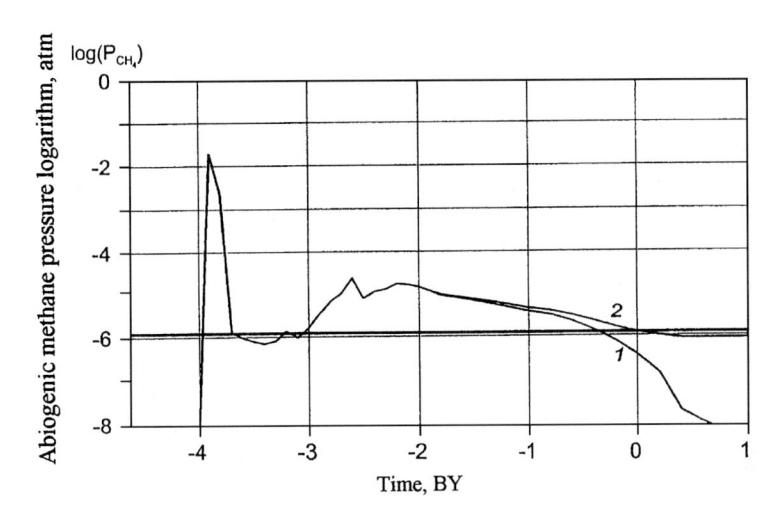

Figure 2.22: The Evolution of Partial Pressure of Methane in the Earth's Atmosphere: (1) Generation of Methane in the Oceanic Crust; and (2) Total Generation of Methane in the Oceanic Crust and on the Continents.

atmosphere and stratosphere and under the solar radiation disintegrates; thus, carbon returns again into the CO_2 reservoir of the Earth's atmosphere:

$$CH_4 + H_2O + h\nu \rightarrow CO_2 + 4H_2 \tag{2.30}$$

$$CH_4 + 2OH + h\nu \rightarrow CO_2 + 3H_2 \tag{2.31}$$

In the oceanic water, methane is consumed by bacteria emanating carbon dioxide (especially after extinction of microorganisms):

$$CH_4 + \text{bacteria} \rightarrow CO_2 + [\text{organic matter}] \tag{2.32}$$

During the Early Precambrian time in the zones of active metamorphism, methane reduction could have probably occurred, leading to the formation of oil shale, shungite, and even graphite. In the Late Precambrian and, especially, in the Phanerozoic time the direct reactions of methane oxidation could have occurred:

$$CH_4 + 2O_2 \rightarrow 2H_2O + CO_2 \tag{2.33}$$

In spite of all these losses, the methane, remaining in the sedimentary deposits, could have formed hydrocarbon deposits, in the form of gas hydrates, for example. One should also consider the possibility of a synthesis of more complex hydrocarbons from methane (Rudenko and Kulakova, 1986).

2.9. Origin of the "Ozone Holes"

Due to ultraviolet solar radiation, oxygen is partially transformed to ozone (O_3) in the stratosphere. This stratospheric ozone layer serves as a protective shield against the solar ultraviolet radiation, which is ruinous for the life species. Destruction of this ozone layer is the greatest peril for the Earth's life. Therefore, understanding of the origin of ozone holes is one of the most important scientific problems.

The ozone holes are the regions in stratosphere in the polar and temperate latitudes with a reduced (by approximately 20–30%) concentration of ozone. As a rule, they emerge during the winter–spring time above the areas of stable anticyclones in Antarctica and Yacutia, for example. The latter is caused by a pronounced decrease (or disappearance in the polar latitudes) in the insolation during the winter time and uplifting of the air masses, which overflow into the stratosphere. As a result, the ozone layer above these areas is rarefied. In the summer time, the ozone holes shrink or totally disappear.

This phenomenon was brought to the attention of scientists only at the end of 1950s when the researchers began measuring the ozone concentration in the stratosphere. The first ozone hole was discovered in Antarctica and, shortly after, speculations on their anthropogenic origin had appeared. The freon gas was named as the most probable culprit. However, one important factor remained unexplained: why the deepest and widest holes were observed in Antarctica (in the Southern Hemisphere), whereas freon was released mostly in the Northern Hemisphere. In addition, during volcanic eruptions the natural freons are released into the atmosphere in considerably greater quantities than industrial freons. Both energetically and quantitatively, the role of industrial freons is small compared to the role played by natural gases (methane, hydrogen, and natural freons) of volcanic origin.

General chemical reaction of methane with ozone occurs as a multi-step sequence:

$$CH_4 + 4O_3 \rightarrow CO_2 + 2H_2O + 4O_2 + 64.96 \text{ kcal/mole} \quad (2.34)$$

$$H_2 + O_3 \rightarrow H_2O + O_2 + 25.1 \text{ kcal/mole} \quad (2.35)$$

The amount of heat generated by ozone–freon reaction is considerably less (no greater than several kcal/mol).[2] Thus, the reaction of ozone destruction by methane generates a great amount of energy. According to Eq. (2.34), 4.062 kcal/g of energy is generated as a result of reaction of 1 g of methane

---

[2]Heat of formation: $CH_4 \rightarrow -17.89$ kcal/mole; $O_3 \rightarrow +34.50$ kcal/mole; $O_2 \rightarrow -2.80$ kcal/mole; $H_2 \rightarrow -1.00$ kcal/mole; $H_2O \rightarrow -57.80$ kcal/mole; and $CO_2 \rightarrow -94.05$ kcal/mole.

with ozone in the stratosphere. As a comparison, burning of 1 g of methane in the air according to the reaction $CH_4 + 2O_2 \rightarrow CO_2 + 2H_2O$ produces 11.98 kcal/g.

The formation of serpentines in the process of oceanic crust hydration releases about 6–10 million tons/year of CH_4 and H_2, whereas the technogenic release of freons does not exceed 100,000 tons/year. To this amount one should add many tens of millions of tons of methane and hydrogen entering the atmosphere in the tectonically active regions and tropical forests, and also the methane emanated by the marshes of the northern regions of Canada and Eurasia. The total mass of natural gases entering the atmosphere reaches hundreds of millions of tons annually. Thus, the amount of natural methane, hydrogen, and volcanic freons entering the atmosphere is almost 4 orders of magnitude higher than the amount of technogenic freons. In addition, their heating effect is considerably higher than that of freons.

This leads one to a conclusion that the anthropogenic effect on the ozone layer in the stratosphere is negligible in comparison with the effect of natural factors – approximately three to four orders of magnitude lower. Consequently, all fluctuations in the ozone concentration are natural and cannot be attributed to human activities.

Kapitza and Gavrilov (1996) showed that the variations in ozone concentration in stratosphere are solely natural and occur with the seasonal periodicity. Moreover, they discovered that the ozone concentration in the equatorial and tropical areas was lower than that within the deepest ozone holes at the near-polar areas, and there was no danger to life.

Chapter 3

The Adiabatic Theory of Greenhouse Effect

3.1. Introduction

The Earth's climate is mainly determined by the intensity of solar radiation, density of the atmosphere, and the Earth's capacity to absorb, accumulate, and disperse this radiation. During the 20th century, the global average surface temperature has increased by approximately 1 °F ($\approx 0.6\ 1$ °C) (Fig. 3.1). "The 20th century warmest years all occurred in the last 15 years of the century. Of these, 1998 was the warmest year on record.[3] The snow and ice cover in the Northern Hemisphere has decreased. Globally, sea level has risen 4–8 inches (10–20 cm) over the past century. Worldwide precipitation over land has increased by about 1%. The frequency of extreme rainfall events has increased around much of the United States" (EPA). Over the same period of time, the concentrations of water vapor, carbon dioxide, methane and other greenhouse gases in the troposphere increased, in small part as a result of human utilization of coal, oil, and natural gas. Increasing concentration of carbon dioxide, for example, is illustrated in Fig. 3.2, which is very popular.

3.2. Greenhouse Effect

Energy radiated from the Sun heats up the Earth's surface and determines the Earth's climate. In turn, heated Earth reradiates (at a longer wave length) part of this energy back into space. Some of this reradiated energy is absorbed and reemitted in all directions by the molecules of greenhouse gases in the troposphere. When the concentration of greenhouse gases in the atmosphere increases, the amount of trapped energy increases, heating the Earth's atmosphere and changing the Earth's climate. A schematic diagram of interaction between the solar radiation and the molecules of greenhouse gases is shown in Fig. 3.3. A generic term for this phenomenon is the "greenhouse effect".

---

[3]On the other hand, the coldest winter on record in Russia occurred in 2005–2006.

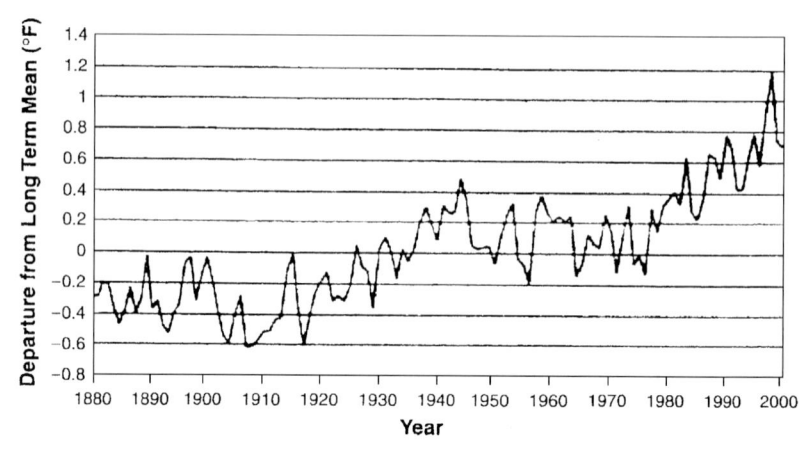

Figure 3.1: Global Average Temperature Changes in the 20th Century. (Modified from the EPA Global Warming Site: *Climate.* U.S. National Climatic Data Center, 2001.)

According to many scientists, without the greenhouse effect, the present-day Earth's global temperature would be 5 °F instead of 59 °F. This phenomenon has been going on throughout the Earth's history over the past 4 billion years and has been known for about 200 years. The great French mathematician Fourier described it first in 1822 (Mester, 1996). At the end of 19th century, Arrhenius (1896) presented a hypothesis on heating the atmosphere by the increasing content of carbon dioxide. For a long time this hypothesis was accepted as absolutely true and was used to explain the greenhouse effect practically without verification (Greehouse Effect, 1989; Green Peace Report, 1993; Budyco, 1997).

Increasing concentration of greenhouse gases may accelerate climatic changes. Environmentalists predict that the global surface temperature could rise 1–4.5 °F over the next 50 years, and 2.2–10 °F in the next century (Kyoto Protocol, 1997). Rising temperature will increase evaporation, which, in turn, will increase global precipitation. The intense rainstorms around the world will become more frequent. Ice melting will elevate the global sea level, which will be likely to rise 0.6 m in the next 50 years around the world. These disastrous climatic changes are usually attributed to the greenhouse effect. The latter is traditionally blamed on the increase in anthropogenic carbon dioxide emission.

The above-described changes in climate (if they occur) would profoundly affect the human environment, making people's life uncomfortable and in many cases unbearable. That is why environmentalists called for large-scale actions aimed at reducing the emission of greenhouse gases (especially

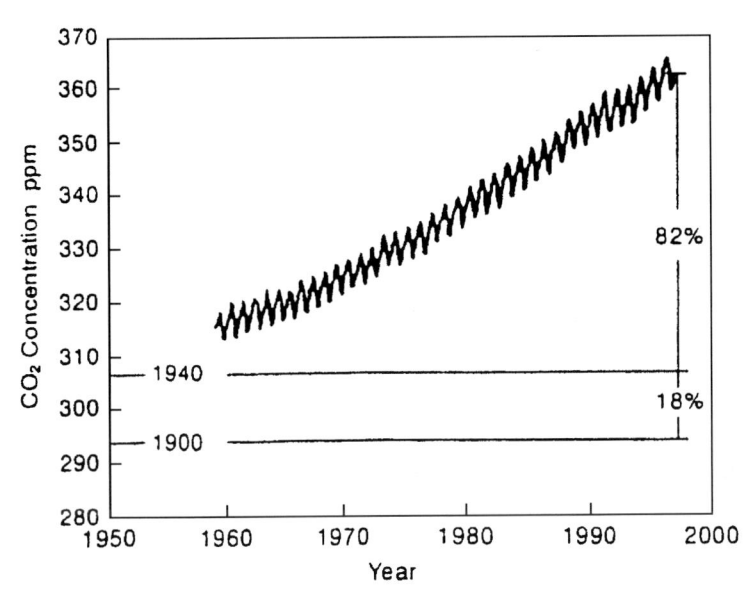

Figure 3.2: Changes in Atmospheric CO_2 Concentration in Parts Per Million (ppm) at Mauna Loa, Hawaii (Modified from Keeling & Whorf, 1997). Approximate Global Levels of Atmospheric CO_2 Concentration in 1900 and 1940 are also shown.

carbon dioxide) into the atmosphere. Joint evaluation of Figs. 3.1 and 3.2 creates a mesmerizing effect for the believers of the Arrhenius hypothesis. Their explanation is very simple and powerful: (1) the atmospheric temperature is rising because of the increasing greenhouse effect, (2) the increasing greenhouse effect is due to the rising concentration of carbon dioxide (and other greenhouse gases) in the atmosphere, and (3) the rising concentration of carbon dioxide is due to an increase in the burning of fossil fuels by humans to generate power. Their recipe for cooling is also simple and powerful: considerable cuts in human-induced CO_2 emission, which translates into drastic cuts in energy production and consumption for developed countries. Unfortunately, this recipe is incorrect because forces of nature leading to atmospheric warming (and cooling) are much stronger than current human influence (either positive or negative).

The rates and levels of the admissible emission for various countries were defined in the frame of the Kyoto Protocol of 1997, which required a worldwide 5% cut in the carbon dioxide emission. This directly translates into corresponding cuts in energy production and consumption. The latter will inevitably seriously damage the economies of developed countries. The greatest damage would be done to the United States economy. To achieve

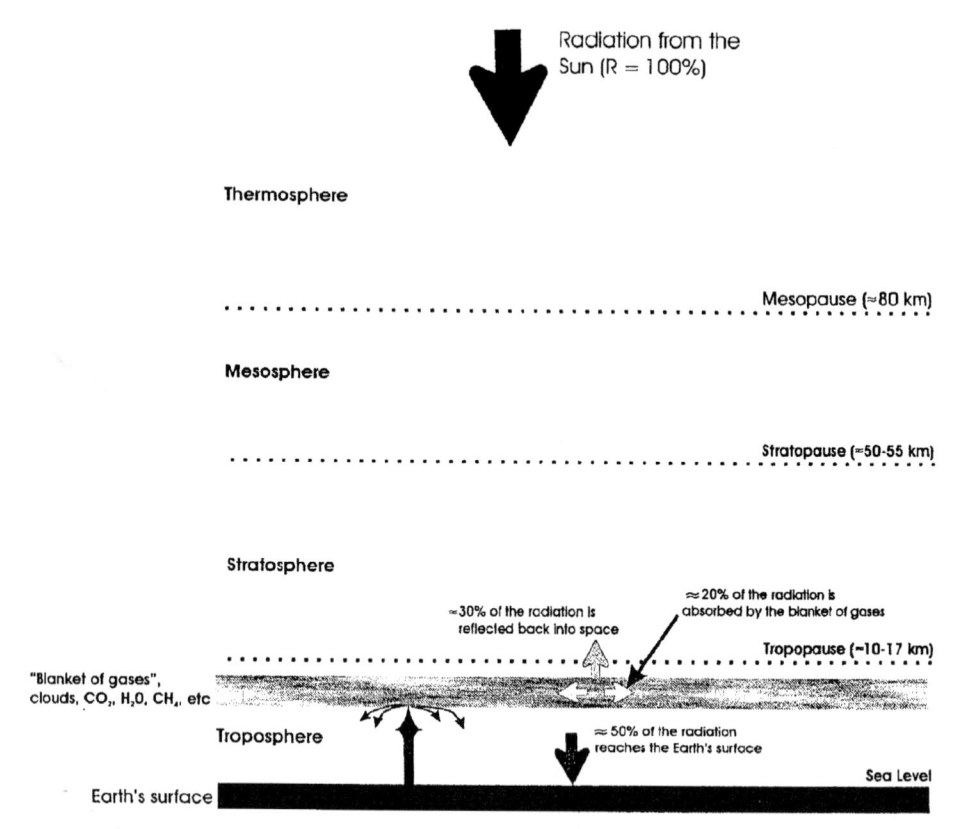

Figure 3.3: Interaction between the Solar Radiation and the Greenhouse Gases. (Modified from the World Resources Institute, *Changing Climate: A Guide to the Greenhouse Effect*, 1989.)

the required carbon dioxide emission cuts by the year 2012, the USA would have to reduce its projected 2012 energy use by 25%. "Most economic studies indicate that the cost of the Kyoto carbon dioxide emission cuts to the US would amount to between $100 billion and $400 billion per year" (Baliunas, 2002). The covenants of the Kyoto Protocol will be especially damaging for the oil- and gas-producing countries, such as USA, Canada, and Russia, the economies of which are founded on production and consumption of hydrocarbon fuels.

Serious analyses and critique of the basic assumptions made in the traditional studies on global warming were made by Robinson et al. (1998), Sorokhtin (2001a, b), Baliunas (2002), and Khilyuk and Chilingar (2003, 2004, 2006). Thorough examination showed that until recently a sound physical theory of the greenhouse effect did not exist and all the numerical calculations and predictions were based on intuitive models using numerous, poorly defined parameters (Greenhouse Effect, 1989). This examination revealed that inherent uncertainties in the estimates of model parameters (models contained at least 30 such parameters) make in fact the numerical solution of the problem incorrect. Therefore, we decided to investigate the greenhouse phenomenon based on the well-established relationships among physical fields describing the mass and heat transfer in the atmosphere, i.e. synergetic approach (Haken, 1980, 1983; Prigozhin and Stengers, 2003).

The Earth's atmosphere presents a typical example of an open dissipative system, which can be described by nonlinear equations of mathematical physics. These circumstances allow one to assume that the self-organization of physical fields and the formation of stable thermodynamic structures are possible in the atmosphere on the time–space scale determined by the parameters of phenomenon. In this approach, one may use the most significant and reliable parameters of the medium and the most general characteristics of the processes driving the Earth's climate. For example, one may use only such general parameters as the total mass of atmosphere, its thermal capacity, and the averaged energy of solar irradiation. In addition, one has to take into consideration the existence of strong negative feedback between the spherical albedo[4] of Earth and its averaged surface temperature. In such a general approach, however, the local details in the description of greenhouse effect are lost, because the constructed model becomes one-dimensional and averaged over the entire Earth.

In many cases, however, this general approach possesses definite advantages, because it allows one to obtain an analytical and unambiguous solution of such global problems as, for example, the influence of the composition of atmosphere on the total magnitude of greenhouse effect (for the entire planet). In addition, the constructed general model can be further specified by adding additional parameters and local variables. In the general model, one can incorporate, for example, the latitude of a certain locale, inclination of the axis of Earth's rotation to the ecliptic plane and the precession of axis, inflow of additional heat with the air flows (cyclones), the reflection capacity of snow cover, etc. In such a way, one can construct a three-dimensional or even four-dimensional model (the fourth dimension is the time) of the greenhouse effect. We will attempt to show that the main factors determining the climate on the Earth are the amount of solar radiation and the composition, pressure, and thermal capacity of Earth's atmosphere (Sorokhtin, 1990, 2001a, b; Sorokhtin and Ushakov, 2002).

---

[4]Albedo: the fraction of solar radiation that is reflected back into space.

3.3. Principal Characteristics of Present-day Atmosphere

The mass of atmosphere is equal to about 5.15×10^{21} g, the average pressure at sea level p_0 is equal to one physical atmosphere or $1.0132\,\text{bar} = 1{,}013.2\,\text{mbar}$ (760 mm of mercury), and the density ρ_0 is about $1.27 \times 10^{-3}\,\text{g/cm}^3$. The atmospheric pressure and air density decrease exponentially with the elevation (Fig. 3.4):

$$p = p_0 \exp\left\{\frac{-g\mu h}{RT}\right\} \tag{3.1}$$

where g is the gravitational acceleration (981 cm/s^2); μ is the average molecular weight of atmospheric gases (for the Earth $\mu = 28.97$ at $p = p_0$); the universal gas constant $R = 1.987\,\text{cal/}^\circ\text{C}\cdot\text{mole} = 8.314 \times 10^7\,\text{erg/}^\circ\text{C}\cdot\text{mole}$; T is the absolute temperature in K; and h is the elevation above sea level.

Nitrogen–oxygen content of the Earth's atmosphere is unique for the planets of Solar System. The dry air contains 75.51% (by mass) of nitrogen, 23.15% of oxygen, 1.28% of argon, 0.046% of carbon dioxide, 0.00125% of neon, and about 0.0007% of all the other gases. Water vapor is a very important active component of atmosphere (also water present as droplets in the clouds). Total average content of water in the atmosphere reaches 0.12–0.13×10^{20} g, which, if condensed, is equivalent to a layer of 2.5 cm in thickness or an average of 2.5 g/cm^2 of the Earth's surface. Taking into consideration the average annual quantity of evaporation and precipitation

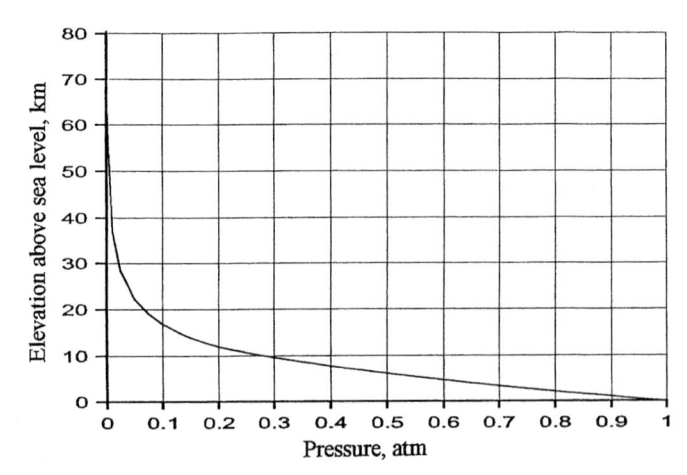

Figure 3.4: Atmospheric Pressure versus Elevation, Computed According to Eq. (3.1).

(≈ 780 mm of the water column), the water vapor in atmosphere is renewed about 30 times per year, or approximately every 12 days. The content of ozone in the atmosphere is approximately equal to 3.1×10^{15} g, whereas the content of oxygen is approximately equal to 1.192×10^{21} g.

Considering the temperature distribution in the Earth's atmosphere (Fig. 3.5), one can identify three characteristic layers: (1) *troposphere* (the lower and densest layer) extends up to an elevation of 8–10 km at high latitudes and 16–18 km in the equatorial belt (average of 12 km). Troposphere layer contains up to 75% of the entire mass of atmosphere and is characterized by a linear distribution of temperature. (2) The middle, significantly rarefied, layer includes the *stratosphere* and *mesosphere*. In this layer the temperature distribution exhibits a clear maximum of 270 K at elevations of about 50 km. (3) The highest layer of atmosphere is *thermosphere*, where the temperature of ionized gases increases with elevation up to 1,000 K and higher. At elevations above 1,000 km, the thermosphere grades into the *exosphere* and then further into the open space.

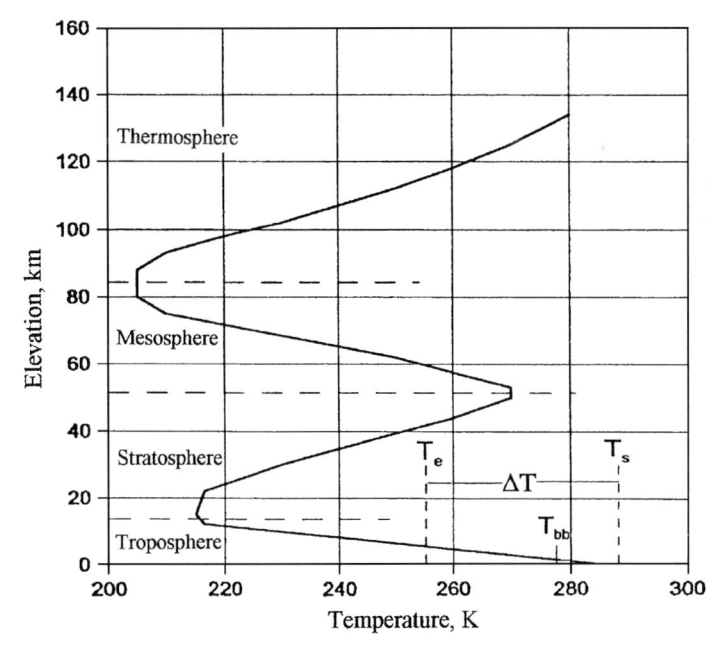

Figure 3.5: Temperature Distribution in the Earth's Atmosphere (After Sorokhtin and Ushakov, 2002). T_e is the Effective (Radiation) Temperature of Earth; T_s is the Average Earth's Surface Temperature; ΔT is the Value of "Greenhouse Effect"; T_{bb} is the Temperature of Absolutely Black Body at the Distance of Earth from the Sun.

The absorption of ultraviolet radiation in the stratosphere and mesosphere occurs mostly because of dissociation of oxygen molecules:

$$O_2 + hv \to 2O. \tag{3.2}$$

The temperature maximum at the boundary between stratosphere and mesosphere is due to the energy generation in the process of ozone formation:

$$O_2 + O \to O_3 + 31.1 \text{ kcal/mole} \tag{3.3}$$

and also in the process of ultraviolet radiation absorption by ozone:

$$O_3 + hv \to O_3 + \text{heat} \tag{3.4}$$

There are transitional layers between the troposphere and stratosphere (*tropopause*, with temperatures of about 190–220 K) and between the mesosphere and thermosphere (*mesopause*, with temperatures of about 180–190 K).

The distribution of average temperature in the troposphere differs from the temperature distribution in the stratosphere, mesosphere, and thermosphere. In the troposphere, this distribution is almost linear, whereas in the upper two layers of atmosphere the distribution is nonlinear with the characteristic maximum at the elevation of about 50 km and increasing temperature above the elevation of 90 km. Thus, the heat transfer is determined mostly by radiation in the stratosphere and mesosphere, whereas in the dense troposphere the heat transfer occurs mainly by convection.

3.4. Key Points of the Adiabatic Theory of Greenhouse Effect

By definition, the greenhouse effect is the difference ΔT between the average temperature of planet surface T_s and its effective temperature T_e, which is determined by the solar radiation:

$$\Delta T = T_s - T_e \tag{3.5}$$

The average surface temperature of the Earth is about 288 K (or 15 °C), and its effective radiation temperature is determined by the classic Stefan–Boltzmann law:

$$T_e^4 = \frac{(1-A)S}{4\sigma} \tag{3.6}$$

where $\sigma = 5.67 \times 10^{-5} \text{ erg/cm}^2 \text{ s } °C^4$ is the Stefan–Boltzmann constant; S is the solar constant at the distance of Earth from the Sun ($S = 1.367 \times 10^6 \text{ erg/cm}^2 \text{ s}$); A is the albedo, which is determined mostly by the cloud cover (for the Earth $A \approx 0.3$). According to Eq. (3.6), the effective temperature T_e is equal to 255 K (or −18 °C). Therefore, the present-day greenhouse effect for the Earth should be equal to 33 °C.

Applying the Stefan–Boltzmann law for the evaluation of heat transfer in the atmosphere, one has to realize that the heat transfer by radiation dominates only in the upper diffuse layers of stratosphere, mesosphere, and thermosphere. The heat transfer in the lower, denser layer (troposphere) occurs mostly by convection (Sorokhtin, 2001a, b). In the troposphere (with the air pressure exceeding 0.2 atm) the heat transfer by convection is dominant. When the temperature of a given mass of air increases, its volume increases proportionally. As the hot air expands, it becomes less dense and rises. In turn, the denser cooler air drops down and replaces the warmer air. This system works in the Earth's atmospheric system as a continuous surface cooler. The cooling effect of the convection in the lower layer of troposphere can considerably exceed the warming of air by radiation.

The most important conclusion from this observation is that the temperature distribution in the troposphere has to be close to adiabatic, because the air mass expands and cools while rising and compresses and heats while moving down. This does not necessarily imply that at any particular instant, distribution of temperature has to be adiabatic. One should consider some averaged distribution over the time intervals of the order of a month.

The adiabatic temperature distribution is determined by the atmospheric pressure p and by the effective thermal capacity of air, which takes into account its additional heating as a result of absorption of infrared radiation of the Earth's surface by the greenhouse gases and heat of condensation of water vapor in troposphere. For adiabatic process, the temperature of gas is a function of pressure that can be presented in the following form (Landau and Lifshits, 1979):

$$T = Cp^\alpha \tag{3.7}$$

where C is a constant, which can be estimated using experimental data; $\alpha = (\gamma - 1)/\gamma$; $\gamma = c_p/c_v$; and c_p and c_v are the specific heats of gas at constant pressure and constant volume, respectively. For tri-atomic gases (CO_2 and H_2O), $\gamma = 1.3$ and $\alpha = 0.2308$, whereas for the diatomic gases (N_2 and O_2) $\gamma = 1.4$ and $\alpha = 0.2857$. As a result of condensation of water vapor in troposphere, which emits heat, the adiabatic exponent α decreases. For example, the average value of this parameter for the humid and heat-absorbing troposphere $\alpha = 0.1905$, whereas for the dry air, $\alpha = 0.2846$ (Sorokhtin and Ushakov, 1999).

Condensation of water vapor in troposphere begets cloudiness, which is the main factor that determines the Earth's albedo. This process gives rise to strong negative feedback between the near-surface and radiation temperatures of the Earth, which stabilizes the temperature regime of troposphere. The rising surface temperature increases the water evaporation and the cloudiness of Earth, which, in turn, increases the albedo of planet and the reflection capacity of Earth's atmosphere. This increases the portion of solar radiation reflected back to space, decreasing the heat supply to the Earth. As a result, the average temperature of Earth's surface decreases to the previous level.

In any system, a negative feedback leads to a linear dependence of system's output on the system's input. This is a universal property of the systems with the negative feedbacks. This property manifests itself in various systems, e.g. the centrifugal regulator of James Watt in a steam engine or the thermal (self-organizing) system of atmosphere.

In the atmospheric thermal system, the input is the temperature T_{gb}, which is determined by the solar radiation at the distance of Earth from the Sun, whereas the output is the average surface temperature T_s. For the Earth, the temperature $T_{gb} = 288.2\,\text{K} = +15\,°\text{C}$. Thus, one arrives at the second important conclusion that the average surface temperature T_s is a linear function of T_{gb}, which characterizes the solar radiation at the distance from Earth to Sun. This enables one to determine the average temperature, T, at any point in the troposphere with pressure p: $T = T_{gb}(p/p_0)^\alpha$.

In Eq. (3.7), C and α are some constants that are defined below. For the temperature $T_{gb} = 288.2\,\text{K}$ and the atmospheric pressure at sea level, one can rewrite Eq. (3.7) in the following form:

$$T_{gb} = C p_0^\alpha \tag{3.8}$$

In the above equation, the constant C is equal to $288.2/(p_0)^\alpha$. Thus, one can define the average temperature at any elevation in the troposphere by the following exponential function (at $p > 0.2\,\text{atm}$):

$$T = 288.2\,(p/p_0)^\alpha \tag{3.9}$$

It is noteworthy that in the formula

$$T_{bb} = (S/4\sigma)^{1/4} \tag{3.10}$$

for computing the temperature of "absolutely black body" the solar constant S is divided by 4, because the area of Earth's disk insolation is four times lower than the total illuminated area of Earth. Equation (3.10) is valid only if the axis of rotation of planet is strictly perpendicular to the ecliptic plane, i.e. the angle of precession ψ is equal to zero.

The angle of inclination of the equatorial plane to the ecliptic plane is not equal to zero and is changing in time. Therefore, each of the Earth's polar regions is insolated during half a year only. Another half a year it is deprived of the influx of solar energy. When one of the polar regions is insolated, the other is situated in the shadow of Earth's body and does not receive the solar energy. The rest of the Earth's surface receives its portion of solar energy on a regular basis and, consequently, Eq. (3.8) is valid for calculation of temperature. Therefore, in computing the average temperature of "inclined" planet at high latitudes (polar regions), one needs to divide the solar constant by 2 (not by 4). In addition, one has to take into consideration the spherical shape of polar region. As a result, the solar constant in Eq. (3.10) has to be divided by a number N, which lies between 2 and 4. Taking all of the above conditions into account and considering that the precession angle is relatively small, one can derive the following equation for the distribution of average temperature in the troposphere:

$$T = \{S \div [\sigma \times \{4(\pi/2 - \psi)/(\pi/2) + 2\psi(\pi/2) \times 2/(1 + \cos\psi)\}]\}^{1/4}(p/p_0)^\alpha \quad (3.11)$$

where ψ is the Earth's precession angle; p is the atmospheric pressure at a given altitude ($0.2\,\text{atm} < p < p_0$), $p_0 = 1\,\text{atm}$; $\alpha = (c_p - c_v)/c_p$; where c_p and c_v are the specific heats of atmosphere at a constant pressure and a constant volume, respectively.

For the present-day Earth's precession angle ($\psi \approx 23.44°$), the average surface temperature at sea level $T_s \approx 288.2\,\text{K}$. This implies, in particular, that the average surface temperature of Earth (considering the Earth's precession) is equal to the efficient temperature of effectively "gray" body at the average distance of Earth to Sun ($T_s = T_{gb}$). Thus, the average temperature of contemporary troposphere at any altitude ($p > 0.2\,\text{atm}$) is equal to:

$$T = T_{gb}(p/p_0)^\alpha = 288.2(p/p_0)^\alpha \quad (3.12)$$

If the axis of Earth's rotation were perpendicular to the ecliptic plane, then, at $p = 1\,\text{atm}$, the average surface temperature would be equal to $278.6\,\text{K}$ (the temperature of absolutely black body at a distance from Earth to Sun). The difference in temperature for $\psi \approx 23.44°$ and $\psi = 0°$ reaches 9.6–10 °C. Consequently, the present-day radiation temperature (considering the precession angle) at the distance from Earth to Sun is equal to 263.6 K and not 255 K.

If the specific heat of gas c_p is measured in cal/g °C, and the gas constant $R = 1.987\,\text{cal/mol\,°C}$, then the adiabatic exponent α can be found according to the following formula:

$$\alpha = R/\mu(c_p + C_w + C_r) \quad (3.13)$$

and

$$c_p = [p(N_2) \cdot c_p(N_2) + p(O_2) \cdot c_p(O_2) + p(CO_2) \cdot c_p(CO_2)$$
$$+ p(Ar) \cdot c_p(Ar)]/p \quad (3.14)$$

where μ is the molecular weight of air; $p(N_2) = 0.7649\,\text{bar}$, $p(O_2) = 0.2345\,\text{bar}$, $p(CO_2) = 0.00046\,\text{bar}$, and $p(Ar) = 0.01297\,\text{bar}$ are the partial pressures of nitrogen, oxygen, carbon dioxide, and argon, respectively; $p = 1.013\,\text{bar}$ is the total atmospheric pressure; $c_p(N_2) = 0.248$, $c_p(O_2) = 0.218$, $c_p(CO_2) = 0.197$, and $c_p(Ar) = 0.124\,cal/g\,°C$ are the specific heats of nitrogen, oxygen, carbon dioxide, and argon at a constant pressure, respectively (Handbook on Geochemistry, 1990); C_w and C_r are corrective coefficients, considering the total heating caused by water condensation C_w (in the wet atmosphere) and absorption of radiation heat of Earth and Sun C_r. For the dry gaseous mixture of Earth's atmosphere $c_p = 0.2394\,cal/g\,°C$.

The adiabatic exponent α can be determined as the parameter of best fit of theoretical model (Eq. (3.12)) to the experimental data (Table 3.1). For the Earth's atmosphere, $\alpha = 0.1905$. In this case, $C_w + C_r = 0.1203$.

Using the radiation ($T_e = 263.6\,K$) and average surface temperature ($T_s = 288.2\,K$), one can determine the corrective coefficients C_w and C_r (components of specific heat) attributed to the water vapor condensation and absorption of infrared radiation of Sun and Earth by the Earth's atmosphere, respectively (Sorokhtin, 2001a, b). On introducing the notations Q_a (total effective heat resource) and m_a (total effective mass of Earth's atmosphere), one can present the radiation component of specific heat C_r as a function of radiation temperature T_e in the following form:

$$C_r = \frac{Q_a}{m_a T_e} \tag{3.15}$$

By analogy, one can consider that additional heating of atmosphere from the radiation temperature T_e to the average surface temperature T_s is determined by the corrected specific heat considering the water vapor condensation:

$$c_p + C_w = \frac{Q_a}{m_a(T_s - T_e)} \tag{3.16}$$

Combining Eqs. (3.15) and (3.16), one can obtain the following equation for C_r:

$$C_r = \frac{(c_p + C_w)(T_s - T_e)}{T_e} \tag{3.17}$$

Together with Eq. (3.13), Eq. (3.17) implies that:

$$C_r = \frac{R(T_s - T_e)}{\mu \alpha T_s} \tag{3.18}$$

$$C_w = \left\{\frac{RT_e}{\mu \alpha T_s}\right\} - c_p \tag{3.19}$$

where the specific heat of dry air c_p is determined by using Eq. (3.14).

Substituting the values of parameters of the Earth's atmosphere ($\alpha = 0.1905$, $\mu = 29$, $c_p = 0.2394\,cal/g\,°C$, $T_s = 288.2\,K$, $T_e = 263.6\,K$, and

Table 3.1: Comparison of the Theoretical Temperature Distribution (Eq. (3.11)) in Troposphere with that based on the Standard Model of Earth's Atmosphere.

Elevation, h (km)	Standard Model of Atmosphere			Adiabatic Theoretical Model of Atmosphere (Eq. (3.11))		
	Pressure, p (mm of mercury)	Temperature, T (°C)	Temperature, T (K)	Pressure, p (atm)	Temperature, T (K)	Temperature, T (°C)
0.0	760.00	15.00	288.2	1.00	288.20	15.00
0.5	716.01	11.75	284.95	0.9421	284.95	11.77
1.0	674.11	8.50	281.70	0.8870	281.70	8.50
1.5	634.21	5.25	278.45	0.8345	278.44	5.24
2.0	596.26	2.00	275.20	0.7846	275.20	2.00
2.5	260.16	−1.25	271.95	0.6084	262.20	−11.00
3.0	525.87	−4.50	268.70	0.6919	268.59	−4.51
3.5	493.30	−7.75	265.45	0.6491	265.45	−7.76
4.0	462.40	−11.00	262.20	0.4660	249.21	−23.99
4.5	433.10	−14.25	258.95	0.5699	258.95	−14.25
5.0	405.33	−17.50	255.70	0.5333	255.69	−17.51
5.5	379.04	−20.75	252.45	0.4987	252.45	−20.75
6.0	354.16	−24.00	249.20	0.7371	271.94	−1.26
6.5	330.72	−27.25	245.95	0.4345	245.91	−27.29
7.0	308.52	−30.50	242.70	0.4059	242.75	−30.45
7.5	287.55	−33.75	239.45	0.3784	239.53	−33.67
8.0	267.79	−37.00	236.20	0.3524	236.30	−36.90
8.5	249.16	−40.25	232.95	0.3278	233.07	−40.13
9.0	231.62	−43.50	229.70	0.3048	229.87	−43.33
9.5	215.09	−46.75	226.45	0.2830	226.64	−46.56
10.0	199.60	−50.00	223.20	0.2626	223.44	−49.78

Table 3.1: (*Continued*).

	Standard Model of Atmosphere				Adiabatic Theoretical Model of Atmosphere (Eq. (3.11))		
Elevation, h (km)	Pressure, p (mm of mercury)	Temperature, T (°C)	Temperature, T (K)		Pressure, p (atm)	Temperature, T (K)	Temperature, T (°C)
10.5	185.01	−53.25	219.95		0.2434	220.23	−52.97
11.0	171.34	−56.50	216.70		0.2254	217.04	−56.16
11.5	160.11	−56.50	216.70	Tropopause	0.2107	214.27	−58.93
12.0	149.64	−56.50	216.70		0.1969	211.52	−61.68

$R = 1.987\,\text{cal/mol}\,°C$), one obtains: $C_r = 0.0307\,\text{cal/g}\,°C$, $C_w = 0.0896\,\text{cal/g}\,°C$, and $C_w + C_r = 0.1203\,\text{cal/g}\,°C$, i.e. exactly the same value as obtained using the theoretical model of tropospheric temperature distribution (Eq. (3.12)). The latter presents additional evidence supporting the adiabatic theory of "greenhouse" effect. The comparison of components of specific heat of the air mass participating in the heat transfer shows that convection accounts for approximately 67%, radiation accounts for about 8%, and the condensation of water vapor accounts for about 25% (Sorokhtin, 2001a, b) of the total amount of heat transfer from the Earth's surface to troposphere.

The convective component of heat transfer dominates in the troposphere. When infrared radiation is absorbed by the greenhouse gases, the radiation energy is transformed into the oscillations of gas molecules, i.e. in heating of the exposed volume of gaseous mixture. Then, the further heat transfer can occur either due to diffusion or by convective transfer of expanded volumes of gas. Inasmuch as the specific heats of air are very small (about $5.3 \times 10^{-5}\,\text{cal/cm s}\,°C$), the rates of heat transfer by diffusion do not exceed several cm/s, whereas the rates of heat transfer by convection in the troposphere can reach many meters per second. Analogous situation occurs upon heating of air as a result of water vapor condensation: the rates of convective transfer of heated volumes of air in the troposphere are many orders of magnitude higher than the rates of heat transfer by diffusion.

Equation (3.11) (the adiabatic model of atmospheric temperature) can be applied for computation of atmospheric temperature distribution for any planet possessing a dense atmosphere (with atmospheric pressure higher than 0.2 atm) and also for various geologic periods of Earth's development. To modify the adiabatic model for different conditions, one needs to specify the value of solar constant S, the angle of precession of planet ψ, and the value of adiabatic exponent α.

The adiabatic model of greenhouse effect can be verified by comparison of the theoretical temperature distribution in the troposphere of Earth (constructed based on Eq. (3.11)) with the standard model based on experimental data. For the Earth, the parameters of adiabatic model were chosen as follows: $S = S_0 = 1.367\,\text{erg/cm s}$; $\psi = 23.44°$; and $\alpha = 0.1905$. Data in Table 3.1 show that the theoretical temperature distribution based on Eq. (3.11) is identical to the standard temperature distribution of Earth (Handbook, 1951) with precision of 0.1%. It is noteworthy that the standard model of Earth's atmosphere presents averaged (over the Earth's surface) values of temperature and pressure as the functions of elevation above the sea level. This model (with the temperature gradient of 6.5 K/km) is applied worldwide for calibration of aircraft gauges and also for barometers, which are used for weather observations.

The adiabatic model can be further verified by comparison of the theoretical temperature distribution in the dense (consisting mostly of carbon dioxide) troposphere of Venus with the experimental data. For Venus, $\psi \approx 3°$, $\alpha = 0.179$, $\mu = 43.5$, $c_p = 0.2015 \, \text{cal/g} \, °C$; $T_s = 735.3 \, \text{K}$, $T_e = 228 \, \text{K}$, and, therefore, $C_r = 0.177 \, \text{cal/g} \, °C$, $C_w = -0.122 \, \text{cal/g} \, °C$, and $C_w + C_r = 0.055 \, \text{cal/g} \, °C$. The increased value of parameter C_r, which is a measure of the radiation component of heat transfer, most probably can be explained by the extremely hot condition of the troposphere of Venus. The fact that $C_w < 0$ means that in the troposphere of Venus (especially in its lower and middle layers) the endothermic reactions of dissociation of some compounds dominate (e.g. dissociation of sulfuric acid H_2SO_4 into SO_3 and H_2O). Meantime, in the upper layers of the troposphere of Venus, at the altitudes of 40–50 km and above the altitude of 60 km, the parameter $C_w > 0$, and the exothermic reactions of formation of chemical compounds (sulfuric acid, for example) dominate there. In addition, the water vapor condenses in the clouds of Venus heating the atmosphere.

For the Venus atmosphere, $p_s = 90.9 \, \text{atm}$ and $S = 2.62 \times 10^6 \, \text{erg/cm}^2 \, \text{s}$ (Marov, 1986; Planet Venus, 1989). Substituting all these values into Eq. (3.11), one can construct the temperature distribution for the atmosphere of Venus. The results of testing of the adiabatic model (3.11) by comparison with the experimental data are presented in Table 3.2 and Fig. 3.6.

As shown in Table 3.2 and Fig. 3.6, the theoretical temperature distribution in the troposphere of Venus (which is absolutely different than that of Earth) has a good fit to the experimental data. Below an elevation of 40 km, the empirical data are approximated by the theoretical model with precision of 0.5–1.0%, whereas at the elevation of 40–60 km the theoretical temperatures lie between two sets of data obtained at high and low latitudes of atmosphere of Venus. Above the elevation with the pressure of 0.2 atm the Venus tropopause begins, and the temperature model presented here does not apply.

The above-presented analysis indicates that the average surface temperature of a planet is determined by the (1) solar constant, (2) mass (pressure) of atmosphere, (3) specific heat of atmospheric mixture of gases, and (4) precession angle of planet.

3.5. Possible Directions of Generalizing the Adiabatic Theory

The above-described model of greenhouse effect is one-dimensional. In this model, the planet is considered as a dimensionless point, and the only dimension is the elevation "above" this point. Such a synergetic model is most accurate when one needs to determine the global characteristics of

Table 3.2: Comparison of the Theoretical Temperature Distribution in the Venus Troposphere (Eq. (3.11)) with the Empirical Data (Planet Venus, 1989).

	Experimental Data				Adiabatic Theoretical Model (Eq. (3.11)) of Venus Atmosphere		
Elevation, h (km)	Temperature, T (K)	Pressure, p (bar)	Temperature, T (K)	Pressure, p (bar)	Elevation, h (km)	Pressure, p (atm)	Temperature, T (K)
0	735.3	92.10			0	90.92	737.7
1	727.7	86.45			1	85.34	729.3
2	720.2	81.09			2	80.05	721.0
3	712.4	76.01			3	75.03	712.7
4	704.6	71.20			4	70.29	704.4
5	696.8	66.65			5	65.79	696.2
6	688.8	62.35			6	61.55	687.9
7	681.1	58.28			7	57.53	679.6
8	673.6	54.44			8	53.74	671.4
9	665.8	50.81			9	50.16	663.2
10	658.2	47.39			10	46.78	654.9
12	643.2	41.12			12	40.59	638.5
14	628.1	35.57			14	35.11	622.1
16	613.3	30.66			16	30.27	605.8
18	597.1	26.33			18	25.99	589.5
20	580.7	22.52			20	22.23	573.3
22	564.3	19.17			22	18.92	557
24	547.5	16.25			24	16.04	540.7
26	530.7	13.70			26	13.52	524.4
28	513.8	11.49			28	11.34	508.2
30	496.9	9.581			30	9.458	492

Table 3.2: (*Continued*).

	Experimental Data				Adiabatic Theoretical Model (Eq. (3.11)) of Venus Atmosphere		
Elevation, h (km)	Temperature, T (K)	Pressure, p (bar)	Temperature, T (K)	Pressure, p (bar)	Elevation, h (km)	Pressure, p (atm)	Temperature, T (K)
Latitude of Measurements 0–30°			Latitude of Measurements of 75°			Theoretical Calculations	
33	417.7	7.211	471.7	7.211	33	7.118	467.5
36	448.0	5.346	446.5	5.345	36	5.277	443.2
39	425.1	3.903	420.5	3.894	39	3.848	418.8
42	403.5	2.802	394.5	2.78	42	2.755	394.5
45	385.4	1.979	368.7	1.941	45	1.935	370.3
48	366.4	1.375	343.5	1.321	48	1.331	346.3
51	342.0	0.9347	318.5	0.8741	51	0.8905	322.3
54	312.8	0.6160	290.0	0.5582	54	0.5796	298.4
57	282.5	0.3891	258.2	0.3392	57	0.3595	274.0
60	262.8	0.2357	237.5	0.1948	60	0.2125	249.4

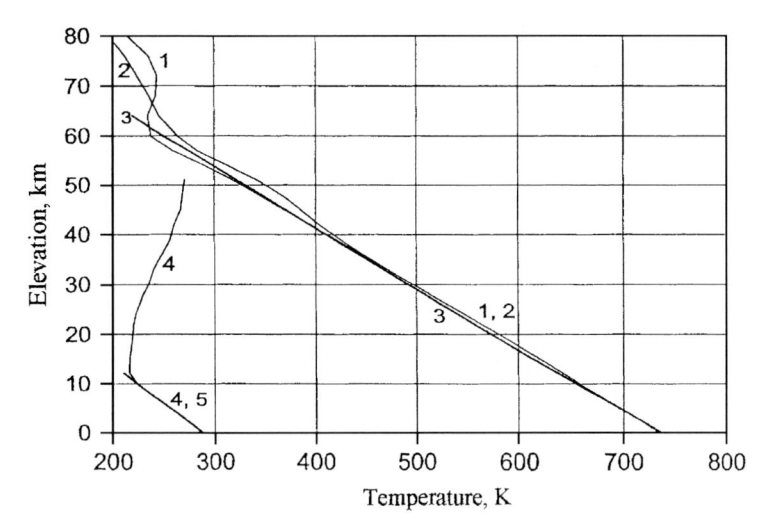

[Figure: plot of Elevation (km) vs Temperature (K)]

Figure 3.6: The Experimental Temperature Distributions in the Troposphere and Stratosphere of Earth (Curve 4) and in the Troposphere of Venus (Curves 1 and 2; Planet Venus, 1989) in Comparison with the Theoretical Distributions (Curves 5 and 3) Constructed According to the Adiabatic Theory of Greenhouse Effect.

troposphere: for example, the greenhouse effect, the average temperature distribution, and the average characteristics of heat transfer. Using the Lambert's law of a sphere insolation and introducing the local latitude, one can convert this model into two-dimensional. Introducing, in addition, the local longitude and seasonal deviations of planet's insolation, the model can be converted into three- and four-dimensional (the fourth dimension is time). The precision of determining the dependence of greenhouse effect on the composition of atmosphere for these models, however, becomes considerably lower.

In this case, the physical concept of "black body" temperature should be replaced by the concept of "gray body" temperature.

$$T_{gb}(\varphi) = \{S \div [\sigma \times 4(\pi/2 - \psi)/(\pi/2) + 2\psi(\pi/2) \\ \times 2/(1 + \cos\psi)]\}^{1/4} \cos\varphi \qquad (3.20)$$

where φ is the local latitude.

If, in addition, one takes into consideration the convective heat transfer in the atmosphere, then the temperature of simulating Earth gray body can be written in the followiing form:

$$T_{gbc}(\varphi) = [(T_{gb}(\varphi))^4 + Q^v/\sigma]^{1/4} \qquad (3.21)$$

where Q^v is the rate of heat transfer with the moving air mass (the cyclones, for example). Considering this process, however, one needs to account for the transfer of air masses, which can alter the adiabatic temperature distribution in troposphere (though the energy redistribution attributed to this process is insignificant). At the night time, when $S = 0$, in addition to the heat transfer by air fluxes, the rate of heat transfer can be determined by the heat emission from the Earth's surface warmed during the day.

Under the above-considered conditions, the Earth's surface temperature $T_s(\varphi)$ at the latitude of φ is equal to:

$$T_s(\varphi) = [(T_{gb}(\varphi))^4 + Q^v/\sigma]^{1/4} \cdot (p_s/p_0)^\alpha \qquad (3.22)$$

Equation (3.22) allows one to determine the latitudes for surface temperatures T_s. If the latitudinal distribution of empirical averaged temperatures is known, then it is possible to determine the average speed of influx of heat by air masses Q^v at a given latitude (Fig. 3.7). Figure 3.7 shows also the graph of averaged air temperature at the surface versus the geographic latitude (Khromov and Petrosyants, 1994). In the same figure, the theoretical distribution of T_s is presented, which was constructed as the best fit to the experimental data. Figure 3.8 shows the relationship between Q^v and geographic latitude, constructed as the best fit of Eq. (3.22) to the empirical data given in

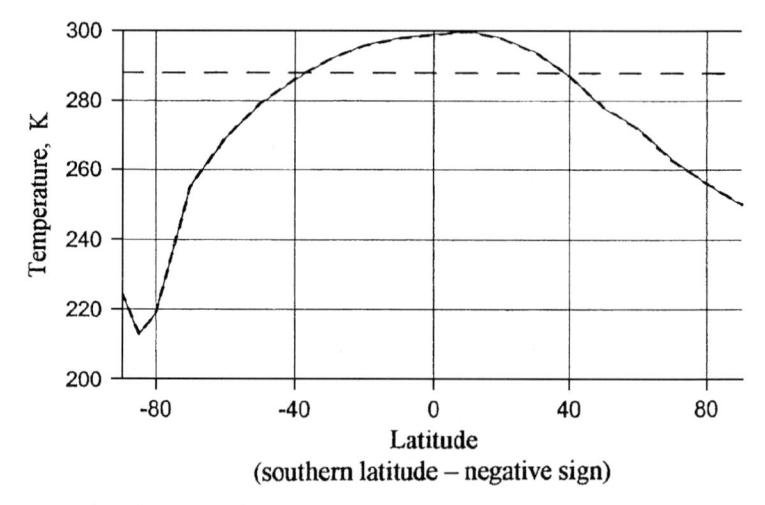

Figure 3.7: The Averaged Yearly Air Temperature at the Earth's Surface versus the Geographic Latitude (Modified after Khromov and Petrosyants, 1994). The Thin Solid Curve Represents the Empirical Distribution. The Dashed Curve Represents the Theoretical Distribution (Eq. (3.22)) Constructed According to the Data of Fig. 3.8. The Horizontal Dashed Line Shows the Average Earth's Surface Temperature (= 288 K).

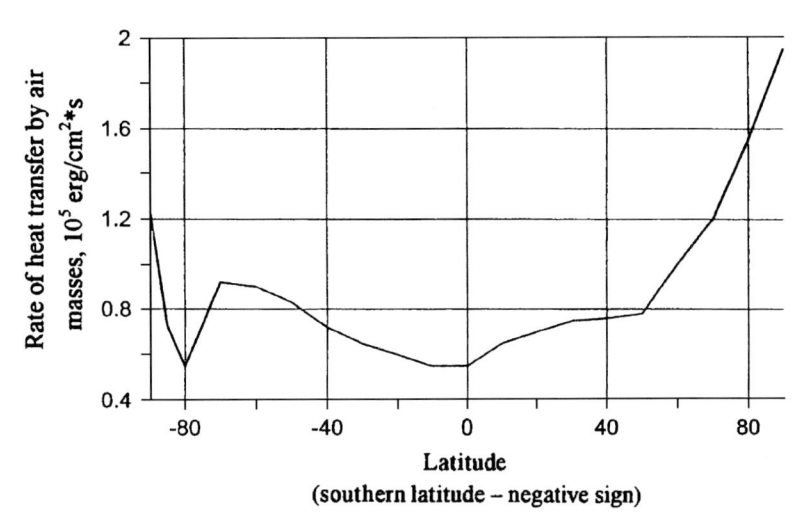

Figure 3.8: The Rate of Release of Energy of Synoptic Processes in the Earth's Troposphere, Q_v (Average Yearly Rates of the Heat Transfer by Air Flows), versus the Geographic Latitude. The Graph was Constructed by Equating the Theoretical Values of Surface Temperatures (Eq. (3.22)) to Empirical Data.

Fig. 3.7. Figure 3.8 shows that the rate of heat influx with the trade winds reaching $5.5 \times 10^4 \, \text{erg/cm}^2\,\text{s}$. At high latitudes, the rate of energy liberation with cyclonic activity is gradually increasing. Thus, one can reach the conclusion that the kinetic energy of air masses is transferring from the tropical belt to the boreal and polar zones. The local minimum of the energy transfer over Antarctica ($\varphi \approx -80°$ SL) is explained by the high position of cupola of its glacial cover (about 3.5–4 km above the sea level) and the stable anticyclonic conditions above it. The surface elevation at the vicinity of South Pole is lower (about 2.8 km) and, consequently, some cyclones can penetrate this region from the South Ocean. To construct the distributions at various latitudes of average yearly total intensity of synoptic processes, one needs to multiply their specific values (determined above) by the areas of latitude zones. For the 10°-latitude belt, the intensity of energetic synoptic processes is shown in Fig. 3.9. For comparison, in the same figure the intensity of insolation versus latitude is also shown. The intensity of insolation recalculated for the entire Earth's surface is approximately equal to $1.11 \times 10^{24} \, \text{erg/s}$. Taking the Earth's albedo ≈ 0.3 into consideration, the corrected intensity is about $7.76 \times 10^{23} \, \text{erg/s}$. The total average power of tropospheric synoptic processes in the Earth's atmosphere is about $3.79 \times 10^{23} \, \text{erg/s}$, which, according to the virial theorem, is almost 50% of the power of Earth's insolation. In

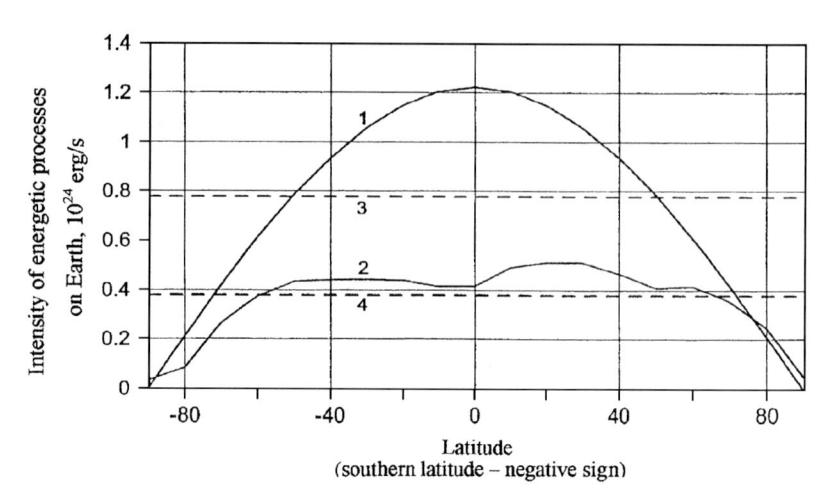

154 Global Warming and Global Cooling: Evolution of Climate on Earth

Figure 3.9: The Intensity of Average Earth's Insolation (Curve 1) and Synoptic Processes in the Earth's Troposphere (Curve 2) versus the Geographic Latitude. Curve 3 Represents the Average Value of Earth's Insolation, and Curve 4 shows the Average Intensity of Earth's Synoptic Processes.

comparison with this huge amount of energy, the total amount of anthropogenic energy production (1.3×10^{13} watts = 1.3×10^{20} erg/s) is negligible. Therefore, the anthropogenic energy production cannot noticeably affect the Earth's global climate. It is important to note that the atmospheric heat content is about 1.3×10^{31} erg and the oceanic heat content is 1.6×10^{34} erg (the heat content in the "solid" Earth is about 1.6×10^{38} erg), whereas the total anthropogenic generation of heat is only 4.1×10^{27} erg/year.

The two-dimensional model allows one to construct the latitudinal surface temperature distributions and describe the energetic synoptic processes, which contribute enormously to the understanding of seasonal and spatial climatic variability. To construct a three-dimensional model, one also needs to introduce the longitudinal coordinates and determine the areas of oceans and continents. The four-dimensional model would require introduction of the seasonal and daily variations in insolation, the precession angle, orbital deviations, etc.

The adiabatic theory considered here provides a good explanation for the various weather phenomena, such as intense cooling of near-surface air layers during the cloudless nights under the anticyclones, for example. For these conditions, $S \approx 0$ and the heat influx Q^v is small. In the cyclonic regions, the convective air mass exchange is slowing down and, consequently, the convective heat influx is decreasing. Meantime, the Earth's surface (heated during the day) continues emanating the accumulated heat. Under these conditions, the factor Q^v in Eq. (3.22) decreases and, as a result, at night (when $S\approx 0$) the near-surface temperature decreases considerably. At high latitudes in the winter time, when the Earth's surface is covered by the snow (with high Earth's albedo) and its heating by the solar radiation is insignificant, these processes lead to the air overcooling and cause extremely cold weather. For the long-lasting stable anticyclones ($Q^v \approx 0$) in the snow-covered regions, general overcooling of the troposphere occurs, and the tropopause drops almost to the Earth's surface. Revealing examples of such overcooling are regularly observed in the central regions of Antarctica and in the Siberian regions of Verkhoyansk and Yacutia. However, as soon as the anticyclonic regime is changed for cyclonic activity, the convective stirring of air masses is renewed, weather warms up, and the approximately adiabatic distribution of atmospheric temperature is restored. As mentioned above, in the snow-covered anticyclonic regions in the winter time the near-surface temperature drops almost to the temperature of tropopause (i.e. the lower layer of stratosphere). In the summer time, however, in such anticyclonic regions with dry air the near-surface layer of atmosphere experiences overheating of about 4–5 °C and higher (Fig. 3.10).

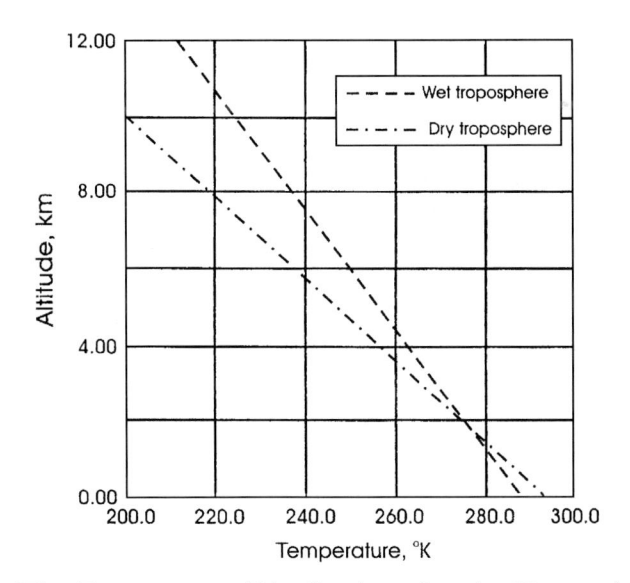

Figure 3.10: The Temperature Distributions for the Dry and Transparent (Eq. (3.23)) and Wet, Absorbing Infrared Radiation (Eq. (3.24)) Earth's Troposphere. Comparison of these Distributions Shows that (At the Constant other Conditions) the Surface Temperature of Wet and Heat Absorbing Troposphere is Always Lower than that of Dry and Transparent Troposphere (the Differences Reaches 3.8 °C).

3.6. Prognostic Atmospheric Temperature Estimates

Applying Eq. (3.12), one can construct the temperature distribution and determine the temperature gradient in completely dry and transparent Earth's troposphere. For this case $C_w + C_r = 0$ and, using Eqs. (3.13) and (3.14), one obtains $c_{pd} = 0.2394\,\text{cal/g}\,°C \approx 1.0023 \times 10^7\,\text{erg/g}\,°C$ (the value of specific heat for dry troposphere) and $\alpha = 0.286$. Then, for dry troposphere, the temperature gradient grad T_{dry} is equal to:

$$\text{grad}\,T_{dry} = g/c_{pd} \approx 9.8\,\text{K/km} \tag{3.23}$$

where g is the gravitational acceleration ($981\,\text{cm/s}^2$) of Earth. For the wet and heat-absorbing troposphere, the temperature gradient grad T_{wet} is determined by the following equation:

$$\text{grad}\,T_{wet} = g/(c_{pd} + C_w + C_r) \approx 6.5\,\text{K/km} \tag{3.24}$$

where $C_w + C_r \approx 0.123\,\text{cal/g}\,°C = 0.504 \times 10^7\,\text{erg/g}\,°C$ is an additional component of the effective specific heat of air attributed to the water vapor condensation and absorption of infrared radiation of Sun and Earth (Eqs. (3.18) and (3.19)). The computation of gradient of temperature for wet and infrared radiation-absorbing atmosphere using Eq. (3.24) indirectly verifies validity of calculations of C_w and C_r using Eqs. (3.18) and (3.19).

As a rule, anticyclones contain dry air, whereas cyclones contain wet air. The pressures in these two different regimes of air mass movement can differ by 60 mbar; therefore, the near-surface temperatures under anticyclones can exceed the values of those under cyclones by about 4 °C. Figure 3.10 shows the temperature distributions for the wet and dry atmosphere. The graphs of Fig. 3.10 demonstrate that *the near-surface temperature of dry and transparent troposphere is somewhat higher than that of wet and radiation-absorbing troposphere*. For the dry but infrared radiation-absorbing troposphere the temperature gradient is somewhat lower – about 8.1 K/km. In the above example, the average temperature difference reaches +3.8 °C. These computations provide a sound explanation to overheating of near-surface atmospheric layers and draughts in deserts and arid zones and also in the regions where together with anticyclones these dry masses of air penetrate from the arid zones (steppes of Zavolzh'e in Southern Russia, for example).

It is very important to consider the influence of so-called "greenhouse" gases on the temperature regime of troposphere. One can analyze the situation by conducting a conceptual experiment in which the nitrogen–oxygen atmosphere of the Earth is replaced by the carbon dioxide atmosphere having the same pressure. Inasmuch as the pressure of atmosphere p at sea level is equal to 1 atm and the normalization coefficient p_0 in Eq. (3.12) is also equal to 1 atm, this equation cannot be used for analysis of atmospheric temperature changes caused by the above-considered replacement. For this purpose, one can use the temperature distribution in different forms using the classic definition of the temperature of absolutely black body:

$$T = b^\alpha T_{bb}(p/p_0)^\alpha = 1.194^\alpha \times 278.6(p/p_0)^\alpha \qquad (3.25)$$

The temperature distributions constructed according to Eqs. (3.12) and (3.25) (at the fixed values of precession angle ($\psi = 23.44°$) and adiabatic exponent $\alpha = 0.1905$) coincide. Therefore, the near-surface temperature change will be determined by the coefficient b only.

Thus, if the nitrogen–oxygen atmosphere of Earth would be replaced by the carbon dioxide one with the same pressure of 1 atm, then the average near-surface temperature would decrease (not increases as commonly assumed) by approximately 2.5 °C. As shown in Fig. 3.11, the temperature decreases throughout the entire layer of troposphere. Both curves of Fig. 3.11 were constructed using the same equations (Eqs. (3.1) and (3.25)) in which only parameters depending on the atmospheric composition were subject to change ($\mu_1 = 29$, $\mu_2 = 44$, $\alpha_1 = 0.1905$, $\alpha_2 = 0.1423$), with all other parameters remaining the same ($C_w + C_r \approx 0.12\,\text{cal/g}\,°\text{C}$, $b = 1.194\,\text{atm}^{-1}$, $T_{bb} = 278.6\,\text{K}$, and $p_s = 1\,\text{atm}$).

By analogy, if the carbon dioxide atmosphere of Venus would be replaced by the nitrogen–oxygen one with the same pressure of 90.9 atm, then its average near-surface temperature would increase from 735 to 930 K (from 462 to 657 °C). Considered examples show that the saturation of atmosphere with the carbon dioxide gas (in spite of the fact that this gas absorbs the infrared radiation) always leads not to an increase but to a decrease in the "greenhouse" effect and the average temperature throughout the entire layer of the planetary troposphere (assuming that all other conditions remain the same). There is a simple explanation for the above conclusion: the molecular weight of CO_2 is about 1.5 times higher and its specific heat is about 1.2 times lower than those of Earth's air. As a result, from Eq. (3.13) the adiabatic exponent for the carbon dioxide atmosphere $\alpha(CO_2) = 0.1423$, which is about 1.34 times lower than that for the wet nitrogen–oxygen atmosphere [$\alpha(N_2 + O_2) = 0.1905$]. Additional heat absorption by the molecules of carbon dioxide increases the coefficient C_r and, consequently, decreases the

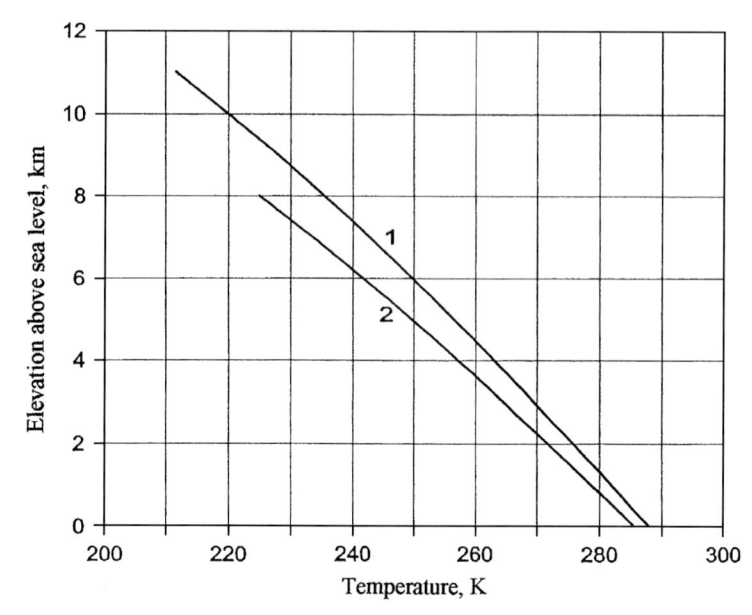

158 Global Warming and Global Cooling: Evolution of Climate on Earth

Figure 3.11: The Distributions of Averaged Temperature in the Earth's Troposphere Constructed using Eqs. (3.1) and (3.25). The Standard Model of the Nitrogen–Oxygen Earth's Atmosphere is Presented by Curve 1 and the Carbon Dioxide Atmosphere Model is Presented by Curve 2 (With Other Conditions Being Constant).

adiabatic exponent $\alpha(CO_2)$. This, in turn, leads to additional decrease in temperature. Clearly, after absorption of radiation heat by greenhouse gases its energy is transformed into the energy of heat oscillations of gas molecules. This, in turn, leads to the expansion of gaseous mixture with its consequent rapid rise to the stratosphere.

Thus, *the heat transfer in troposphere occurs mostly by convection, which is significantly more efficient than radiation.* The convective heat transfer occurs up to the lower layers of stratosphere, where the surplus of transported heat further dissipates by radiation. With increasing carbon dioxide concentration and by its intense absorption of infrared radiation, the convective heat transfer intensifies to an even greater degree, which accelerates the heat transfer beyond the troposphere (Fig. 3.5). This implies that with the increased carbon dioxide content in the troposphere, the convective mass transfer is also intensified. The intensification of synoptic processes in the Earth's troposphere, which has been observed during the last several years, may be attributed to the accumulation of carbon dioxide in troposphere.

A similar situation should occur in the process of saturation of troposphere with methane gas. In the denominator of Eq. (3.13), the product μc_p for the dry air is equal to 6.94, for methane it is 8.45, and for carbon dioxide, 8.67. This implies that $\alpha(\text{air}) > \alpha(CH_4) > \alpha(CO_2)$ and, consequently, *the maximal cooling should occur with the accumulation of carbon dioxide in the troposphere and somewhat lower cooling effect upon accumulation of methane gas.*

It is important to emphasize that accumulation of carbon dioxide and methane in the troposphere intensifies the convective processes of heat and mass transfer and leads to subsequent cooling, but not warming of climate. At the same mass of atmosphere, the total specific heat of carbon dioxide atmosphere is always lower than that of nitrogen–oxygen atmosphere. In addition, because of the higher density of carbon dioxide than that of Earth's air, the carbon dioxide atmosphere must be thinner and will retain the surface heat less in comparison with the thicker layer of the nitrogen–oxygen atmosphere, which also has a higher specific heat.

Considering the temperature distribution for a cloudy troposphere, one has to use the adiabatic distribution for the wet air below clouds and that for the dry one above them. The temperature distribution in the cloud blanket itself should depend on the heat generated as a result of water vapor condensation. The average mass of water in the atmosphere can be estimated by using Eq. (3.19) and considering that the average specific heat (C_w) is equal to 0.0791 cal/g °C. Taking into account the fact that the specific heat of 1 g of vapor is equal to 0.49 cal/°C, one can determine the water content over 1 cm^2 of the Earth's surface: $m_w = 0.49/0.0791 \approx 6.2$ g/cm^2. Therefore, the average water content in the near-surface layers of Earth's atmosphere is approximately equal to 62%. The specific heat of wet and heat-absorbing air is equal to 0.3597 cal/g °C $[(C_w)_\Sigma = c_p + C_w + C_r = 0.2394 + 0.0791 + 0.0412]$; whereas the specific heat of dry, but absorbing infrared radiation, air is equal to 0.2806 cal/g °C $[(C_r)_\Sigma = c_p + C_r = 0.2394 + 0.0412]$.

As an example, let us consider a cloud blanket at an elevation of 1–2.5 km. The difference in pressures between the upper and lower boundaries of the cloud blanket is equal to $\Delta p = 887 - 737 = 149.9$ g/cm^2 ($M_{cl} \approx \Delta p$) (see Table 3.1), and the water mass in the cloud layer is equal to $(m_w)_{cl} = 149.9 \times 0.0062/2 \approx 465$ g/cm^2. Using the value of 595.8 cal/g for the heat of water condensation (q), one can estimate the average heating of air in the cloud layer, ΔT_{av}:

$$\Delta T_{av} = q(m_w)/M_{cl}(C_w)_\Sigma \approx 5.14\ °C \qquad (3.26)$$

For the above example, Fig. 3.12 shows the temperature distribution in the cloudy troposphere. It is noteworthy that the inversions of air temperature in this case can occur without the horizontal air mass transfer.

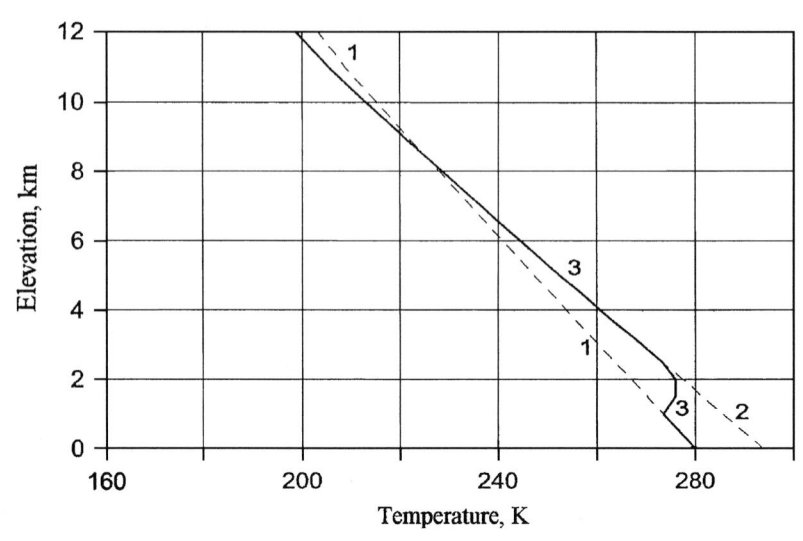

Figure 3.12: The Temperature Distribution in the Cloudy Troposphere at an Elevation of 1–2.5 km: (1) Temperature Distribution in the Wet and Heat Absorbing Troposphere ($\text{grad}\,T_w \approx 6.5\,°\text{C/km}$); (2) Temperature Distribution in the Dry but Heat Absorbing Troposphere ($\text{grad}\,T_d \approx 8.1\,°\text{C/km}$); and (3) Theoretical Temperature Distribution in the Cloudy Troposphere (at the Dew Point of $0\,°\text{C}$).

3.7. Impact of Anthropogenic Factor on the Earth's Climate

The adiabatic theory allows one to evaluate quantitatively the influence of anthropogenic emission of carbon dioxide on the Earth's climate. The carbon content in the atmosphere was increasing by approximately 3 billion tons per year at the end of the century. The rate of the total human-induced CO_2 emission to the Earth's atmosphere is currently about 5–7 billion tons per year (Schimel, 1995; Robinson et al., 1998), or about 1.4–1.9 tons of carbon per year. This amount of carbon dioxide not only changes the gas composition of atmosphere and its specific heat, but also slightly increases the atmospheric pressure.

One can use the adiabatic model together with the sensitivity analysis (Sorokhtin, 2001a, b; Khilyuk and Chilingar, 2002, 2004) to evaluate the effect of anthropogenic emission of carbon dioxide on global temperature. Taking logarithms of both sides of Eq. (3.9) results in:

$$\ln T = \ln 288.2 + \alpha \ln p \qquad (3.27)$$

Introducing notation $A(\alpha, p)$ for the right part of Eq. (3.27), one can find the sensitivity functions of $\ln T$ as partial derivatives of $A(\alpha, p)$ with respect to parameters α and p. Differentiating Eq. (3.27), one obtains:

$$\frac{1}{T}dT = \frac{\partial A(\alpha,p)}{\partial \alpha}d\alpha + \frac{\partial A(\alpha,p)}{\partial p}dp \tag{3.28}$$

Substituting partial derivatives of $A(\alpha, p)$ and multiplying both sides of Eq. (3.28) by T, the following equation is obtained:

$$dT = T(\ln p\, d\alpha + \frac{\alpha}{p}d\alpha) \tag{3.29}$$

Transition to finite differences in differential Eq. (3.29) yields:

$$\Delta T \approx T \ln p \Delta \alpha + T \frac{\alpha}{p} \Delta p \tag{3.30}$$

Equation (3.30) is convenient for analysis of the effect of anthropogenic emission of greenhouse gases on global temperature. If, for example, the concentration of CO_2 in the atmosphere increases two times (from 0.035% to 0.07%), which is expected by the year 2100, then the pressure will increase by $\Delta p = 1.48 \times 10^{-4}$ atm, and $\Delta \alpha = -4 \times 10^{-6}$ (Sorokhtin, 2001a, b). At the sea level, if the pressure is measured in atmospheres, then $p = 1$ and $\ln p = 0$. Thus,

$$\Delta T \approx T \alpha \Delta p \tag{3.31}$$

After substitution of $T = 288$ K, $\alpha = 0.1905$, and $\Delta p = 1.48 \times 10^{-4}$ atm into Eq. (3.31), one obtains $\Delta T \approx 8.12 \times 10^{-3}$ °C. Thus, the increase in the surface temperature at the sea level caused by the doubling of human-induced CO_2 emission will be less than 0.01 °C, which is negligible compared to the natural temporal fluctuations of global temperature.

This increase will be slightly higher at the higher altitudes, where the pressure is less than one atmosphere and its logarithm is negative. Multiplication of $\ln p$ by the negative value of $\Delta \alpha$ will increase ΔT (the global temperature change) (Eq. (3.30)). If, for example, $h = 10$ km, then the barometric pressure $p \approx 0.24$ atm and the temperature is about 200–220 K (FMH-1, 1995). Substituting all necessary data ($T = 220$ K, $p \approx 0.24$ atm, $\ln p \approx -1.4286$, $\Delta p = 1.48 \times 10^{-4}$ atm, $\alpha = 0.1905$, $\Delta \alpha = -4 \times 10^{-6}$) into Eq. (3.30), one obtains $\Delta T \approx 2.710 \times 10^{-2}$ °C, which is less than 0.03 °C.

From these estimates, one can deduce a very important conclusion that even considerable increase in the anthropogenic emission of carbon dioxide and other greenhouse gases practically does not change the global atmospheric temperature. Thus, the hypothesis of current global warming resulting from the increased emission of greenhouse gases into the atmosphere is a myth. Humans are not responsible for the increase in the global surface temperature of 1 °F ≈ 0.56 °C during the last century and one should explain this increase by the natural forces heating the atmosphere.

Evaluating the climatic consequences of anthropogenic CO_2 emission, one has to take into consideration that the emitted carbon dioxide dissolves mostly in oceanic water (according to the Henry Law) and then is fixed into carbonates. In this process, together with carbon, a part of atmospheric oxygen is also transferred into the carbonates. Therefore, instead of a slight increase in the atmospheric pressure one should expect its slight decrease with a corresponding insignificant climate cooling. In addition, part of the mass of carbon dioxide is reduced to methane in the process of hydration of oceanic crust rocks. Because during formation of carbonates and generation of methane, about 2.3×10^8 tons/year of CO_2 are removed from the atmosphere; the potential consumption of CO_2 in the process of hydration, however, is considerably higher. Although the period of this geochemical cycle is over 100 years, the effect of CO_2 consumption is additive.

Together with the anthropogenic CO_2, part of O_2 is also removed from the atmosphere (about 2.3 g per 1 g of carbon). If the ocean and plants consume all this additional CO_2, after the year of 2100 this would lead to a reduction in atmospheric pressure by approximately 0.34 mbar and cooling of the climate by -8.2×10^{-2} °C ≈ -0.01 °C. In reality, however, the metabolism of plants should almost completely compensate for the disruption of equilibrium by mankind and restore the climatic balance.

3.8. Influence of Oceans on the Atmospheric Content of Carbon Dioxide

The oceanic water contains carbon dioxide (in the form of HCO_3^- ions) about 92 times more than that in the atmosphere (Handbook on Geochemistry, 1990). The equilibrium distribution of this gas between the ocean and atmosphere is modeled by the Henry's Law, according to which the concentration of CO_2 in the oceanic water $C(CO_2)_{oc}$ is proportional to its partial pressure in the atmosphere, $p(CO_2)_{atm}$:

$$C(CO_2)_{oc} = A p(CO_2)_{atm} \tag{3.32}$$

where A is the Henry's coefficient, depending significantly on temperature T.

The present-day ocean contains about 2.14×10^{20} g of CO_2 and the atmosphere contains 2.33×10^{18} g of CO_2 (Handbook on Geochemistry, 1990), which corresponds to the partial pressure of CO_2 [$p(CO_2)_{atm}$] of

4.6×10^{-4} atm. The average mass concentration of CO_2 in the oceanic water is equal to 1.56×10^{-4}. Therefore, according to Eq. (3.32) the present-day Henry's coefficient $A \approx 0.34$ atm^{-1}. Also,

$$A = A_0 \exp\{-\Delta H^0/RT\} \tag{3.33}$$

where ΔH^0 is the change in enthalpy (heat generation with negative sign) caused by dissolving 1 mole of CO_2 in the oceanic water; $R = 1.987$ cal/mol °C is the universal gas constant; A_0 is the limit value of Henry's coefficient when $\Delta H^0 \to 0$. As a result of dissolution of CO_2 in the oceanic water, carbonic acid is formed according to the following reaction:

$$CO_2 + H_2O \to H_2CO_3 \tag{3.34}$$

The values of enthalpy for these substances are equal to -94.05, -68.32, and -167 kcal/mol, respectively. Therefore, the change in enthalpy caused by the dissolution of one mole of CO_2 in the oceanic water is equal to $-167-(-94.05-68.32) = -4.63$ kcal/mol, or $-4,630$ cal/mol. Taking into consideration that the present-day average oceanic water temperature is about 280 K and using Eq. (3.33), one obtains:

$$A_0 = 0.34/\exp\{4630/1.987 \times 280\} \approx 8.27 \text{ atm}^{-1} \tag{3.35}$$

Equations (3.32) and (3.33) imply that the increase in temperature of present-day ocean water by 1 °C leads to the change in partial CO_2 pressure by 13.8×10^{-6} atm, i.e. by 13.5 ppm. During glacial periods, the average temperature of oceanic water could drop by 4 °C (to 276 K). Then, the carbon dioxide partial pressure should be reduced by 52–54 mass ppm (or 79–82 volume ppmv). The Antarctic ice core measurements (Fig. 3.13) show that the deviations in concentration of CO_2 are of the same magnitude (about 80–90 ppmv). During the warm Cretaceous period, when the average temperature of oceanic water could have risen to $+17$ °C (290 K), the partial pressure of carbon dioxide could have increased up to about 740 ppm (i.e. about 1.6 times higher than contemporary partial pressure of CO_2).

Considering the causes of greenhouse effect, one should discuss the classic Arrhenius hypothesis on direct influence of changing concentration of atmospheric CO_2 on the temperature of troposphere. In support of this hypothesis, the proponents present the data on CO_2 content in ancient ice cores of Greenland and Antarctica, which indicate that the CO_2 content always rose during the interglacial warming periods (Fig. 3.13). On a larger scale this effect occurred during the warm climatic periods (Cretaceous Period, for example). Proponents of the greenhouse global warming theory usually point out that the increased emission of greenhouse gases is accompanied by an increase in global temperature, i.e. the increased emission of these gases causes the global

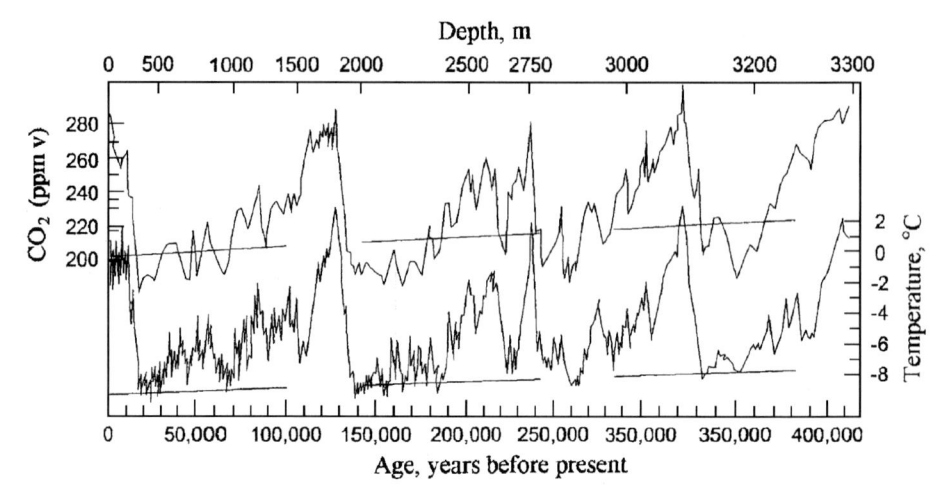

164 Global Warming and Global Cooling: Evolution of Climate on Earth

Figure 3.13: The Correlation between Changes in Concentration of Carbon Dioxide and Temperature Deviations during the last 420,000 years at the Antarctic Station Vostok. On the Graph, Time Dimension is Directed from Right to Left. The Temperature Data and Carbon Dioxide Concentrations are Obtained from the Ice Core Measurements (Petit et al., 1999). The Comparison of these Graphs shows that the Temperature Deviations Lead the Changes in Carbon Dioxide Concentration by Approximately 500–600 Years (Fischer et al., 1999; Mokhov et al., 2003).

warming. Yet, increase in the carbon dioxide concentration closely follows a long-term warming trend. The adiabatic theory, however, shows that the orthodox proponents of the theory of greenhouse effect as a result of increase in anthropogenic emission of carbon dioxide are confusing the cause and effect: the changes in partial CO_2 pressure occur as a result of corresponding deviations in temperature and not vice versa. Figure 3.13 shows that the temperature deviations lead the corresponding changes in the concentration of atmospheric CO_2. Mohov et al. (2003) determined that the changes in the atmospheric CO_2 concentration are delayed by about 500 years with respect to the corresponding changes in atmospheric temperature. Moreover, when the temperature decreases, this delay becomes more significant and easily observed. Judging from these data, *the temperature deviations drive the changes in atmospheric CO_2 concentration.*

One can explain this situation by the inverse dependence of CO_2 solubility in oceanic water on temperature (Eq. (3.33)) and the Henry's Law (Eq. (3.3)), which describes the dynamic equilibrium between the partial atmospheric CO_2 pressure and its content in the hydrosphere. The increase in temperature of oceanic water drives a part of CO_2 content from the ocean into the atmosphere. On the contrary, cooling increases solubility of CO_2 in the oceanic water and a part of atmospheric CO_2 dissolves in the ocean. Thus, the global warming always leads the increase in atmospheric concentration of CO_2, whereas the global cooling causes a corresponding drop in the CO_2 concentration. It is noteworthy that the delay in changes of the CO_2 concentration in comparison with the changes in the global temperature is close to the time of complete stirring of water in the World Ocean (from several hundred to thousand years, Fig. 3.13). As shown in Fig. 3.13, at the present time, we live in the cooling geologic epoch.

In order to understand the climatic changes attributed to anthropogenic CO_2 emission into atmosphere, one can study climates of the past epochs, which obviously were not influenced by the human activity. One of these observations relates to the persistent periodic oscillations of temperature throughout the Pleistocene Epoch (Raymo, 1998), during which variations of (a) 3–4.5 °C occurred during the glacial periods and (b) 0.5–1 °C occurred during the interglacial periods (Oppo et al., 1998). Pagani and Freeman (1999) reconstructed the history of atmospheric CO_2 concentration throughout the Miocene Epoch (25–9 million years ago). They found that CO_2 concentrations during the Miocene were similar to those observed during the Pleistocene Epoch (180–290 ppm). In the middle of Miocene Epoch, however, deep water and high-latitude surface ocean waters were 6 °C warmer than they are today. These authors stated that the "uniformly low" CO_2 concentration during the Miocene time "appears in conflict with greenhouse theories of climate change".

3.9. Methane Emission Effect

The concentration of methane in the atmosphere is continuously increasing (Fig. 3.14). Due to high volatility and low density, the methane released at the Earth's surface rapidly rises into the higher layers of troposphere and further to stratosphere. In the presence of ozone, methane is transformed into hydrogen, water vapor, and carbon dioxide (possible reaction: $CH_4 + O_3 +$ solar radiation $\rightarrow CO_2 + H_2O + H_2$). The resulting hydrogen gas rises further upward into stratosphere, where it combines with ozone: $H_2 + O_3 \rightarrow H_2O + O_2$. Formed in the upper layers of troposphere, carbon dioxide can contribute to the greenhouse effect. The life span of CO_2 (several hundred years) is much longer than that of CH_4 (7–10 years).

The main sources of methane on the Earth are: (1) volcanic activity, (2) marshes and tundra, (3) rice paddies, (4) cattle, (5) oil and gas production,

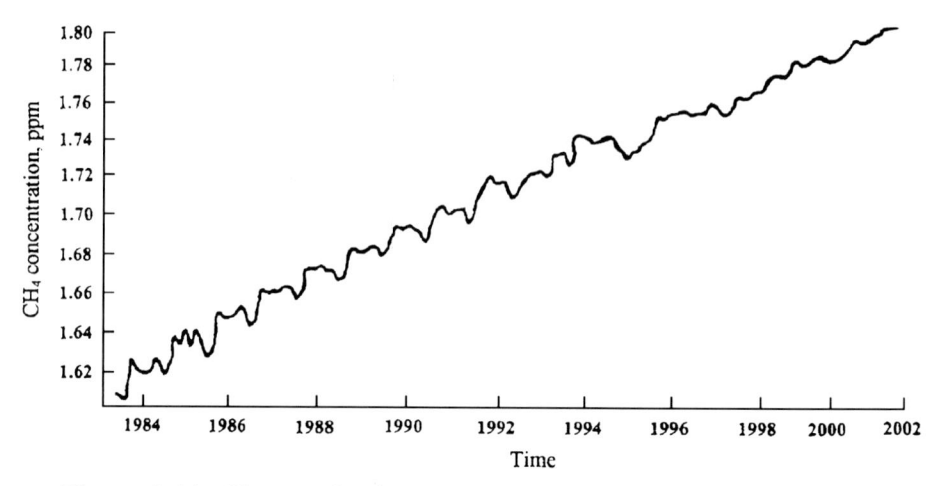

Figure 3.14: Changes in the Atmospheric Methane Concentration.

and (6) natural gas migration (through faults and fractures) to the Earth's surface (Khilyuk et al., 2000).

A great volume of methane is emanated as a result of volcanic eruptions. Yasamanov (2003) estimated that about 5×10^{15} g of CH_4 are released yearly to oceanic water at the spreading zones of mid-ocean ridges. The amount of methane released to the atmosphere from marshes and tundra is estimated between 5×10^{13} and 7×10^{14} g annually. From tundra, the amount of CH_4 released is about 4×10^{13} g/year. Enteric fermentation contributes 2×10^{13} to 2×10^{14} g/year. Thus, the total amount of methane released annually into the atmosphere from the natural sources is estimated at the level of $2-3 \times 10^{15}$ g. Methane is also generated as a result of anthropogenic activities. Rice paddies produce about 5×10^{14} g of methane annually. Oil and gas production and related operations add up to 9×10^{14} g of methane annually, which is one order of magnitude lower than the amount of methane released from the natural sources. Great amount of methane is released to the atmosphere as a result of gas migration to the Earth's surface from coal, oil, and gas deposits and the Earth's mantle (Khilyuk et al., 2000). Total amount from the latter source has not been estimated at the present time. Probably, it is around 5×10^{15} g/year, whereas the burning of fossil fuels contributes only 1×10^{13} g/year of methane to the atmosphere. Thus, the amount of methane released from natural sources is at least two orders of magnitude higher than that released as a result of the fossil fuels utilization.

Chapter 4

Evolution of Earth's Climates

4.1. Introduction

The adiabatic theory of heat transfer in the Earth's atmosphere allows one to construct the temperature distribution for various geologic epochs and periods. We live in the cooling geologic time when, *beginning from mid-Mesozoic time (about 150–100 million years ago), the Earth's climate is gradually cooling.* There are several causing factors, including the nitrogen removal from the atmosphere and its further transfer into the soil as nitrites and nitrates (Sorokhtin and Sorokhtin, 2002; Sorokhtin and Ushakov, 2002). As a result, the mass of atmosphere (and its pressure) has been gradually decreasing, which led to the global climate cooling by approximately 2.3–2.5 °C and, at the present time, this cooling is not even compensated by a continuous increase in solar radiation. The drift of continents resulting in the relocation of continents to high latitudes during the Cenozoic Era, also led to the global cooling. There are numerous geological data that corroborate this conclusion, among them: (1) no traces of glaciations in the Mesozoic Era, (2) emergence of the first glacial covers in Antarctica in the Middle Cenozoic time (about 38 million years ago), and (3) occurrence of periodic glaciations in the Northern Hemisphere during the last 2 million years. We are now living in the interglacial period and the next ice age probably will be a bleak one.

4.2. Climatic Cooling during Cenozoic Era

General cooling of climate is clearly evidenced by the steady decrease in temperature of near-bottom oceanic water throughout the Cenozoic Era (Fig. 4.1). The pronounced cooling of oceanic water occurred at the boundary between the Eocene and Oligocene epochs, which was caused by the emergence of Antarctic glaciation (other data indicate that this event occurred 38 million years ago). During the Miocene Epoch, however, one could observe an increase in temperature, which could be attributed to the disappearance of Tethys Ocean and drift of some of the Gondwana

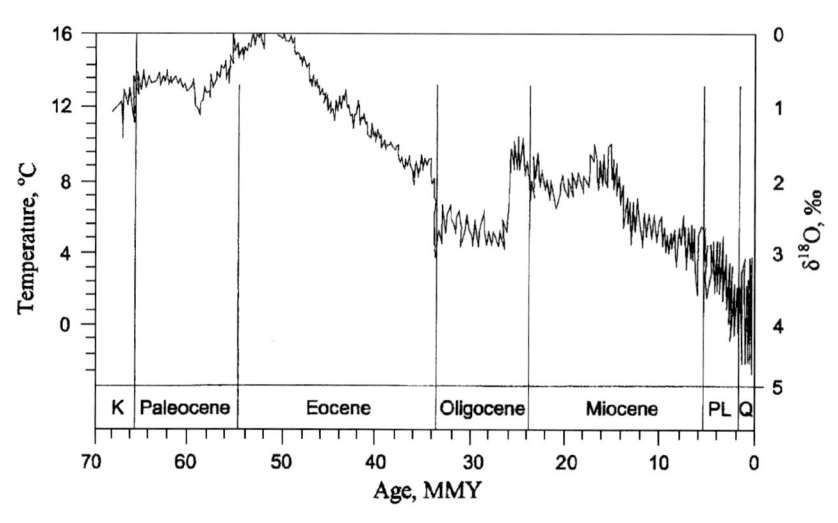

168   Global Warming and Global Cooling: Evolution of Climate on Earth

Figure 4.1: Decreasing Temperature of Near-Bottom Ocean Water during Cenozoic Era. The Temperature was Determined based on Shifts in Oxygen Isotope Ratios $\delta^{18}O$ in Remains of Benthos Fauna (Zachos et al., 2001).

continents to equatorial regions. In the Paleocene–Quaternary time, new glaciations were initiated. At that time, the glaciations affected also the Northern regions, which led to rapid drop in the near-bottom temperature to almost 0 °C. General temperature reduction during the last 70–60 million years, most probably was caused by the partial removal of nitrogen from the atmosphere by bacteria and its transfer into the soils, sediments, and organic matter. As a result of microbial activity, the atmospheric pressure during the Cenozoic Era was gradually reduced, leading, in turn, to the climate cooling.

Various geologic data indicate considerable oscillations in the Earth's surface temperature with steady atmospheric cooling during the last several thousand years. For example, using the information on oxygen isotope ratios in the remains of foraminifera and other microflora in the Sargasso Sea, Keigwin (1996) concluded that there was a steady reduction in surface temperature of seawater in this area (Fig. 4.2). Figure 4.2 indicates that we are approaching another peak in the temperature increase, which had already started in the 17th century. Friedman (2005) reported that the oxygen isotopic composition of Gulf of Aqaba (Red Sea) beachrock reflected a temperature decrease of ambient seawater for approximately half of the Holocene Epoch. There appears to be a temperature decrease from 33 °C to 17 °C between 7.07 and 2.62 thousand years ago, a time interval of approximately 4,500 years. If one combines these findings with a general cooling trend (≈ 2 °C) over the last 3,000 years (Fig. 4.2), then one could infer that the latest cooling period in

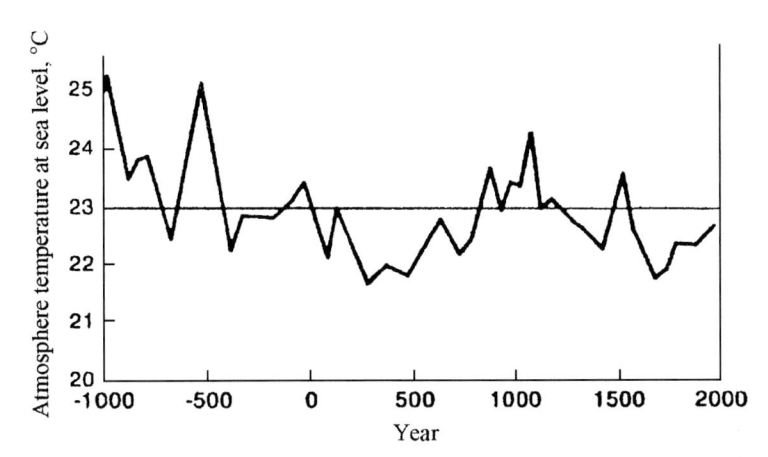

Figure 4.2: The Near-Surface Temperatures of Sargasso Sea (Averaged over the Moving 50-Year Time Interval) (Modified after Keigwin, 1996). Horizontal line Indicates the Average Temperature over the 3,000-Year Period.

geologic history comprises most of the Holocene Epoch. Thus, we live in the cooling geologic time and the currently observed global atmospheric warming is an insignificant episode in the geologic history.

The global warming, which was observed during the last three centuries, most probably is a temporary phenomenon caused by variations in the solar magnetic activity. The instrumental observations of temperature were carried out in South England since 1749, whereas those of solar magnetic activity were started in France in 1850s. Comparison of these data indicates a good correlation between the Earth's surface temperature and the solar magnetic activity (Fig. 4.3). The Sun is a major supplier of energy to the Earth. Therefore, it is an absolute must to use data on variations in solar activity in explaining the evolution and forecasting of the Earth's climate.

Analysis of the varying solar irradiation during the past two millennia indicates that within the next hundred years a long period of climate cooling should be expected with its coolest phase around the year of 2030 (Landscheidt, 2001a, b). The editors of the *Science* journal (2002) commented on the gradually increasing number of publications linking the Earth's climatic changes to the varying solar activity: "As more and more wiggles matching the waxing and waning of the Sun show up in records of past climate, researches grudgingly take the Sun seriously as a factor in climate change… the Sun seems to play a pivotal role in triggering droughts and cold snaps."

Landscheidt (2002) pointed out that the Sun's eruptional activity (energetic flares, coronal mass ejections, eruptive prominences) "heavily affecting the

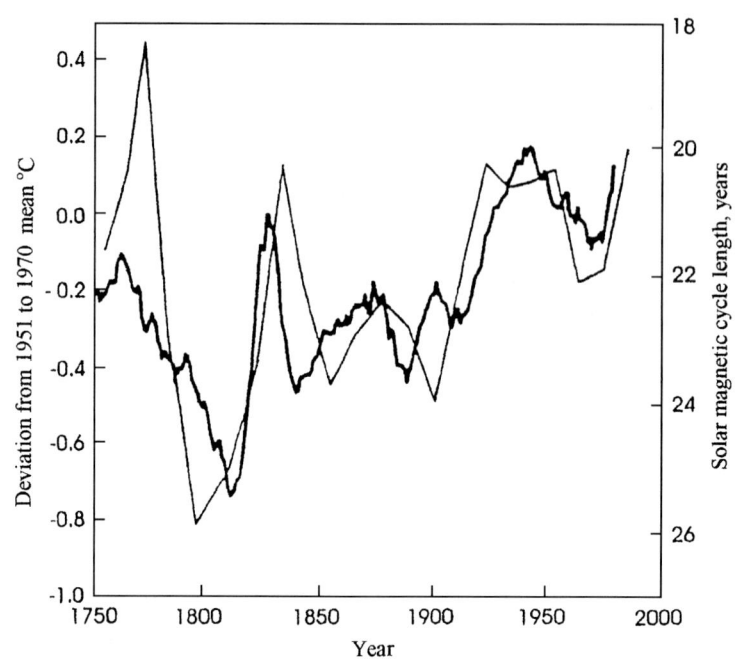

Figure 4.3: The Correlation of Temperature Deviations in the Northern Hemisphere (Left Vertical Axis and Thick Curve) with the Solar Magnetic Cycle Lengths (the Wolf's Numbers) (Right Vertical Axis and Thin Curve). (Modified after Grovesman and Landsberg, 1979; Jones, 1986; Baliunas and Soon, 1995; Robinson et al., 1998.)

solar wind as well as softer solar wind contributions by coronal holes have much stronger effect than the total irradiance". The total magnetic energy flux from the Sun dragged out by the solar wind has risen 2.3 times since 1901 (Lockwood et al., 1999). During the same time interval, the global average surface temperature of Earth has increased by 0.6 °C. The energy of solar irradiation is transferred to the near-surface environment of Earth by magnetic reconnection and directly into the atmosphere by the charged particles (Landscheidt, 2002). Egorova and associates (2000) carried out the surface temperature and atmospheric pressure measurements at the altitude of 10 km during the years from 1981 to 1991 at the Antarctic Station Vostok. The collected experimental data show a strong impact of solar eruptions on the atmospheric temperature and pressure at the elevations close to tropopause. In addition, precise satellite measurements (Fig. 4.4) indicate a slight but steady cooling during the last 25 years (Christy et al., 1996).

In summary, the global climatic evolution should be studied as a system problem based on a strict physical theory, taking into account the evolution of

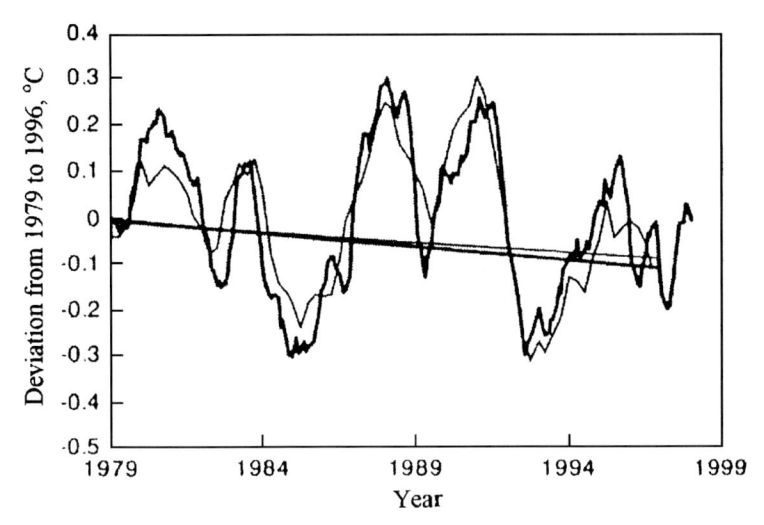

Figure 4.4: Gradual Atmospheric Cooling between 1979 and 1996 (Modified after Christy et al., 1996). The Temperatures are Averaged Over 6-Month Moving Interval. Thick Curve Presents the Satellite Measurements, Whereas the Thin Curve Represents the Measurements Obtained using Radio Probe Balloons. Both Curves Present the Temperature Deviations from the Average (Over the Period of Observations) Values.

atmospheric gas content, geological environment, changing solar irradiation, varying conditions of the Earth's rotation and revolution, and evolution of oceans and continents. One must also understand all reverse compendency in this extremely complex system. The preceding discussion persuasively shows that one cannot explain global climatic changes by a single and imaginary dependence of climate on atmospheric concentration of so-called "greenhouse gases".

4.3. Bacterial Nature of the Earth's Glaciations

Regional glaciations on the Earth occur only if the Earth's surface temperature in the region drops below the freezing point of water. According to Chumakov (2001, 2004), the emergence of ice ages was accelerating throughout the geological history. The Early Proterozoic (Huron) glaciations occurred in the mountains only, similarly to the Archaean ice age (Sorokhtin and Sorokhtin, 2002). Until the Late Riphaean time (about 1 BY ago), the normal (not mountains) continental glaciations did not occur at all.

To understand the nature of continental glaciations, one needs to study the causes leading to the long-term drop in surface temperature for vast areas of the continents. One of the possible causes of persistent cooling could be reduction in the solar irradiation. The evolution theory of stars, however, indicates that after reaching the time of beginning of its development, the star's luminosity could only progressively increase in time. For example, during the time of Sun's existence, its luminosity has increased by 30–37% (Bachall et al., 1982). A strong correlation between the magnetic solar activity and the Earth's surface temperature during the last 250 years has been established. Changes in the luminosity of Sun over this interval of time occur, as a rule, with periodicity of about 11 and 60 years (Fig. 4.3). However, one can probably detect a component of a longer periodicity causing some temperature rise over the interval of observations.

Similar temperature changes in the Earth's climate with a dominating periodic component of longer periodicity (up to 1,500 years) were discovered using the paleontology records (Fig. 4.2). As shown in this figure, during the last 3,000 years of geologic history the temperature deviations have been reaching 3 °C. In addition, this graph indicates a negative temperature trend over the same time interval.

Another possible cause of climatic oscillations could be the changing Earth's insolation due to precession of Earth's axis and eccentricity of its orbit around the Sun. The corresponding changes in temperature (judging from the isotopic hydrogen ratios in the ice cores of Antarctica) could reach almost 10 °C, whereas the periods of their oscillations, as a rule, do not exceed 110,000–120,000 years (Petit et al., 1999). The same data indicate also that the changes in concentration of carbon dioxide in atmosphere do not influence the Earth's climate, because they always occur after the changes in temperature and, therefore, represent the effect (not the cause) of these changes (e.g. see Hoyt and Schatten, 1997).

Looking for the global cause of long-term geologic cooling one has to exclude also volcanic eruptions, because the warmest Cretaceous Period of the Mesozoic Era was accompanied by a significant volcanic activity. The question remains: what is the natural cause of evolutionary (on a scale of billions of years) changes in the average temperature of Earth's surface? *Could it be the atmosphere itself, with changes in temperature due to changes in the atmospheric content of gases and pressure?*

The adiabatic theory of greenhouse effect developed in Chapter 3 implies that the average surface Earth's temperature T_s expressed in K is determined completely by the atmospheric pressure p_s and adiabatic exponent α, which, in turn, is determined by the atmospheric content of gases and the thermal capacity of air. Experimental verifications (see Chapter 3) proved a high degree of precision of this model. Any changes in the atmospheric pressure lead directly to corresponding changes in the surface temperature.

In Chapter 2, a model of Earth's degassing was described. Based on this model, the evolution of content and pressure of atmosphere was determined. This model was constructed taking into account the removal of a part of atmospheric nitrogen by the nitrogen-consuming bacteria, which reduced considerably the partial pressure of nitrogen and total atmospheric pressure by the present time. When the evolution of atmospheric pressure and content of various gases in atmosphere in time are determined, the adiabatic theory allows one to establish the corresponding temperature regimes of Earth's climates. The evolution of temperature regimes throughout the Earth's geologic history with the forecast for the future 1 billion years is shown in Fig. 4.5.

Experimental temperature data for the Venus troposphere at low (0–30°) and high (75°) latitudes at various elevations (i.e. at various atmospheric pressures) indicate that the climatic contrast (determined by the difference in temperatures at these latitudes) varies approximately inversely with the atmospheric pressure (Fig. 4.6). This regularity allowed us to determine the evolution of latitudinal temperature distribution in the Earth's atmosphere. In addition to the average Earth's surface temperature, we determined

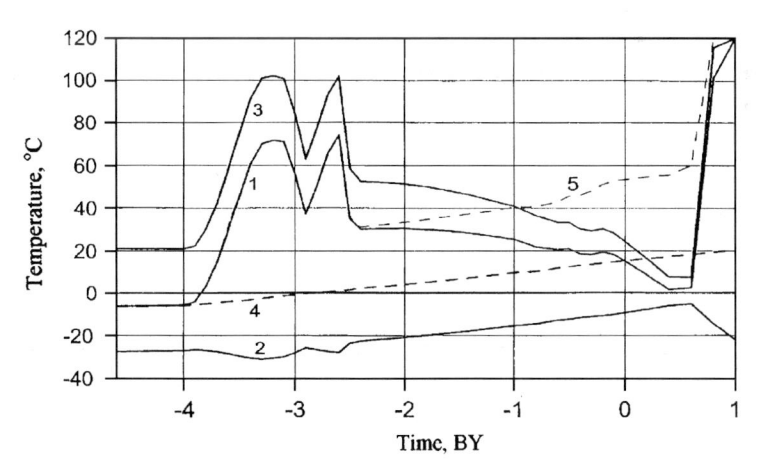

Figure 4.5: Evolution of Earth's Climates at a Constant Precession Angle ψ of 24°. (1) Averaged (over the Earth's Surface) Near-Surface Temperature at Sea Level; (2) Radiation (Effective) Temperature of Earth; (3) "Greenhouse Effect" of Earth's Atmosphere; (4) Temperature of "Gray Body" at a Distance of Earth from Sun, Characterizing Increase in the Sun's Luminosity with Time; (5) Average Temperature of Earth Assuming that Nitrogen is not Consumed by Bacteria.

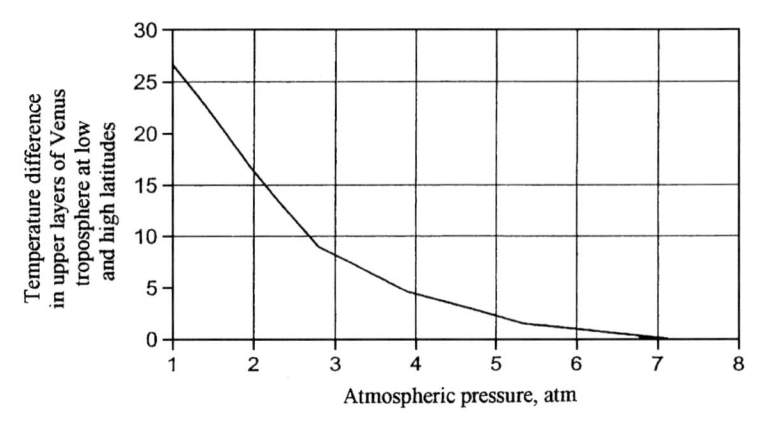

Figure 4.6: Climatic Contrast – Difference in Temperatures at Low (0–30°) and High (75°) Latitudes at an Elevation of 33–50 km (After Planet Venus, 1989).

also the equatorial (T_{eq}) and polar (T_{pl}) temperatures using the following equations:

$$T_{eq} \approx T_s + \delta T \cdot 0.36; \quad T_{pl} \approx T_s - \delta T \cdot 0.64 \qquad (4.1)$$

where $\delta T \approx \Delta T_0/p_s$; $\Delta T_0 \approx 30\,°C$ (difference in temperatures at the equator and poles of the present-day Earth without taking into account the influence of albedo of snow cover); and p_s is the current value of near-surface pressure. Evolution of equatorial and polar temperatures is shown in Fig. 4.7.

According to our model, the first time when the polar oceans were covered by ice occurred only in the second half of Cenozoic Era – during the Oligocene or in Middle Miocene epochs. Before that time, even during the Precambrian and Paleozoic continental glaciations, the temperature of oceanic water at the Earth's poles, most likely, remained above the freezing point, and the polar oceans were not covered with ice. In spite of the fact that in the Early Archaean time the temperatures at the Earth's poles were below the freezing point, the ancient oceans remained warm because at that time they were located only in the equatorial belt and at low latitudes (Sorokhtin, 2004).

If the temperature at the sea level is known (Fig. 4.7), then the adiabatic theory allows one to determine the temperature at the average elevation of the continents (Fig. 4.8). The vast, snow-covered, polar territories with high albedo (about 0.7) should have temperatures close to the radiation temperature of Earth. Because the polar regions receive twice as little insolation as the other regions of Earth (Chapter 3), the temperature T_{es} for

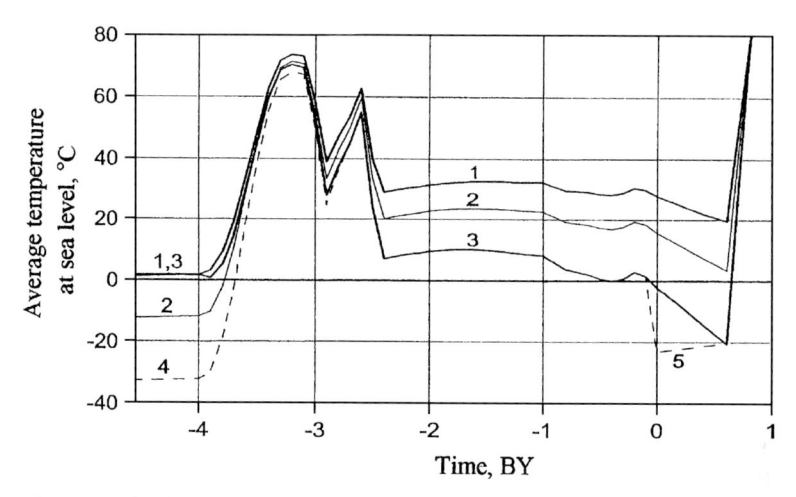

Figure 4.7: Evolution of Averaged Temperatures of Ocean Surface at the Constant Precession Angle ψ of 24°: (1) Temperature at the Earth's Equator; (2) Average Temperature of Earth; (3) Temperature at the Poles (in the Early Archaean Time this is the Temperature at Low Latitudes, because Oceans did not Exist at the Polar Regions at that Time); (4) Temperature at the Poles during the Archaean Time; (5) Temperature of Surface of Ice Cover of Polar Oceans (Taking into Account the Albedo of Snow Cover).

such snow-covered territories can be expressed in the following form (using Eq. (3.21)):

$$T_{es} = [(1 - A)(1 + \cos\psi)S/4\sigma]^{1/4} \cdot (p/p_0)^\alpha \qquad (4.2)$$

Using Eqs. (3.21) and (4.2), one can estimate the temperature of ice covers in the polar regions of Earth (Fig. 4.9). In the same graph, the correlation of theoretical calculations with the geologic data at times of Earth's glaciations is shown (Chumakov, 2004).

In any observational science, the theoretical estimates should be verified by comparison with the corresponding experimental data. To determine the theoretical estimates, the following numerical values of model parameters were assigned: precession angle $\psi = 23.44°$; average albedo for the North Polar Ocean $A_N = 0.66$; average albedo for the central regions of Antarctica $A_A = 0.69$; average temperature gradient over Antarctica $\text{grad}\, T = 8.1$ K/km. Using Eqs. (4.2) and (4.1), one can obtain the average present-day temperature at the North Pole of $-22.8\,°C$, whereas according to the experimental data this temperature is $-23\,°C$ (Khromov and Petrosyants, 2001). By analogous theoretical computations for the central regions of Antarctica, we obtained: -36.6,

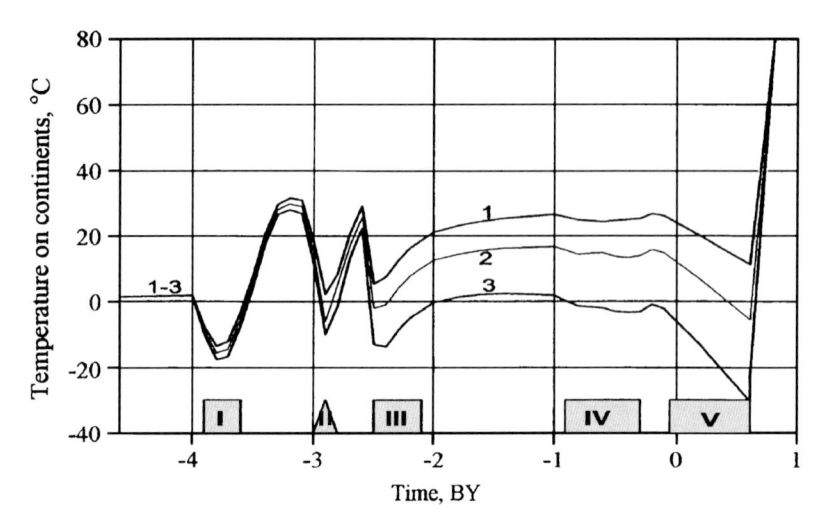

176   Global Warming and Global Cooling: Evolution of Climate on Earth

Figure 4.8: Evolution of Averaged Temperatures on the Surface of Continents at a Constant Precession Angle ψ of 24°. (1) Temperature at the Earth's Equator; (2) Average Temperature of Earth; (3) Temperature at the Poles (without Counting on the Albedo of Snow Cover); I through V are the Periods of Possible Glaciations on Continents, located at high Latitudes.

−44.7, −52.8, and −60.9 °C at the elevations of 1, 2, 3, and 4 km above the sea level, respectively.

To verify the obtained theoretical estimates of average temperatures on the surface of Antarctic ice cover, we compared them with the experimental data (using the ice cores) on the average yearly temperatures (T_{av}) in the central regions of Antarctica, which practically are not penetrated by the sea cyclones. Kotlyakov (2000) provided the following experimental data: (1) T_{av} = −53 °C at the Cupola-C (with the elevation of 3.24 km above the sea level) and (2) −55 °C at the Komsomol'skaya Station; and (3) −55.5 °C at the Vostok Station (with the elevations of 3.24 and 3.5 km above the sea level, respectively). According to observations of O. Sorokhtin and A. Kapitsa, the average temperature (T_{av}) at the top of main Cupola of Antarctica (elevation of about 4 km), is −60 °C. According to our model, for these elevations the theoretical temperature estimates are: −54.9 °C, −56.9 °C, and −61 °C, respectively, which showed a good fit of the theoretical to experimental data.

Our theoretical computations are valid with the proviso that they are conducted for the time intervals, which are sufficiently long for the emergence of distinct ice ages. To detect such ice ages, one needs to consider jointly such global climatic drivers as (1) continental drift, (2) changes in the Earth's precession angle, and (3) Milankovitch cycles. To cause the

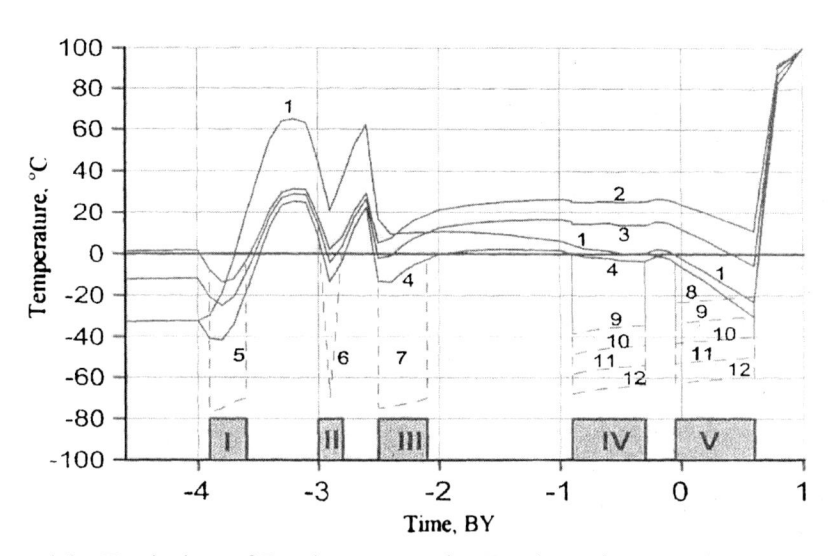

Figure 4.9: Evolution of Ice Ages over the Earth's History. (1) Temperature of Ocean at the Poles; (2) Average Temperature of Continental Surface at the Equator; (3) Average Temperature at a Mean Elevation of Continents (4) Average Surface Temperature of Continents at the Poles; (5)–(12) Temperatures of Surface of Ice Covers (taking into Account the Albedo of Snow Cover); (5) Early Archaean Equatorial Continental Glaciations; (6) and (7) Temperatures of Continental Glaciations during the Middle Archaean and Early Proterozoic Time at the High Mountains at Middle Latitudes; (8) Surface Temperature of Polar Ocean (taking into Account the Albedo of Snow Cover); (9)–(12) Temperature of Continental Glaciations at the Polar Latitudes at the Elevations of 1, 2, 3, and 4 km, respectively; (I) Early Archaean High Mountain Continental Glaciations at the Equator; (II) Middle Archaean (2.9 Billion Years Ago) High Mountain Glaciations at an Elevation of about 6 km above the Sea Level; (III) Early Proterozoic (Huronian) High Mountain Glaciations at an Elevation of about 3 km; (IV) Late Riphaean–Paleozoic Era of Glaciations at the High Latitudes; (V) Cenozoic and Future Glaciations of Continents and Sea Basins at High Latitudes.

continental glaciations, the climatic cooling alone is not sufficient. It is also necessary for the continents to be situated at polar regions and, in conformity with precession cycles, for the Sun's radiation to be minimal.

It is important to consider the causes of ancient ice ages. The oldest Early Archaean ice age (group I, Fig. 4.9) did not leave any visible geologic records. *At that time (about 3.9–3.7 billion years ago) the juvenile sea basins and the nuclei of future continental shields were located within a narrow equatorial Earth's belt.* Therefore, the sea basins were warm with the temperatures ranging from 3 to 20 °C (Fig. 4.7). The peaks of juvenile continental formations (possibly reaching a height of 3–4 km) were located at the equatorial belt and, despite a general climatic warming, could reach the elevations having negative temperatures and experience glaciation (Fig. 4.8). If the glaciers emerged at that time, they were local and occurred in the mountains only. It is very likely that glaciers of long duration did not emerge at all at that time, because the volume of Earth's surface water was insignificant.

The middle-Archaean glaciation (group II, Fig. 4.8), which occurred about 2.9 billion years ago, could develop not only at low but also at moderate latitudes. By that time, the zone of Earth's tectonic activity was extended to the latitudes of 40–50° and the juvenile continental shields could drift to these latitudes. In addition, their average surface elevation rose to 5.5–6 km. In the middle-Archaean time, the water volume in sea basins increased considerably and the water temperature rose to 30–40 °C, which facilitated evaporation. Therefore, the *Alpine glaciations of the middle-Archaean time could be accompanied by the formation of small glaciers, which were recorded in the geological chronicle of Earth* (Figs. 4.5 and 4.9).

In the Early Proterozoic Era large but not very deep oceans emerged with average depths of about 1 km (Sorokhtin and Ushakov, 2002), whereas the surface elevation of continents remained comparatively high (average elevation of 2–4 km) (Sorokhtin and Sorokhtin, 1997). At the beginning of the third group of glaciations (Huron Ice Age), the ice covers could develop only in the middle latitudes (Figs. 4.8 and 4.9). After breaking up of the first supercontinent Monogaea (when its fragments were scattered around the entire Earth's surface), in the middle and at the end of third glaciation's period (2.4–2.1 billion years ago) ice covers could have started forming on many continental shields, which were located in high latitudes at that time.

The ice covers of that age were identified on the four continents: (1) in North America (including the Huronian system), (2) in Europe (on the Baltic Shield), (3) in South Africa (Transvaal Formation), and (4) in Western Australia (the Tury Formation). The age of glaciations for all of them is about the same: between 2.4 and 2.2 billion years ago (Chumakov, 2001).

As shown in Figs. 4.8 and 4.9, during the time interval between 2 and 1 billion years ago, the negative average yearly temperatures on the continents did not exist. Consequently, the continental glaciations could not occur at that time. According to Chumakov (2004), in the upper layers of the Early Proterozoic, in the Early Riphaean time, and during the larger part of the Middle Riphaean time the traces of glaciations were not discovered.

During the late Riphaean and Paleozoic time, the glaciations of the fourth group (Figs. 4.8 and 4.9) occurred due to significant cooling of the Earth's climate at that time. Stable negative average yearly temperatures dominated on the continents situated in the polar regions, resulting in the formation of vast ice covers. If the area of snow-covered glaciers became significant, then a high albedo led to the temperature reduction to −40 to −65 °C. The stable oceanic glaciers similar to that of contemporary North Polar Ocean probably were not formed during either the Late Riphaean or the Paleozoic time, because the temperature of polar regions of oceans was close to zero or even slightly positive at that time (Fig. 4.7).

In the Middle Cenozoic time, about 40 million years ago, the last (fifth) glaciation period had begun, which was characterized by the widest ice covers of Earth's surface. This glaciation period will last into the future, but will be the last one in the Earth's history (group V, Figs. 4.8 and 4.9). In the Oligocene or Miocene epochs, the first (in the Earth's history) vast oceanic glaciations occurred in the North Polar Ocean and also in the South Ocean (shelf glacial cover of Ross Sea), which, in turn, led to significant drop in temperature of near-bottom waters in the oceans (Fig. 4.1). The glaciation periods shall stop occurring in about 600 million years from now, because at that time the intensive degassing of endogenic oxygen should start due to complete oxidation of mantle iron to magnetite. The oxygen is liberated as a result of formation of Earth's core matter (Fe_2O or smelts of $Fe \cdot FeO$) (Sorokhtin, 1971, 1974, 2004), according to the following reaction: $2Fe_3O_4 \rightarrow 3Fe_2O + 5O$. Consequently, the partial pressure of oxygen in the Earth's atmosphere should rise up to several tens of atmospheres and the Earth's surface temperature will eventually exceed 400 °C. As a result, the Earth's life will cease to exist in about 500–600 million years from now.

How could one then explain a warm climatic period at the end of Mesozoic Era, which occurred between the two severe ice ages: the Riphaean-Paleozoic and Late Cenozoic? The emergence of this warm period was mainly caused by the formation of Pangaea and intense generation of oxygen. When the free iron disappeared completely from the mantle (Sorokhtin, 1974; Sorokhtin and Ushalov, 2002), the generation of biogenic oxygen was intensified. After the emergence and proliferation of blossoming plants during the Jurassic and Cretaceous periods, this process reached its maximum. The increase in partial pressure of oxygen at the end of Mesozoic Era temporarily compensated by the reduction in atmospheric nitrogen. As a result, the total atmospheric pressure increased slightly at that time. In turn, the pressure increase led to significant warming of Earth's climate (Eq. (3.11)). At the end of Cretaceous Period when the partial pressure of oxygen reached equilibrium (about 0.231 atm), the further reduction in partial pressure of nitrogen stopped to be compensated by the increase in partial pressure of oxygen and the total atmospheric pressure began decreasing. In addition, the supercontinent Pangaea began breaking apart, which led to the emergence of fifth glaciation period at the high latitudes of Earth (about 38 million years ago).

Thus, the living organisms played a significant role in the process of formation of Earth's climate. The nitrogen-consuming bacteria proved to be especially influential – they reduced the partial pressure of nitrogen more than half during the Late Proterozoic and Phanerozoic time. This led to considerable cooling of atmosphere and formation of extensive continental glaciers in the moderate latitudes. The generation of biogenic oxygen by phytoplankton and land plants during the Phanerozoic time not only facilitated the development of highly organized life on Earth, but also partially compensated for the reduction in partial pressure of nitrogen. As a result, the climate became significantly warmer at the end of Mesozoic Era. However, when the partial pressure of oxygen had reached its equilibrium level of 0.231 atm, the atmospheric pressure began dropping again, and a new cooling period (leading to glaciation) commenced.

4.4. Precession Cycles and the Earth's Climate

The precession phenomenon is caused by the deviation of Earth's mass distribution from the spherical symmetry. This deviation in symmetry is mainly caused by the nonuniformity of Earth's crust in its continental and oceanic regions and also by the possible heterogeneity in mantle matter density. The average period of the complete cycle of Earth's axis precession is about 26,000 years (Zommerfeld, 1947; Kaula, 1971). In addition, weaker and of shorter-period nutation oscillations are superposed on the precession movement of planet. These shorter and weaker oscillations manifest themselves as the migration of Earth's poles, which most probably is caused by the tidal perturbations, movement of Earth's core, and rotation of the Earth–Moon mechanical system around a common center of gravity. As a result, the overall movement of precession axis becomes rather complex.

The shape of the Earth is close to the shape of ellipsoid of rotation of liquid body with inertial equatorial bulging. In addition, the equatorial radius ($R_e = 6,378.2$ km) is greater than the polar radius ($R_p = 6,356.8$ km) by 21.4 km, which corresponds to the polar compression $e \, [= (R_e - R_p)/R_e = 1/298.3]$. Consequently, the large excessive masses of matter are concentrated at the equatorial region of Earth interacting gravitationally with other celestial bodies. Such an interaction has a tendency to turn the Earth in a manner that put its axis of rotation in the direction perpendicular to the direction of interacting body and the equatorial plane tends to coincide with

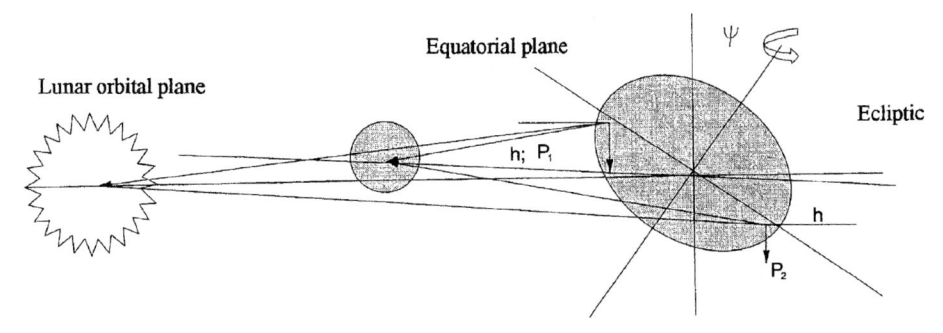

Figure 4.10: Schematic Diagram of Influence of Sun–Moon Attraction on the Equatorial Inertial Bulging of Earth (not to scale).

this direction. A leading role in the reduction of precession angle, however, belongs to the mutual attraction of the Moon and Sun (Fig. 4.10).

The gravitational attraction of the Moon and Sun acts simultaneously on both sides of the equatorial bulges of Earth trying to turn its rotation axis in opposite directions. The force of attraction of the Moon and Sun on the side of Earth, which faces the Moon or the Sun, is somewhat greater than that applied to the opposite side (Fig. 4.10).

In order to determine the mass of equatorial bulging of Earth, at first one needs to estimate its volume. This volume, V_b, can be computed as the difference between the volume of ellipsoid of rotation of Earth and the sphere inscribed in it:

$$V_b = 4/3\pi R_e^2 R_p - 4/3\pi R_p^3 = 7.1 \times 10^{24} \text{ cm}^3 \qquad (4.3)$$

Then, the mass of equatorial bulging of Earth is equal to:

$$m_b = \rho_b \cdot V_b \approx 2 \times 10^{25} \text{ g}$$

where ρ_b ($\approx 2.8 \text{ g/cm}^3$) is the average density of bulge taking into account the layer of ocean water (depth of ≈ 3–4 km) and the underlying layer of oceanic crust and upper mantle (about 17 km). One half of this mass (m_b) faces the Moon, whereas the other half is situated at the opposite side of Earth (Fig. 4.10). Therefore, the effective gravitational mass of each half of the bulge is about two times less (10^{25} g). The difference in the gravitational forces of Moon applied to the two sides (ΔP_l) can be estimated using the following expression:

$$\Delta P_m = 1/2 \gamma m_m m_b [1/(L_m - R)^2 - 1/(L_m + R)^2] \qquad (4.4)$$

where γ ($= 6.67 \times 10^{-8} \text{ cm}^3/\text{g s}^2$) is the gravitational constant; m_m ($= 7.35 \times 10^{25}$ g) is the mass of Moon; L_m ($= 3.844 \times 10^{10}$ cm) is the distance between the centers of gravity of Earth and Moon; and R ($= 6.371 \times 10^8$ cm) is the average radius of Earth. On substituting these values into Eq. (4.4), one obtains $\Delta P_m \approx 2.2 \times 10^{21}$ g cm/s^2.

By analogy, one can estimate the difference in the forces of Sun gravitation applied to the equatorial bulges ΔP_s:

$$\Delta P_s = 1/2 \gamma m_s m_b [1/(L_s - R)^2 - 1/(L_s + R)^2] \qquad (4.5)$$

where $m_s (= 1.99 \times 10^{33}$ g) is the mass of Sun; $L_s (= 1.5 \times 10^{13}$ cm) is the distance between the centers of gravity of Earth and Sun; and $R (= 6.371 \times 10^8$ cm) is the average radius of Earth. On substituting the given information into Eq. (4.5), one obtains $\Delta P_s \approx 1.02 \times 10^{21}$ g cm/s^2.

The forces of the Moon and Sun attraction are applied to the centers of mass of each half of equatorial bulge. The shape of each half of the bulge resembles a convex dome with the center of gravity located at a distance of about 2/3 of the Earth's radius from the geometric center of Earth (Fig. 4.11). In this case,

$$h \approx 2/3 R \sin \psi \qquad (4.6)$$

and the difference in momentums of gravitational forces applied by the Moon on the equatorial bulges of Earth (Fig. 4.11) is equal to $\Delta M_m = \Delta P_m \cdot h = 3.71 \times 10^{29}$ g cm^2/s^2 and by the Sun, $\Delta M_s = \Delta P_s \cdot h = 1.72 \times 10^{29}$ g cm^2/s^2. Then, the total momentum of gravitational forces applied by the Moon and Sun on the Earth's bulges is equal to $\Delta M_m + \Delta M_s = 5.43 \times 10^{29}$ g cm^2/s^2.

In addition to gravitational influence of the Moon and Sun, the precession of Earth's axis is affected by the asymmetry of location of continents on the Earth's surface. The total mass of all continents m_{con} is approximately equal to 2.25×10^{25} g (Ronov and Yaroshevskiy, 1978), the average thickness of continental crust H is equal to 40 km $= 4 \times 10^6$ cm and the average level of continental surface (Δh) is equal to 875 m $= 8.75 \times 10^4$ cm above the sea level. The continents influence the Earth's mass asymmetry because the center of their mass is situated slightly higher than that of the expelled mantle matter. At the isostatic equilibrium of continents, the mass of expelled

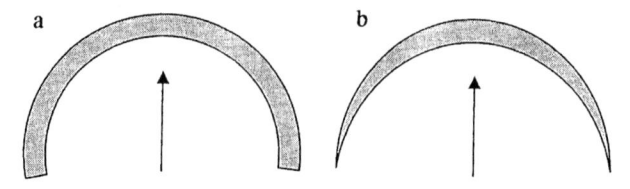

Figure 4.11: The Center of Gravity of Equatorial Bulging of Earth in the Section Facing the Moon: (a) in the Equatorial Plane and (b) in the Meridian Plane.

mantle is equal to the mass of continents. Judging from the elevation of continental surface, center of continental mass is positioned higher that of the ousted mantle ($\delta h \approx 500$–600 m). Then, the excessive mass of continental crust (Δm_{cont}) is equal to 3.38×10^{23} g. The centrifugal acceleration g_{cf} of continents can be defined by the equation $g_{cf} = \Omega^2 R \cos\varphi$, where $\Omega = 7.27 \times 10^{-5}$ radians/s is the angular speed of Earth's rotation and φ is the latitude of center of continental mass. For example, if φ is $30°$, then $g_{cf} = 2.9$ cm/s^2 and $\Delta P_{cont} = \Delta m_{cont} \cdot g_{cf} \approx 9.8 \times 10^{23}$ g cm/s^2. Depending on the distance of center of total continental mass from the center of Earth, the values of ΔM_{cont+m} lie between 10^{29} and 10^{31} g cm^2/s^2. In addition, one has to take into account the non-uniformity of Earth's mantle, especially the locations of light, ascending mantle fluxes, the heavier, subsiding lithospheric plates, and the rugged surface of Earth's core (Fig. 1.11). Thus, one can conclude that the resulting influence of discussed factors on the precession angle becomes rather uncertain. Using our indirect estimates based on comparison of theoretical temperature deviations in the Pleistocene Epoch with the isotopic temperatures of glacier in Antarctica (Fig. 4.16), we arrived at the contemporary estimate of continental momentum (ΔM_{cont+m}) of about 0.6×10^{29} g cm^2/s^2.

Applying the theory of free gyroscopes, one can determine the average angular speed of turning of axis of rotation of Earth ω (around the line of intersection of equatorial plane with the plane of revolution of Moon around the Earth) and of the Earth around the Sun:

$$\omega = (\Delta M_l + \Delta M_s - \Delta M_{cont+m})/I\Omega \tag{4.7}$$

where $I (= 8.04 \times 10^{29}$ g cm^2) is the moment of inertia of Earth.

At the precession angle of $23.44°$, the present-day angular speed of Earth (ω) is equal to about 1×10^{-11} rad/s or 3.3×10^{-4} rad/year. The time necessary for the turning of Earth by $1°$ is approximately equal to 855 years. Turning of the Earth from the present-day precession angle of $23.44°$ to the angle of inclination of Moon orbit to the ecliptic plane is an asymptotic process in which the precession angle gradually approaches a value of $5.1°$. The process takes millions of years. Thus, the dependence of angular speed of turning of Earth on the precession angle is essentially nonlinear. The time of Earth's turning is determined by the following equation:

$$\tau = 2\pi/\omega \tag{4.8}$$

Inasmuch as $\omega = d\psi/dt$, one can find the dependence of precession angle on time (Fig. 4.12):

$$\psi = \int_0^t \omega ds \tag{4.9}$$

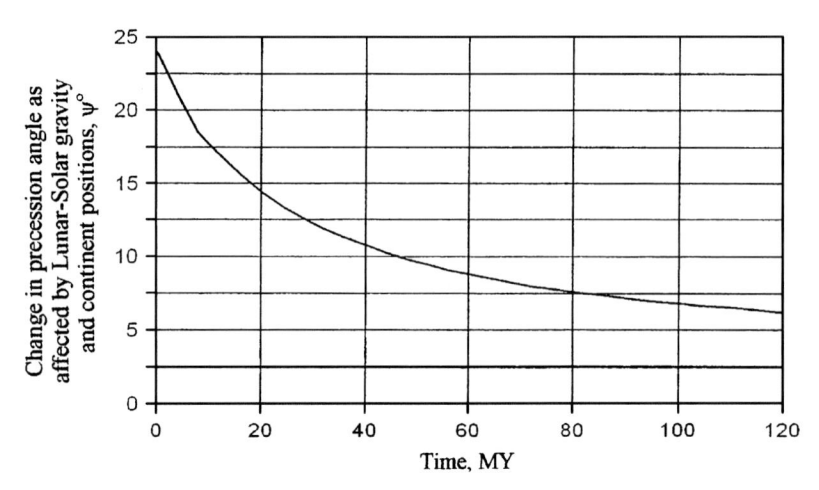

184   Global Warming and Global Cooling: Evolution of Climate on Earth

Figure 4.12: Changes in the Precession Angle of Earth in Time under Influence of Sun–Moon Tides and Asymmetric Disposition of Earth's Continents (at the Initial Precession Angle of 24°). The Minimal Possible Value of Precession Angle is Determined by the Geometry of Disposition of Continents on the Earth's Surface, which at the Present Time is Approximately Equal to 2.5°.

If the forces of Moon–Sun gravitational attraction of equatorial bulges of Earth are in equilibrium with the influence of continents and mantle on the precession axis, then:

$$\Delta M_m + \Delta M_s = \Delta M_{cont+m} \qquad (4.10)$$

For $\omega = 0$, the precession angle reaches the equilibrium value when $t \to \infty$. For the present-day positions of continents, this equilibrium value is $\approx 2.5°$ (Fig. 4.12). Using Eq. (3.11), one can evaluate the change in Earth's surface temperature as a function of time (Fig. 4.13).

As the equatorial plane approaches the plane of revolution of Moon around the Earth and the ecliptic plane, the external influence of Moon on equatorial Earth's bulge decreases significantly. This process (according to Eq. (3.11)) leads to a considerable cooling of climate. As soon as the average near-surface temperature reaches a certain critical value, glaciations occur (for Northern regions, this critical value of average near-surface temperature is close to 9–10 °C). In turn, the formation and growth of glaciers at the polar regions leads to imbalance of equilibrium rotation of Earth and to rather rapid increase in the precession angle.

To determine the resulting influence of glaciers on the rotational regimes of Earth, one needs to take into account the glaciations of both Northern

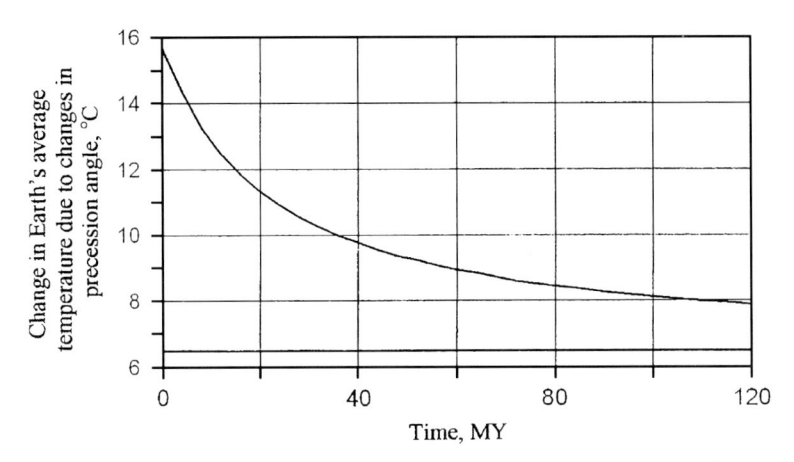

Figure 4.13: Changes in the Average Surface Temperature of Earth with Time under Influence of Sun–Moon Tides and Asymmetric Disposition of Earth's Continents. When the Precession Angle is Monotonously Decreasing Starting at 24°, the Temperature Curve is Asymptotically Approaching its Minimal Value of 6.5 °C (the Classic Value of Temperature of Absolutely Black Body at a Distance of Earth from Sun is Equal to 5.5 °C).

and Southern hemispheres (Fig. 4.14). The growth of Antarctic ice mass, however, was considerably lower than that of the Northern hemisphere because (1) it was limited by the finite size of underlying Antarctic continent and (2) it occurred mostly at the coastal regions and in Western Antarctica with frequent wet cyclones. Above the Eastern Antarctica, with high position of glacial dome (up to 4 km), the anticyclones usually prevail. The snow mass in the central regions increases mainly due to falling of hoarfrost out of relatively dry air and this increase is almost completely compensated by the plastic spilling of ice from the central to coastal regions.

The locations of glaciations in the Late Pleistocene time are shown in Fig. 4.14 (Kotlyakov, 2002). The dominating mass of ice accumulated on the Canadian Shield of Northern hemisphere and on the Western Antarctica of Southern hemisphere. The mass of ice in the Eastern Antarctica changed to a lower degree. In addition, because of the long existence of Eastern Antarctica glaciers, their ice cover is in isostatic equilibrium. Therefore, during the Late Pleistocene time the two momentums of forces, directed against each other were applied to the Earth's body: one of them was caused by the Canadian Shield, whereas the other by the glaciers of Western Antarctica (Fig. 4.15). However, the momentum, caused by the glaciers of Northern hemisphere, prevailed.

The influence of glaciers (located symmetric with respect to the Earth's poles) on the precession angle of Earth can be ignored because they balance

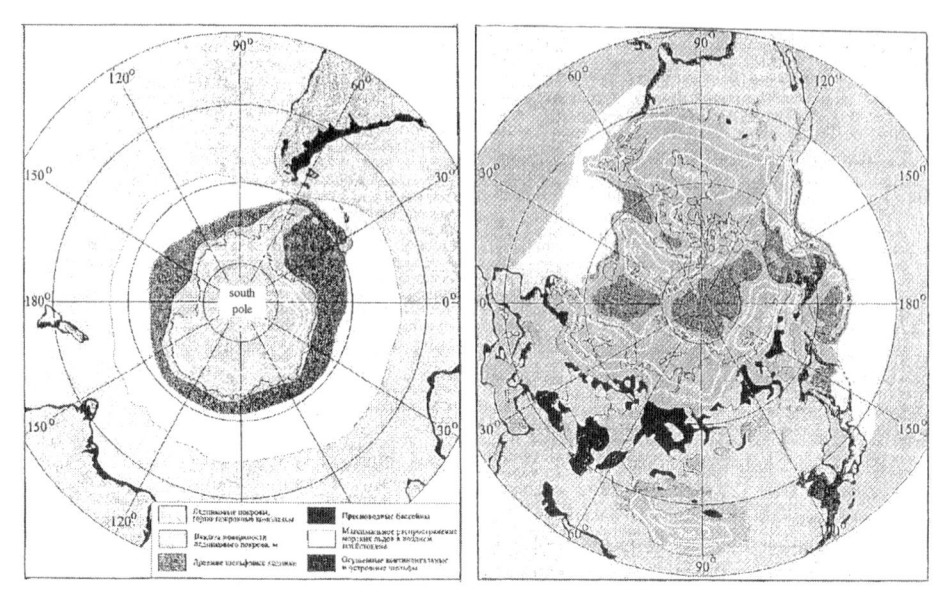

Figure 4.14: Glaciations of Antarctica and Northern Hemisphere during the Late Pleistocene Epoch (Modified after Kotlyakov, 2002). Light Shading around the Polar Regions Indicates the Areas of Maximal Proliferation of Ice during the Late Pleistocene Epoch; Dark-Gray Shading Indicates the Ancient Shelf Glaciers at Antarctica, North Polar Ocean, North Atlantics, and Okhotsk and Bering Seas. In the Central Regions of Antarctica, Canada, North Europe, Northern Regions of Siberia, Shallow Seas of Polar Ocean, and at the High Mountains of Central Asia, the Ice Covers of Thickness up to 2–3 km were Abundant. Left–Southern Hemisphere; Right–Northern Hemisphere.

each other. By our estimates, the excessive mass of ice covers (also considering the Southern glaciers) of 2×10^{22} g was located at about 70° of Northern latitude above Canada. The centripetal acceleration (g_{cp}) at a latitude of 70° is approximately equal to 1.15 cm/s^2. In this case, ΔP_{gl} (the force applied to the excessive mass of glaciers) was about 2.3×10^{22} g cm/s^2, and the distance h_{gl} from the center of excessive mass of ice to the North Pole was about 2.2×10^8 cm = 2,200 km. Then, the additional momentum ΔM applied to the Earth was equal to $\Delta P_{gl} \cdot h_{gl} \approx 5.1 \times 10^{30}$ g cm^2/s^2. According to Eq. (4.7), $\omega = 7.9 \times 10^{-11}$ rad/s and the characteristic warming period $\tau \approx 2,500$ years. The above estimate should be considered as a first approximation. This estimate, however, allows one to obtain the characteristic period of climatic warming and degradation of glaciers. According to Kotlyakov (2002), "the

Lunar orbital plane

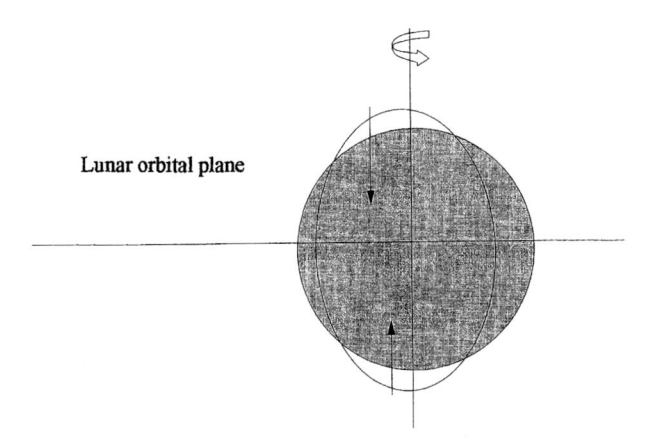

Figure 4.15: The Influence of the Excessive Mass of Glaciers of Northern Hemisphere on the Turning of Earth's Rotation Axis. This Influence Leads to the Increase in Precession Angle. As a Result, the Precession Angle Returns Rapidly (in about Several Thousand Years) to its Initial Value of about 24°, which, in turn, Causes the Climate Warming, Ice Melting, Degradation of Ice Covers, and Advent of Sequential Interglacial Period.

decay of gigantic Pleistocene glaciers in the Northern hemisphere occurred quite rapidly on the geologic timescale – only in several thousand years."

To calculate the climatic temperature deviations, one also needs to take into account the Earth's orbital deviations (Milankovitch cycles). The main harmonic components of these cycles are characterized by the periods of about 40 and 22 thousand years (Milankovitch, 1957; Bol'shakov, 2003). They give rise to the temperature deviations of about $\pm 3\,°C$. Figure 4.16 shows changes in the average Earth's temperature during the Late Pleistocene time (taking such deviations into consideration) in comparison with the isotopic temperature of Antarctic ice at the Vostok Station.

One can observe a strong correlation between the theoretical temperature curves and experimental data, with the theoretical curves being smoother. The final shape of theoretical curve depends considerably on the phase relationships among the main cycles of Moon–Sun and "glacial" deviations of temperature, the cycles of Earth's orbital deviations in the process of revolution of Earth around Sun, and the precession of Earth's rotation. Therefore, by choosing the phase shifts for the components of climatic temperature deviations (the best fit of theoretical curve to experimental data obtained from the ice cores from Antarctica) one can construct a curve that can be used for forecasting the climatic changes in the future. One example of such prediction is shown in Fig. 4.17.

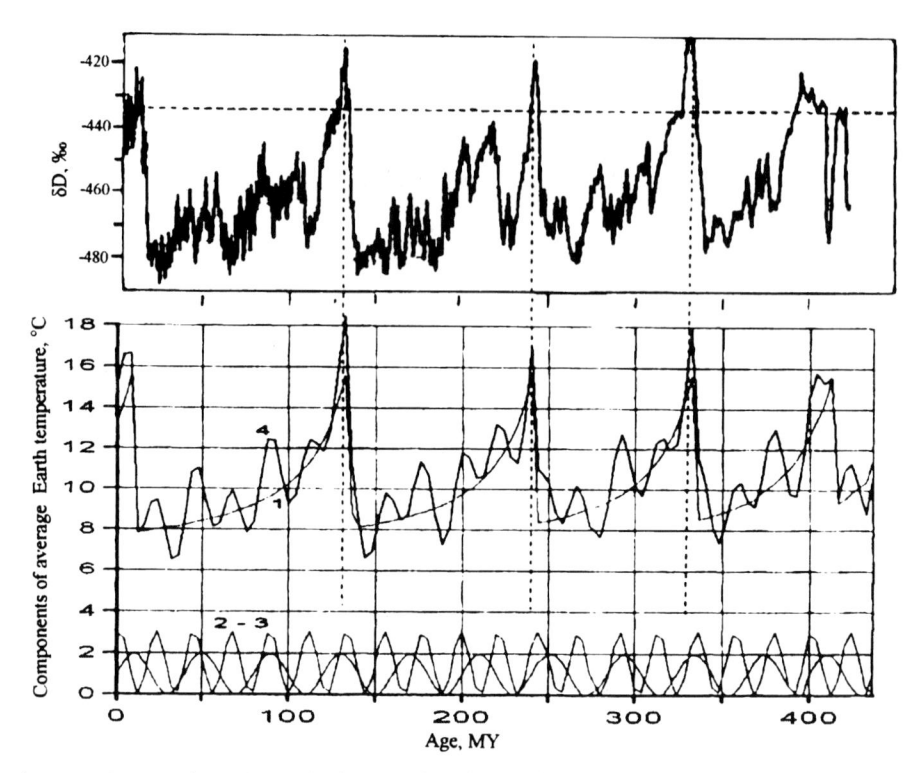

Figure 4.16: The Correlation of Theoretical Temperatures during the Pleistocene Epoch with Experimental Values of Isotope Temperatures of Antarctic Ice Cover Measured at the Vostok Station (Kotlyakov, 2000). The Time Axis is Directed from Right to Left. The Upper Graph Represents the "Isotope Temperature" of Antarctic Cover at the Vostok Station. The Last Peak of "Isotope Temperature" (about 10–20 Thousand Years Ago) Corresponded to 8 °C. The Lower Graph Shows the Theoretical Curve of Average Temperatures on Earth Constructed Considering the Influence of Moon, Sun, and the Positions of Continents on Precession Cycles of Earth. (1) Temperature Deviations Caused by the Changes in Precession Angle (the Maxima of Curve (1) Coincide with the Maxima of Experimental Data); (2) and (3) Temperature Deviations Caused by the Precession of Orbit of Revolution of Earth around the Sun (Two Basic Harmonics of Milankovitch Cycles with the Periods of about 40 and 22 Thousand Years); and (4) Resulting Changes in Temperature due to Combined Influence of all the Considered Causes.

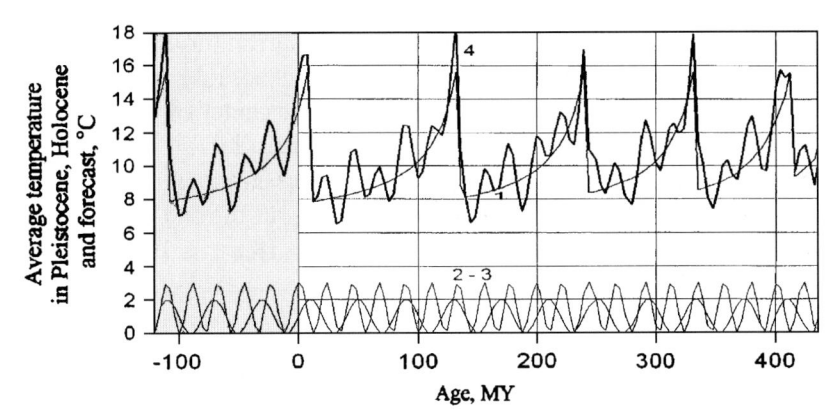

Figure 4.17: Theoretical Prediction of Climate Evolution for the Following 120,000 Years is shown in the Shaded Part of Graph (the Time Axis is Directed from Right to the Left). (1) Theoretical Curve of Changes in Temperature of Earth due to Moon–Sun Influences on the Earth (the Temperature Deviations are Caused by the Changes in the Precession Angle); (2) and (3) Temperature Deviations, Caused by the Precession of Orbit of Revolution of Earth around the Sun (by the Milankovitch Cycles); (4) Resulting Temperature Curve due to Both Causes.

Thus, in the Pleistocene Epoch, as a result of mutual Moon–Earth influences, there were slow but steady periods of climatic cooling by about 8–10 °C, which lasted approximately 100–120 thousand years. After formation of thick ice covers, a rapid warming by the same 8–10 °C occurred that, during only several thousand years, degraded the glaciers completely. Therefore, the Moon–Earth interactions in combination with the Earth's glaciations triggered essentially nonlinear auto-oscillating climatic processes, which were typical for the Late Pleistocene time. In the future, we should expect cooling only and probably the most significant one. Such a general cooling is occurring due to the life-activities of nitrogen-consuming bacteria (see Chapter 3), which are gradually reducing the partial pressure of nitrogen with the consequent gradual, steadily decreasing total atmospheric pressure. The decreasing atmospheric pressure, according to Eq. (3.11), inevitably leads to the acceleration in cooling of climate. This most significant, last and longest geological cooling period had started in the Riphaean time (Figs. 4.7 and 4.8).

The warm climate of second half of Mesozoic Era, which emerged after the vast proliferation of blossoming plants, can be explained by increased generation of oxygen that temporarily compensated for the reduction of partial nitrogen pressure attributed to the nitrogen-consuming bacteria. After establishment of stationary regime of oxygen partial pressure in the Cenozoic

Era, however, the new phase of cooling had begun, even though the luminosity of Sun slightly increased at that time. In addition, because of more compact positioning of continents and the proximity of their center of mass to equator, the momentum of force ΔM_{cont} at that time was considerably higher than that at the present time. This led to an increase in precession angle and, as a result, to the climate warming. According to our estimates, at the end of Mesozoic Era the average near-surface temperature reached $+18 \ldots +19\,°C$ (instead of present-day $+15\,°C$).

4.5. Continental Drift and the Climates of Earth

The position of continents on the Earth's surface can disturb the spherical symmetry of Earth (Eq. (4.7)) and influence considerably the near-surface temperature regimes. At present time, this influence is relatively small ($\Delta M_{cont+m} \approx 0.6 \times 10^{29}\,cm^2\,g/s^2$). Upon formation of compact supercontinents, which are usually positioned at low latitudes with the resulting center of mass at the equator (Monin, 1988, 1999), the magnitude of ΔM_{cont+m} increases pronouncedly. Inasmuch as the direct influence of continental drift on the Earth's precession angle is difficult to evaluate, the authors applied an indirect method. They used the paleotemperatures of ocean to determine the precession angles, which existed in the past.

According to Chumakov (2004), the surface temperatures of Sargasso Sea, which about 100 million years ago was located in the tropical belt at about 20° of Northern latitude, varied from $+25\,°C$ to $+30\,°C$. At that time, the maximum temperatures (at equator) even reached the magnitude of 28–32 °C and positive temperatures dominated at the Earth's poles. Using Eqs. (4.1) and (4.10), one can estimate that about 100 million years ago the equatorial temperature of 32 °C and the positive temperatures at the Earth's poles corresponded to the precession angle of 34°. In addition to the paleotemperatures of Cretaceous Period, one can use the data on distribution of ice ages in the Earth's history for estimation of precession angles.

The stationary precession angle during the formation of Pangaea supercontinent (about 100 million years ago) reached about 34°. Judging from the reconstruction of other supercontinents and considering their smaller masses (Fig. 1.41), our estimates of precession angles at the times of their formation are somewhat lower (Fig. 4.18). During the time intervals between the times of formation of supercontinents, we assumed that the precession angles were close to the stationary values of 6–8°.

Using the model of changes in Earth's precession angle (Fig. 4.18), variation in the average Earth's surface temperature was determined (Fig. 4.19). In

Evolution of Earth's Climates 191

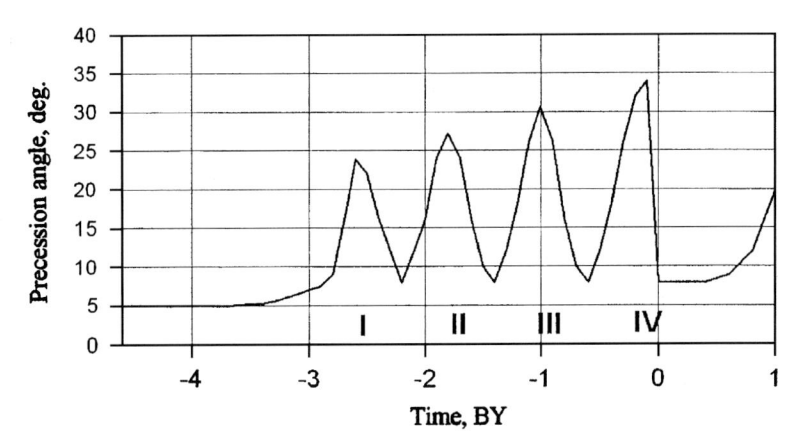

Figure 4.18: Probable Changes of Precession Angle during the Formation of Supercontinents. (I) Monogaea; (II) Megagaea; (III) Mezogaea (Rodinia); (IV) Pangaea. It was Assumed that Excessive Masses of Monogaea, Megagaea, and Mezogaea were 0.7, 0.8, and 0.9 of the Mass of Pangaea, Respectively. It was also assumed that the Future Supercontinent Hypergaea would not be Formed over the Considered Time Interval.

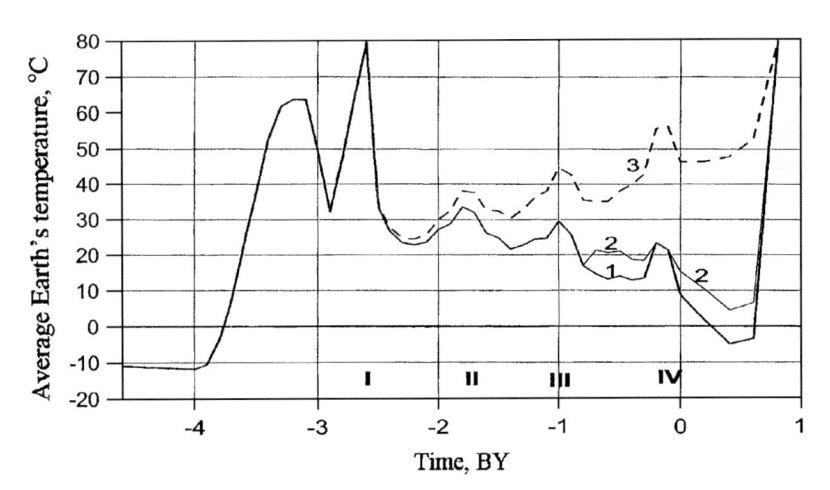

Figure 4.19: Influence of Continental Drift and Formation of Supercontinents on the Average Temperatures on Earth. (1) Average Temperature at Sea Level taking into Account the Changes in Precession Angle of Earth; (2) Average Earth's Temperature during the Interglacial Periods; (3) Average Earth's Temperature with no Nitrogen Consumption by Bacteria; the Roman Numerals Denote the Times of Formation of Supercontinents: (I) Monogaea; (II) Megagaea (Shtille); (III) Mezogaea (Rodinia); (IV) Pangaea (Wegener).

addition, using the same model, the changes in temperature of oceans and continents at the Earth's equator and poles were determined (Figs. 4.20 and 4.21).

As shown in Figs. 4.19–4.21, *the time of formation of supercontinents was accompanied by corresponding increase in the near-surface temperatures.* During the formation of first supercontinent Monogaea in the Earth's history about 2.6 billion years ago, the average temperature at sea level exceeded 70 °C. The temperature was close to 30 °C when the supercontinent Megagaea (Shtille) was formed about 1.8 billion years ago. During the formation of Mezogaea (Rodinia) about 1 billion years ago the temperature decreased to 26 °C. Finally, during the formation of last supercontinent Pangaea (Wegener) the temperature was 24 °C and lower. During the intervals between the formation of supercontinents, the average temperatures at sea level decreased by 7–10 °C.

The formation of supercontinents changes drastically the tectonic structure of Earth. These changes are well correlated with the periods in Earth's

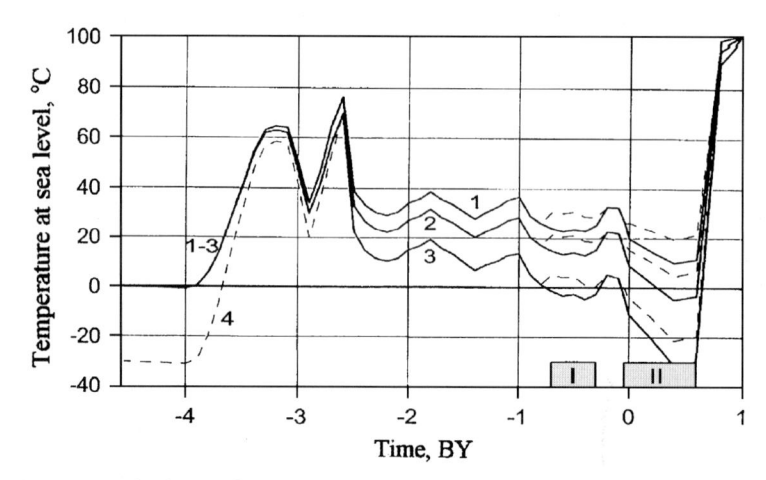

Figure 4.20: Evolution of Temperature of World Ocean due to Joint Influences of Changes in Atmospheric Pressure and the Precession Angle (Fig. 4.18). (1) Temperature of Oceans at the Equator; (2) Average Temperature of Oceans; (3) Temperatures of Oceans at the Earth's Poles during the Proterozoic and Phanerozoic Time (the Temperatures during the Interglacial Periods are Presented by Dashed Curves); (4) Temperature at the Earth's Poles during the Archaean (at that Time the Oceans began Forming at the Equatorial Zones and they moved to the Poles only by the End of Archaean Time). The Shaded Rectangles I and II Indicate Periods of Glaciations of Polar Oceanic Basins in the Riphaean and Phanerozoic Time.

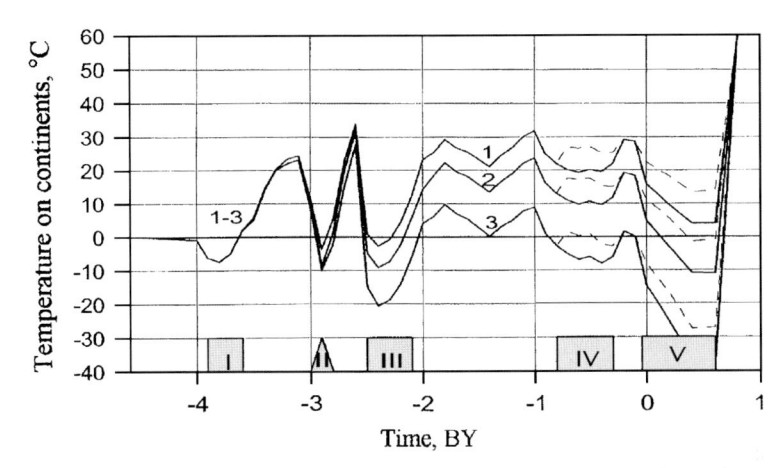

Figure 4.21: Evolution of Temperature of Continental Surface due to Joint Influence of Evolution of Atmospheric Pressure, Changes in the Continental Surface Elevation, and Deviations in Precession Angle of Earth (Fig. 4.19). (1) Temperature of Continents at the Equatorial Zone; (2) Average Temperature on the Continents; (3) Temperature on Continents at the Earth's Poles in the Proterozoic and Phanerozoic Time (Dashed Curves Represent Temperatures during Interglacial Periods); I through V Indicate the Times of Ice Ages at the Polar Regions or High Mountains (Figs. 4.8 and 4.9).

geologic history when major climate changes occurred. This correlation is possibly due not only to changes in insolation of Earth's surface, but also to changes in the directions and intensity of oceanic currents (gyres). When the tectonic geometry is conducive to development of equatorial currents, the events leading to global warming occur, "whereas the physical blockage of equatorial current system appears to coincide with icehouse events" (Gerhard et al., 2001). Thus, one can come to the conclusion that pronounced global climatic changes are caused primarily by changes in insolation and the Earth's tectonics.

Chapter 5

Influence of Climate on Formation of Mineral Deposits on Earth

5.1. Introduction

Distribution of ores in the Earth's crust and their origin are closely linked to global evolution of Earth and climatic changes. On the other hand, the patterns of distributions of ores in space and time may serve as reliable indicators of the Earth's evolution. Consequently, the metallogenic concepts of formation of ores were based on the prevailing theoretical views on development of geotectonic, petrologic and geochemical processes. Despite the multitude of existing concepts on the relationship of ores with the tectonic development of Earth (Rundquist, 1968; Smirnov, 1982; Mitchell and Garson, 1984; Starostin, 1996; Khain, 2000; Sorokhtin et al., 2001), the problem of origin of ores is yet to be solved.

Analysis of observations on the conditions of ore formation and time–spatial relationships between the geologic–petrologic characteristics and the tectonic scenario of formation of mineral deposits shows that metallogenetic periods represent unique "moments" in the history of Earth's evolution. This is absolutely clear from general considerations, because from the thermodynamic point of view the Earth is an open dissipative system irretrievably losing its energy into space. The latter implies that the Earth's evolution is an irreversible process. Using this general approach, Sorokhtin and his associates (Sorokhtin et al., 2001; Sorokhtin and Ushakov, 2002) analyzed the origins of mineral deposits and showed that most of the endogenic deposits formed owing to considerable influence of exogenic factors of Earth's evolution.

5.2. Plate Tectonics and Origins of Endogenic Mineral Deposits

The main difficulty in explaining the causes of formation of large local formations of ores and accumulation of some other scattered elements is the fact that their concentration in the mantle is extremely small, whereas in the places of deposits their concentration increases by hundreds and thousands of times. For example, the concentration of uranium and gold in the present-day mantle does not exceed 2×10^{-9}; mercury and thorium -8×10^{-9}; lead -8×10^{-9}; silver, wolfram, and platinum – about 10^{-7}; and lithium, niobium, and molybdenum – about 10^{-6}. In addition, according to the contemporary concepts of origin and evolution of Earth, the mantle matter (including the upper and lower mantle) was thoroughly stirred during the 4 billion years with tectonic activity by the convective flows and on the average is uniform in its elemental contents. Consequently, one cannot expect any local nonuniformity in the mantle with the increased content of ore elements. Only the elements most abundant in the mantle (chromium, for example) can be separated from the mantle matter into their endogenous deposits by the direct differentiation of mantle smelts. As an example, one can mention the deposits of chromites in the ophiolite belts of Earth.

In addition, it is important to note that, judging from the conditions of smelting out of oceanic basalts and the content of juvenile water in them (the Earth's mantle is practically dry, with the water content not exceeding 0.05%). There are no fluid flows in the mantle that can transport the ore elements into the crust. However, a new geologic theory – the tectonics of lithospheric plates – allows one to develop a sound approach to explanation of enrichment of Earth's crust with ore elements, carbonates, and silica. In particular, it became clear that the main mass of endogenous mineral deposits in the continental crust could have formed only through the multistage process of enrichment of Earth's crust with ore elements.

This multistage process started at the rift zones on oceanic floor. There are extremely powerful hydrothermal systems through which the Earth dissipates over 30% of endogenic heat in the oceanic rift zones and on the slopes of mid-oceanic ridges. By our estimates, the average rate of water transfer in all the oceanic hydrothermal sources of mid-oceanic ridges is equal to $2{,}300\,\text{km}^3/\text{year}$. At such rates, the entire mass of oceanic water (1.37×10^{24} g) transits through the active faults and seeps in the mid-oceanic ridges with the vast and gentle slopes in about 0.6–1.0 million years.

Through the hydrothermal systems of rift zones, huge masses of endogenic matter (including the chalcophilic ore elements and silicon dioxide) are transferred into the oceanic crust and hydrosphere forming the sulfides of zinc, copper, lead, and other metals (Lisitsyn et al., 1990; Lisitsyn, 1993, 2000). Thus, during only the last 150 million years, in the Pacific Ocean alone, about $(10-15) \times 10^{12}$ tons of polymetallic matter entered the Earth's crust from the hydrothermal sources (Gurvich, 1998). In the rift zones of mid-oceanic ridges, there is hydration of near-surface layers of mantle and occurrence of many chemical reactions, which are important for maintaining the stable ecological conditions on the Earth and for the formation of

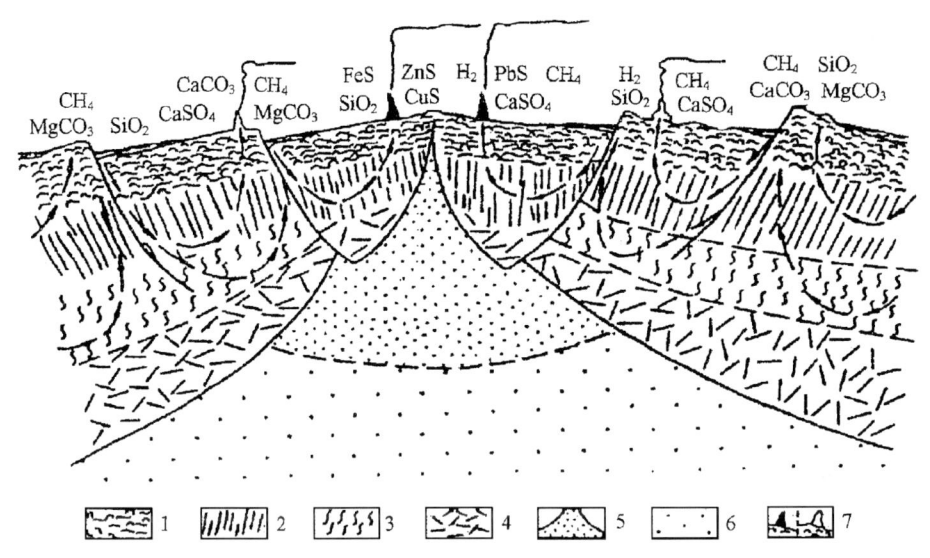

Figure 5.1: The Formation of Oceanic Crust and Geochemistry of Hydrothermal Processes in the Rift Zones of Mid-Oceanic Ridges. (1) Basalts (pillow lavas); (2) Dolerite Dykes; (3) Serpentine Layer; (4) Undercrustal Layer of Lithosphere; (5) Source of Magma under the Mid-Oceanic Ridges; (6) Asthenosphere; (7) Black and White "Smokers". Arrows Indicate the Directions of Water Movement in the Oceanic Crust.

carbonate and silica facies (Fig. 5.1). Moreover, all the reactions are irreversible because they are accompanied by the energy release. The most important are the reactions of hydration of oceanic crust rocks that bind carbon dioxide in carbonates, which are further assimilated by life forms with subsequent burial in sediments:

$$4Mg_2SiO_4 + 4H_2O + 2CO_2 \rightarrow \underset{\text{serpentine}}{Mg_6[Si_4O_{10}](OH)_8}$$
$$\underset{\text{olivine}}{} + \underset{\text{magnesite}}{2MgCO_3} + 72.34 \, \text{kcal/mole} \quad (5.1)$$

$$\underset{\text{anorthosite}}{2CaAl_2Si_2O_8} + 4H_2O + 2CO_2 \rightarrow \underset{\text{kaoline}}{Al_4[Si_4O_{10}](OH)_8}$$
$$+ \underset{\text{calcite}}{2CaCO_3} + 110.54 \, \text{kcal/mole} \quad (5.2)$$

Owing to these reactions, a stable and relatively low carbon dioxide partial pressure in the Earth's atmosphere is maintained. Also, the ocean water is continuously supplied by the primary compounds for the formation of carbonate rocks.

In the process of hydration of pyroxenes, the silicon dioxide is delivered in great amounts from the rift zones according to the following reaction:

$$6MgSiO_3 \underset{\text{enstatite}}{} + 4H_2O \rightarrow \underset{\text{serpentine}}{Mg_6[Si_4O_{10}](OH)_8}$$

$$+ \underset{\text{silicon dioxide}}{2SiO_2} + 27.38 \text{ kcal/mole} \qquad (5.3)$$

Silicon dioxide is necessary for formation of siliceous deposits, chert, and jasper. Upon oxidation of divalent iron to trivalent state in the presence of carbon dioxide, the abiogenic methane forms:

$$4\underset{\text{fayalite}}{Fe_2SiO_4} + 12\underset{\text{forsterite}}{Mg_2SiO_4} + 18H_2O + CO_2 \rightarrow \underset{\text{serpentine}}{Mg_6[Si_4O_{10}](OH)_8}$$

$$+ 4Fe_2O_3 + CH_4$$
$$\text{hematite}$$
$$+ 183.47 \text{ kcal/mole} \qquad (5.4)$$

In the absence of carbon dioxide, hydrogen forms in the process of iron oxidation:

$$\underset{\text{fayalite}}{Fe_2SiO_4} + 3\underset{\text{forsterite}}{Mg_2SiO_4} + 5H_2O \rightarrow \underset{\text{serpentine}}{Mg_6[Si_4O_{10}](OH)_8}$$

$$+ \underset{\text{hematite}}{Fe_2O_3} + H_2 + 30.72 \text{ kcal/mole} \qquad (5.5)$$

According to our estimate, the process of hydration of rocks of the present-day oceanic crust results in the generation of about 9 million tons per year of methane and approximately the same amount of hydrogen. At this rate of methane generation, the amount of hydrocarbons generated on the Earth during each one-million-year period would exceed the entire amount of existing hydrocarbon reserves. This implies that the greater part of methane is oxidized by the methane-consuming bacteria or together with hydrogen is transferred into the atmosphere. A considerable part of these volatile emanations, however, could be conserved in the oceanic sediments as hydrocarbon deposits in the form of gas hydrates, for example.

In the oceanic rift zones, many ore elements (Fe, Zn, Pb, Cu, Mg, etc.) are transferred from the mantle together with sulfur into the oceans. Reacting with sulfur, the ore elements form sulfides. One has to note, however, that a significant part of sulfur in the sulfides of rift zones is due to reduction of sulfates of oceanic water in the presence of methane and hydrogen (Eqs. (5.4) and (5.5)):

$$\underset{\substack{\text{magnesium}\\\text{sulfate}}}{MgSO_4} + CH_4 \rightarrow \underset{\text{magnesite}}{MgCO_3} + \underset{\substack{\text{hydrogen}\\\text{sulfide}}}{H_2S} + H_2O + 11.35 \text{ kcal/mole} \qquad (5.6)$$

The liberated hydrogen sulfide is an extremely aggressive mineralizing agent (during chemical reactions generating a great amount of energy). Therefore, it immediately reduces iron and other ore metals (copper, zinc, lead) to sulfides literally "sucking" them out of basalts and ultrabasic rocks of oceanic crust; for example, according to the following reaction:

$$\underset{\text{fayalite}}{Fe_2SiO_4} + 2H_2S \to 2FeS + SiO_2 + 2H_2O + 56.68 \text{ kcal/mole} \tag{5.7}$$

All these chemical compounds eventually accumulate in sediments, forming (1) deposits of sulfides of ore metals around the geothermal wells ("black smokers") at the rift zones and (2) the layer of metallogenic deposits on the oceanic crust (Lisitsyn et al., 1990). In time, the sulfide deposits are oxidized and destroyed and their ore matter is transferred into the metallogenic deposits.

Besides the hydrothermal enrichment of oceanic crust with ore elements in its lower layers, usually, the deposits of magmatic striped chromite ores form at the contact of gabbro with dunites and periodotites. The origin of chromite ores is linked to direct differentiation of mantle matter in magmatic source under the rift zones. In fact, these ores are the result of separation of a dense phase of basaltic melts, which descended to the bottom of magmatic source, underlain by the restite mantle rocks. This determines the usual position of chromite ores between the layers of gabbro and dunites in the ophiolite complexes. As examples of such deposits, one can include the Kempirsay Paleozoic deposits of chromites in the South Ural Mountains and the Mesozoic deposits at the Balcan Peninsula.

According to the tectonics of lithospheric plates, the second stage of enrichment of continental crust with lithophilic and ore elements occurs at the zones of subduction of continental plates at the expense of transfer of these elements from the oceanic crust (Fig. 5.2). The ore elements are mobilized and transferred into the continental crust due to dehydration processes and smelting of oceanic crust rocks and pelagic and terrigenous sediments at the plate subduction zones. In this process, a great volume of water is liberated from hydrosilicates, creating massive hot and mineralized fluid flows, which transfer large amounts of previously scattered elements into the continental crust.

The chemical reactions of dehydration of rocks of oceanic crust in the subduction zones of lithospheric plates in many aspects are carried out in the opposite (compared to the rift zones) direction – with assimilation of already separated compounds (e.g. SiO_2, $CaCO_3$, and $MgCO_3$) and liberation of new components. In contrast to the rift zones, their liberation

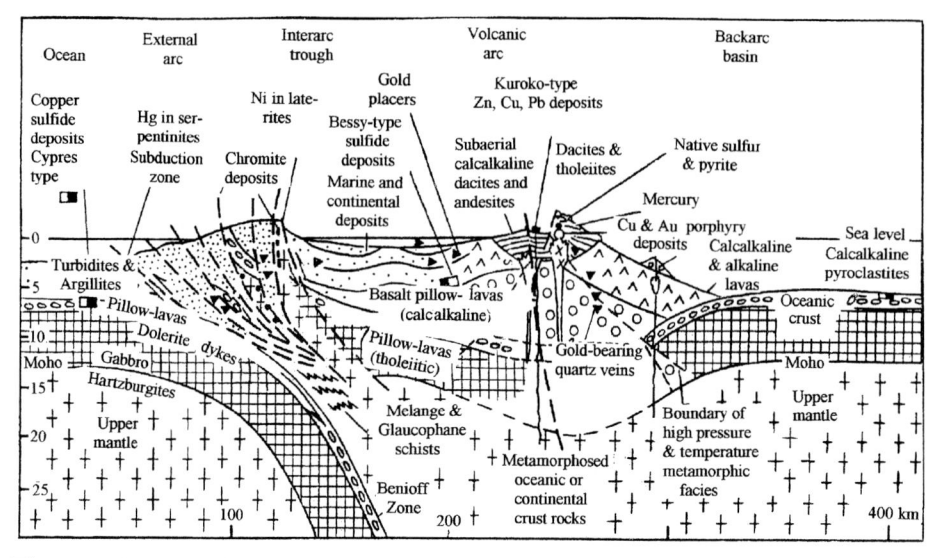

Figure 5.2: Metallogenic Zoning of the Island Arc (Modified after Mitchell and Garson, 1981, 1984).

occurs with consumption of additional heat energy, which is generated there as a result of friction of lithospheric plates at the subduction zones. The main process of liberation of water and silicon dioxide, due to deserpentinization of the third layer of oceanic crust, occurs according to the following reaction:

$$Mg_6[Si_4O_{10}](OH)_8 +38.12 \, kcal/mole \rightarrow 3Mg_2SiO_4 +SiO_2 + 4H_2O \quad (5.8)$$
$$\text{serpentine} \qquad\qquad\qquad\qquad\qquad \text{forsterite}$$

A hard-melting olivine (forsterite) sinks into the mantle, whereas the solutions of silicon dioxide in hot water and alkali and other lithophilic elements move upward and enter the continental crust.

The oceanic sediments are usually sucked into the zones of plate subduction (Sorokhtin and Lobkovskiy, 1976). These sediments, however, are always saturated with seawater and, upon reaching the hot areas of subduction zones, the alkali aluminum silicates are formed, which are also transferred into the continental crust enriching it with the alkali and other lithophilic elements. One of the possible simplified reactions is:

$$Al_4[Si_4O_{10}](OH)_8 +NaCl + 26.98 \, kcal/mole \rightarrow Na[AlSi_3O_8]$$
$$\text{kaoline} \qquad\qquad\qquad\qquad\qquad\qquad \text{albite}$$
$$+HCl + 2H_2O + SiO_2 + 3AlO(OH) \quad (5.9)$$
$$\text{diaspore}$$

At deeper levels, the aluminum silicates are enriched with potassium:

$$Al_4[Si_4O_{10}](OH)_8 + KCl + 23.88 \text{ kcal/mole} \rightarrow K[AlSi_3O_8]$$
$$\text{kaoline} \hspace{4cm} \text{orthoclase}$$

$$+ HCl + 2H_2O + SiO_2 + 3AlO(OH) \hspace{2cm} (5.10)$$
$$\text{diaspore}$$

The important role of chlorine in the hypogenic processes is indicated not only by the formation of some chlorine-containing silicates (skapolites and sodalites), but also by the emanation of large amount of gaseous HCl in the areas of active volcanism. For example, in 1919 in the Ten Thousand Smokes Valley (New Zealand), about 1.25 million tons of HCl and 0.2 million tons of HF were emanated with the water vapor.

In the subduction zones, there are the dissociation of carbonates, with binding of alkali elements in silicates, and formation of free carbon dioxide according to the following possible reaction:

$$Mg_6[Si_4O_{10}](OH)_8 + 4SiO_2 + 2CaCO_3$$
$$\text{serpentine} \hspace{3cm} \text{calcite}$$

$$+ 59.5 \text{ kcal/mole} \rightarrow 2CaMg[Si_2O_8]$$
$$\text{diopside}$$

$$+ 4MgSiO_3 + 2CO_2 \hspace{3cm} (5.11)$$
$$\text{enstatite}$$

The enstatite and diopcide as the hard-melting minerals are removed to the mantle, whereas the water and carbon dioxide return to the hydrosphere and atmosphere.

The oceanic crust is recycled into the continental one using the conveyer-like mechanism, which is cummulative and very powerful. During the time of tectonic activity of lithospheric plates (lasting about 2.6 billion years), the total mass of water–fluid flows in the subduction zones exceeded 8–10 times the mass of entire hydrosphere of Earth and about 13–16 times the mass of water of all oceans and seas! These theoretical estimates deserve a special attention, because they show that these water–fluid flows are millions of times larger than the juvenile ones.

The enrichment of continental crust in ore elements in the subduction zones leads to more or less uniform increase in their content at the fronts of lithospheric plate subduction. In some cases, the commercial accumulations of certain types of mineral deposits (e.g. pyrite) can be formed. This could frequently happen in the long-existing subduction zones because of the erosion of frontal areas of the plate pulling over and the recycling of crustal rocks and sediments overlying the plate. In this way, the unique copper–molybdenum porphyry ores of South American Andes were most probably formed (based on the paleogeodynamic reconstructions) as the oceanic crust of Pacific and Pre-Pacific Ocean subducted continuously for at least the past 1.5 billion years.

The theory of plate tectonics also suggests the third mechanism of enrichment of endogenous mineral deposits with the ore elements. The exogenous processes of erosion and accumulation of sediments, often occurring with active participation of live species, greatly influence the redistribution of elements throughout the Earth's crust. As a rule, the sedimentogenesis is accompanied by an intense differentiation of Earth's matter. This is exemplified by phosphorites, carbonates, and sand–clay deposits with their unique mineral content. In the Russian Platform shales, for example, the tin content as compared to that in the mantle is about 11 times higher, lead – about 20 times, potassium – about 200–250 times, rubidium – about 500–700 times, barium – up to 1,500 times, and uranium – 3,000–3,500 times higher. On the other hand, if the sand–clay sediments were accumulated in the stagnant basins saturated with hydrogen sulfide, which was a typical condition existing in the Precambrian time, then the sulfides of iron, copper, zinc, lead, and molybdenum (and, in some cases, the uranium oxides and the hydroxides of gold and wolfram) could accumulate in the sediments. This is true for other sediments. In the case of carbonates, for example, the strontium content exceeds that in the mantle several times. Sodium, potassium, calcium, sulfur, chlorine, and fluorine are concentrated in the evaporites. The live species accumulate in their organisms many of rare elements, such as uranium, cesium, and neodymium. Therefore, in phosphorites their content is significantly higher than that in the mantle (for uranium – 20–25 thousand times higher and for other rare elements – 500–1,000 times higher).

As a rule, the terrigenous sediments are transported by rivers and material flows to the continental margins and accumulate on the continental slopes as turbidites (Lisitsyn, 1984). In addition, at the base of such deposits, thick layers of evaporites accumulate, which were formed at the early stages of breaking down of original supercontinents (e.g. deposits of evaporites can be observed at the costal zones of Atlantic Ocean, Red Sea, and Gulf of Mexico). When such deposits reach the subduction zones or the places of collision of two continents, the entire spectrum of erupted crustal rocks (granites to syenites and ultrabasic rocks) are smelted out of the sediments, with the specific mineral contents. This observation can explain the scattered occurrence of mineral deposits over the Earth's surface.

If the crustal rocks experience repeated destruction and transfer into the sediments, the fourth stage of enrichment of crustal rocks with ore elements can occur. Therefore, the mineral deposits of this type can be richer in mineral content, because their mineral matter experiences more recycling during the Earth's geologic development. As an example, the concentration of tin in the younger deposits is higher than that in the older ones. For example, the total content of tin in the Mesozoic deposits is more than two orders of magnitude higher than that in the Archaean deposits (Sobolev et al., 2000). One can observe an analogous situation with molybdenum: as a result of magmatic recycling of sediments, the concentration of this metal in the young deposits of wolfram and molybdenum have been continuously increasing in time (Sobolev et al., 1997). This implies that seemingly "endogenic" mineral deposits in fact were formed through the enrichment stage due to destruction of crustal rocks and subsequent sedimentogenesis, i.e. they were formed under exogenic conditions.

The hydrothermal deposits present another example of strong influence of exogenic factors on the formation of mineral deposits. Inasmuch as the water content in the mantle is negligibly small, all hydrothermal mineral deposits are formed as a result of either mobilization of surface or ground water on contact with the hot intrusive bodies or liberation of water from the cooling magma. Even in the latter case, the water-saturated magma could receive water only from either the water-saturated sediments or directly from the hydrosphere.

There are two types of hydrothermal deposits. The first type includes the ore formations in the fumaroles and hot springs of volcanic areas. The sulfide deposits of the "black smokers" in the rift zones of mid-oceanic ridges also belong to this type. The second type, often associated with pegmatite mineralization, comprises the deposits formed over granitoid massifs or due to the rise of hot and mineralized water from the subduction zones. In the Post-Archaean time, the granite magma formed due to remelting of water-saturated sand–clay sediments in the collision zones of continents. Water was thus brought to the subduction zones with the oceanic sediments and pelagic accumulations. In all cases, water in the hydrothermal springs is obtained from the hydrosphere, i.e. it is an exogenic agent.

Influence of exogenic factors on formation of endogenic mineral deposits is well illustrated by tin, rare-metals, and gold deposits of Verkhoyansk-Kolyma foldbelt in East Siberia. In the Early Paleozoic time, the Kolyma Massif broke from the East-Siberian Platform, and the East-Siberian Paleoocean formed between them. After these events, at the eastern passive province of Siberian Platform thick layers of terrigenous sediments began to accumulate. These sediments were brought by the paleorivers from Anabar Archaean Shield at the North and from the Aldan Shield and Vitim-Patom Mountains at the South. The heavy fractions delivered from the ancient shields had been gradually accumulating as nearshore deposits, including cassiterite, gold, minerals of niobium, tantalum and other rare metals. During 200–300 million years of existence of the East-Siberian Ocean, about 12–15 km of terrigenous deposists (including river bed and delta) accumulated at its Western province beginning from the Devonian (but mainly from the Carboniferous) to the Jurassic time.

In the middle of Mesozoic Era, as the Kolyma Mountain Range began approaching the Siberian Platform, the East-Siberian Ocean began closing. Closing down of this paleoocean was accompanied by compression of the entire accumulated mass of oceanic sediments at the edge of continent and pushing it over the Siberian Platform. The Kolyma Mountain Range trampled on the lithosphere of East-Siberian Ocean. Transverse rifts formed in the oceanic lithosphere underlying the marginal continental sediments. The basalt smelts intruded into the layers of sediments through these rifts. The density of basalt smelts is about 2.8 g/cm^3 and is considerably higher than that of sediments (about 2.5–2.7 g/cm^3). Consequently, the magmatic basalts intruded mostly into the lower layers of sediments. The temperature of basaltic magma at a depth of 12–15 km reaches 1,350–1,400 °C, whereas the melting temperature of water-saturated sediments at the same depth range does not exceed 650–700 °C. This implies that the deposits of considered formation should smelt on contact with the hot basaltic magma (with reduction in density and viscosity) and then pierce the upper layers of sediments in the form of granitoid intrusions. The average content of these granitoid rocks (granite-porphyry to lyporites) matches the average contents of the sedimentary rocks at the intrusion zone. In the process of intrusion, the previously accumulated ore elements in the sediments were overworked by hydrothermal processes. This resulted in characteristic mineralizations: in the North region of Verkhoyansk–Kolyma zone, where the sediments were brought mostly from the Anabar Shield, there were tin and tantalum–niobium mineralization, whereas in the South-Eastern part of this region, which received the sediments from the Aldan Shield and Vitim-Patom Mountains, the gold mineralization prevailed.

Another example is the magmatic and metasomatic iron ore formations of Ural Mountains, such as Kachkanar deposits of titanium–magnetite ores of the Magnet and Blagodat' mountains. In all these cases, the formation of iron ore deposits was linked to the closing down of Paleo-Ural Ocean and overthrusting of the Ural island arc over the edge of Russian Platform. In this process, thick sedimentary formations of Riphaean and older age (Taratash type of iron ore) were sucked into the Paleo-Ural subduction zone. Their remelting and hydrothermal activation led to the formation of these deposits. The ore deposits of the Kachkanar Formation, for example, which were formed as a result of intrusion of large bodies of gabbroids, are surrounded by pyroxene minerals. This means that together with iron the basalt smelts also brought silicon dioxide to the surface. The paragenesis of iron and silicon dioxide is observed, for example, in the iron quartzites of Taratash deposits.

The plate tectonics theory opened a new approach in establishing the main regularities in distribution of mineral deposits over the Earth's surface. This theory implies (and the empirical data confirm) that endogenic mineral deposits as a rule form at the edges of lithospheric plates. There is a relationship between the types of plate boundaries and the ore complex types and their location with respect to these boundaries. This allows one to use the geodynamic maps (showing the positions and types of ancient boundaries of lithospheric plates) for forecasting the location (exploration) of mineral resources.

5.3. Earth's Core Separation Determining the Geologic Evolution of Earth

Together with the geologic evolution of Earth, the conditions of Earth's crust formation and concentration of mineral deposits were also changing. One should remember that the Earth's evolutionary processes are irreversible. Consequently, the present-day evolutionary conditions cannot be used directly for description of conditions of formation of mineral deposits in the past. For example, during the Archaean time there were no subduction zones. Therefore, the regimes of formation of continental crust and, consequently, the conditions of formation of mineral deposits were quite different from those of the present-day. According to the generally accepted concept of global Earth's evolution, the Earth's core differentiation is the main energetic process, which determines the development of Earth and its mineral deposits (Chapter 1).

According to Sofronov (1969), like the other planets of Solar System, the Earth was formed owing to the "cold" accretion of gas–dust of protoplanetary cloud. Immediately after its formation, the juvenile Earth was a tectonically passive "cold" planet with uniform distribution of chemical elements. Consequently, the juvenile Earth did not contain any mineral deposits.

The geologic history of the Earth can be divided into three large periods: Pre-Archaean (4.6–3.8 billion years ago), Archaean (3.8–2.6 billion years ago), and Proterozoic together with Phanerozoic (2.6 billion years ago to present). All of them were closely linked to the three periods of Earth's core formation discussed in Chapter 1. Initially, the Earth was heated by the radiogenic and tidal energies. During the Pre-Archaean time (first 800 million years of Earth's history), about 1.1×10^{37} erg of radiogenic and 2.1×10^{37} erg of tidal energy were generated in the bowels of the Earth. Owing to this heating, the initial heat content of Earth increased from 7.12×10^{37} erg to 9.2×10^{37} erg at the beginning of Archaean eon. As a result, about 4 billion years ago, at the equatorial belt of Earth (where the tidal deformations were maximal) the upper mantle matter began melting at a depth of 200–400 km. After that, the strongest source of endogenic energy (process of chemical–gravitational matter differentiation) became operative. This process occurred as a result of separation of smelts of iron and its oxides from the mantle silicates. During the Archaean Eon this process proceeded according to the mechanism of zonal matter differentiation. By the end of Archaean time, this mechanism had led to the "catastrophic" event of Earth's core separation. After occurrence of this event, the core growth continued due to the more moderate process of gravitational mantle matter differentiation. This process played (and continues playing) the main role in originating a large-scale mantle matter convection, which is the major source of tectonic activity.

During the Archaean time, together with iron and its oxides, most of the siderophilic and chalcophilic elements were transferred into the ring layer of smelts. Therefore, in the Archaean time the convective mantle above the sinking ring-shaped layer of zonal differentiation of Earth's matter was impoverished in iron and other siderophilic elements. Probably, this explains why the Archaean continental shields and their greenstone belts do not have a high metallogenic potential.

The transfer of dense iron oxide smelts to the center of Earth and supplanting of primordeal core matter had to be accompanied by generation of huge amount of energy (about 5×10^{37} erg), which had led to the overheating of planet by the end of Archaean time. The intense convective flows in the mantle formed as a result of this process completely changed the existing earlier regime of tectonic development and had led to the formation of the first in geologic history supercontinent Monogaea, which probably occurred at the end of Archaean time. In the same way, one can also explain the separation of Earth's core, which occurred about 2.6×10^9 years ago.

As a supporting evidence, one can consider the paleomagnetic data, which indicate that the dipole magnetic field of the present-day type emerged only at the boundary of Archaean and Proterozoic times about 2.6×10^9 years ago. The analysis of lead isotope ratios shows that the Earth's core was separated without the melting of Earth's silicate matter. Moreover, the process of Earth's core formation could be described by a two-stage model of changes in the lead isotope ratio: (1) prior to the commencement of core separation, the lead isotope ratio evolution occurred in a closed reservoir and (2) after the beginning of this process, the evolution occurred with partial transfer of lead into the Earth's core. By the present time, about 30% of the entire mass of Earth's lead was buried in the Earth's core (Sorokhtin, 1999).

The Earth's core separation process had changed the Earth's tectonic activity. As mentioned before, the juvenile Earth during the almost entire Pre-Archaean time (from 4.6 to 4.0 billion years ago) was a tectonically passive planet. After the initiation of zonal Earth's matter differentiation, which was accompanied by the separation of iron oxides from silicates (Chapter 1), for the first time the convective mantle emerged at the equatorial belt of Earth. Its temperature rapidly exceeded the melting temperature of iron. The first smelts of mantle matter appeared in the upper layers of mantle and the nuclei of oldest continental shields began forming.

The maximal tectonomagmatic activity of Earth was observed in the Late Archaean time when the total heat flux through the Earth's surface exceeded more than ten times its contemporary value (Fig. 1.20, curve 1). If, in addition, one takes into consideration that during the Archaean time the tectonomagmatic activity was concentrated only at the narrow low-latitude belt, then its specific activity would have been even higher (Fig. 1.20, curve 2). It is necessary to emphasize that in the Early Archaean time less than 15% of the mass of continental crust was formed (Taylor and McLennon, 1988), whereas in the Late Archaean time (for the same time interval) about 55% of its mass (or 3.7 times more) was formed. This provides an additional evidence that in the Early Archaean time the tectonomagmatic activity was localized in the narrow equatorial belt, whereas the greater portion of Earth remained cold and tectonically passive.

The evolution of Earth's tectonic activity is shown in Fig. 1.20. The deviations associated with tectonic cycles on this evolution curve are smoothened. The actual tectonic activity curve can be obtained by superposition of quasi-periodic oscillations (with moderate amplitude), characterizing the tectonic cycles, on the evolution curve. As an example, Fig. 5.3 presents the curve of changes in the Earth's tectonic activity during the Phanerozoic time (recalculated to the average rate of movement of lithospheric plates), which was constructed based on the data on sea transgressions and regressions on continents.

Because of the high tectonomagmatic activity during the Archaean time, the overheating of upper mantle occurred over zones of molten iron. The rates of moving apart of oceanic crust in the rift zones were rather high at that time (up to 400–500 cm/year), whereas the life span of the oceanic crust formed at these zones was very short (less than 10–15 million years). Consequently, thick and dense lithospheric plates did not exist in the Archaean time because their formation required a long time (about 150 million years). Instead, only thin basaltic plates (less than 15–30 km in thickness) had been forming.

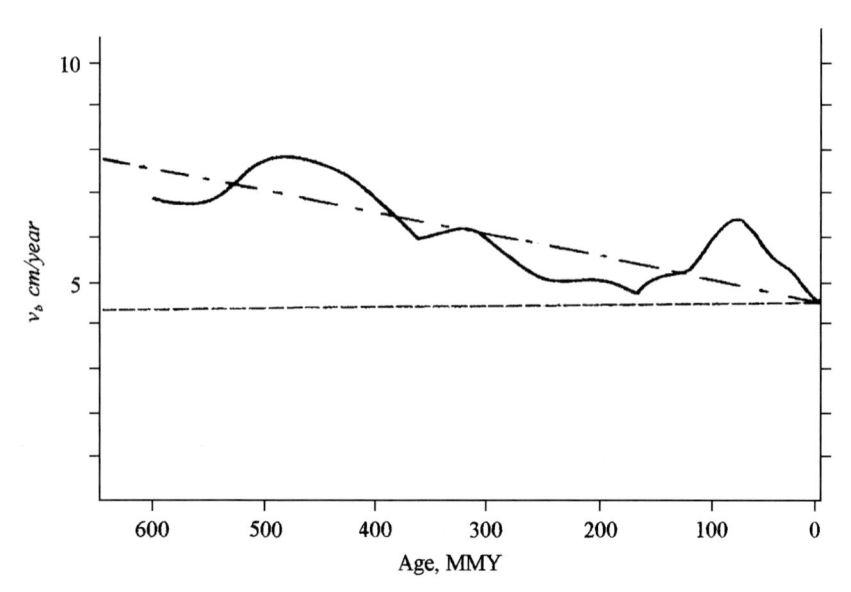

Figure 5.3: The Earth's Tectonic Activity during the Phanerozoic Time Recalculated to the Average Rate of Movement of Lithospheric Plates (Modified after Garkalenko and Ushakov, 1978). Dashed-Dotted Line Represents the Change in Time of Average Tectonic Activity during the Phanerozoic Time; Solid Line Represents the Tectonic Activity during the Phanerozoic Time, Constructed based on the Data on Sea Transgressions and Regressions on the Continents; and Horizontal Axis Represents the Time before Present in Millions of Years.

During the Archaean time, the plate subduction zones did not exist and the compensation of spreading of oceanic floor occurred at the zones of piling-up of thin basaltic plates, which were pulled one over the other above the places of downward flows of matter in the convective mantle. Repeated smelting of these water-saturated basaltic plates (which occurred at the foot of piled-up zones) had eventually led to melting out of the lighter continental magmatic rocks, such as trondemites, tonalites, and plagio-granites, which rose to the upper layers of growing continental crust as diapirs and domes. Judging from the radiolocation images of the surface of Venus, an analogous situation is observed at the present time on that planet: rift zones are present, but there are no subduction zones with their characteristic narrow ridges and deep trenches.

After formation of Earth's core about 2.6 billion years ago, its further growth occurred due to more moderate barodiffusion mechanism. As a result, the Earth's tectonic activity had been reduced. This, in turn, had led to

(1) slowing down of movement of lithospheric plates, (2) increase in the "life span" of oceanic plates, and (3) further cooling down and densification of oceanic plates at the expense of growth of the ultrabasic lithosphere under the oceanic crust. As a result, the first subduction zones were developed during the Early Proterozoic time and the Earth's geologic development proceeded along the pass of tectonics of lithospheric plates.

The separation of Earth's core, which contains about the third of Earth's mass, influenced significantly the content of convective mantle, smelts of which can be observed now at the Earth's surface. In the Archaean time, for example, when the metallic iron was separated in the process of zonal melting, the convective mantle was impoverished in iron and siderophilic and chalcophilic elements. This circumstance can explain the scarcity of metalogenic deposits (except iron) in the Early Archaean time (Smirnoff, 1984; Khain, 2000) and modest deposits of iron and other minerals in the Middle Archaean time.

At the boundary between the Archaean and Proterozoic eons, during the time of Earth's core separation, the content of mantle changed considerably. At that time, the primordeal matter of juvenile Earth (iron concentration of about 13–14% and iron oxides concentration of about 23–24%) was added to the convective mantle. This primordeal matter also contained siderophilic, sulfides, chalcophilic metals, and other ore minerals including platinoids. As a result, at the end of Archaean and during the Early Proterozoic times, favorable conditions were created for the formation of unique deposits of endogenous ores, and the Early Proterozoic time became the most remarkable period of endogenous ore formation.

Unique differentiated intrusions of alkaline and ultrabasic rocks in ancient shields (which occurred in the middle of Early Proterozoic time about 2.3 billion years ago upon the first impulses of stretching and splitting of Archaean supercontinent Monogaea) may be considered as a direct evidence of these events. The Great Dyke of Zimbabwe intrusion, which represents a layered intrusion of a mantle matter of Early Proterozoic time into the Earth's crust, can be considered as the most typical, classical example of this kind. The deposits of chromites in Great Dyke are widespread in its lower layers and are accompanied by dunites and harzburgites, whereas platinum (as sperrylite, $PtAs_2$) and platinoids are encountered in the sulfide layers in the ultrabasic rocks and gabbro–norites. The Early Proterozoic ultrabasic and gabbro–norite intrusions of Bushweld layered magmatic massif (South Africa) have high concentrations of iron, titanium, chromium, and vanadium, and the platinoids reach an industrial concentration in the stratified deposits of copper–nickel sulfides. The Bushweld Pluton intruded into a thick layer of volcanic sediments of Transvaal System of Early Proterozoic age. Thus, the upper (gabbro–norite) part of mantle intrusion came in contact with the crustal granites, which were formed as a result of melting of sedimentary–volcanic rocks enclosing the pluton. Therefore, Bushweld granites are associated only with lithophilic (hydrothermal) mineralization of tin and fluorite.

As additional examples of similar mantle intrusions, one can consider the Sadberry intrusions into Huron (Canada) sedimentary–volcanic deposits of Early Proterozoic time (sulfides of copper, cobalt, nickel, and platinum), the Stillwater deposits in Canada (chromium, titanium–magnetite, platinoids), and Cambalda deposits in Australia (copper, nickel, platinoids). In Russia, examples include (1) the Burakovskiy intrusion at the South-Eastern part of Baltic Shield; (2) gabbro–norite intrusions of Pechenga and Monchegorsk, with high concentration of copper–nickel and cobalt sulfides; (3) Panskaya intrusion at the Kola Peninsula; and (4) magmatic formations of Olong Group in Karelia with platinum mineralization.

It is noteworthy that the intrusions of this kind with high concentrations of ore elements occurred neither before nor after the Early Proterozoic time. This is a strong evidence that at the boundary between the Archaean and Proterozoic eons the convective mantle was enriched in primordeal Earth's matter, which had risen from the central parts of Earth during the formation of Earth's core at the end of Archaean time.

After the formation of dense iron oxide Earth's core, its further growth occurred according to the more moderate barodiffusion mechanism. Consequently, during the Proterozoic time (1) the tectonic activity of Earth was reduced considerably, (2) the mantle chemical–density convection had begun, (3) the rate of spreading of oceanic floor had decreased significantly, (4) the lithospheric plates had become thicker and heavier, (5) the subduction zones emerged, and (6) the tectonics of lithospheric plates had developed.

Chemical–density mantle convection is unstable and its structure changes continuously. This led to periodic formation (with the period of about 800 million years) of mono-cell convective structure with a single upward and single downward mantle flow. In this process, all the existing continents drifted toward the center of downward mantle fluxes, forming gigantic supercontinents such as the Pangaea of Wegener. There were four such supercontinents in the Earth's history: Monogaea (2.6 billion years ago), Megagaea of Shtille (1.84 billion years ago), Mezogaea or Rodinia (1.05 billion years ago, breaking later into two large continents of Laurasia and Gondwana), and Pangaea of Wegener (about 230 million years ago).

Sequential stages of formation and destruction of supercontinents determined the occurrence of different metallogenic scenarios on the Earth. At the time of formation of supercontinents, the ophiolite massifs were formed with the characteristic deposits of chromites, similar to those in the Kempirsay Massif in the Southern Ural Mountains. The continental collisions caused smelting of huge masses of granite, formation of abundant hydrothermal and pegmatite deposits, and polymetallic and pyrite ores.

Usually, the "life span" of supercontinents did not exceed 100–120 million years, which can be explained by specifics of chemical–density matter convection in the mantle. Therefore, the compression of supercontinents was "rapidly" replaced by stretching with formation of (younger than the age of supercontinent) alkali-ultrabasic, syenite, carbonatite, and diamond-bearing kimberlite magmatism. The majority of world kimberlite pipes were formed during the periods of initial stretching of supercontinents. One needs to realize, however, that the open rifts–canals (outlet for deep magma) could sometimes appear at the stage of formation of supercontinent under general compression of continental lithosphere, especially if one of the colliding plates was wedge-like. This, for example, happened during the consolidation of Laurasia and closing of North Paleo-Atlantic Ocean (Yapetus Ocean) during the Devonian time (Fig. 5.4). Apparently, during the Devonian and Carboniferous time, the alkaline-ultrabasic intrusions of Kola Peninsula and diamond-bearing kimberlites of Arkhangel'sk Province (Northern Russia) penetrated into the Earth's crust through rifts. In the process of breaking up, further stretching of supercontinents led to a large-scale intrusion of trap basalts and formation of continental rifts with typical bimodal volcanism. As a result, the supercontinent broke apart into smaller, cenrifugally drifting continents, with formation of young oceans (Atlantic type) between them.

5.4. The Influence of Ocean and Climates of Earth on the Formation of Sedimentary Mineral Deposits

The leading role of atmosphere and hydrosphere in the development of mineral deposits during geologic history (including the Early Proterozoic time) on the Earth's surface was discussed by many geologists (Strakhov, 1963; Voytkevitch and Lebed'ko, 1963; Vinogradov, 1964; Tugarinov and Voitkevitch, 1970). Only during the last years, however, it became possible to evaluate the scope of this phenomenon and clarify and describe the details of such an influence on the endogenic metallogenesis of Earth.

First of all, one needs to take into consideration that the Archaean ocean was hot and, owing to high partial carbon dioxide atmospheric pressure (up to 5 bars), its water was saturated with carbonic acid and was very acidic (pH $\approx$ 3–5). The hot and acid water is an aggressive reagent, which dissolves

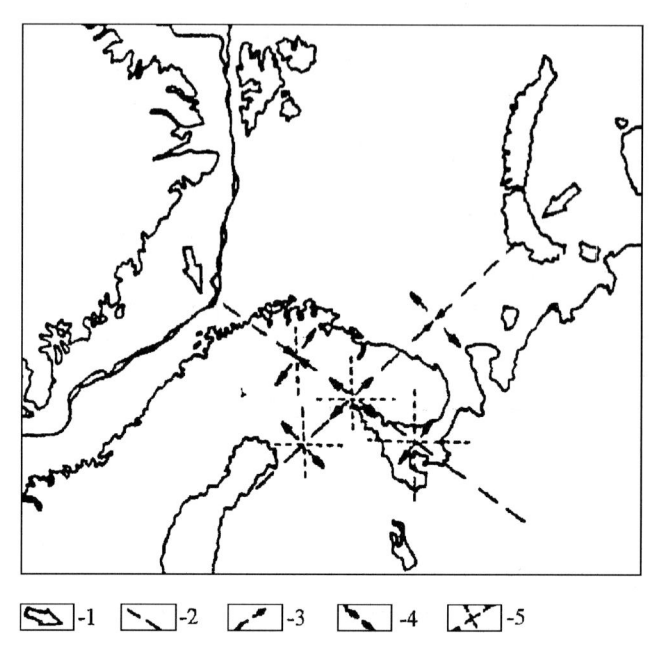

Figure 5.4: Schematic Diagram of Stretching and Compression during the Collision of the American–Greenland and West Siberian Kara Sea Plates with the Baltic–Barents Plate, 400 and 320 Million Years Ago, Respectively (After Sorokhtin et al., 1996.) (1) Direction of Plate Stress; (2) General Extents of Stretching Zones; (3) Directions of Stretching; (4) Directions of Compression; and (5) Extents of Submeridional and Sublatitudinal Rifts, which Determined the Principal Diagonal Stretching Zones.

many ore elements and compounds. Therefore, one should expect that the Archaean water was saturated with many ore elements including gold, uranium, oxides of magnesium and divalent iron, and sulfides of iron, copper, zinc, and lead. Probably, all of these compounds entered the hydrosphere as a result of hydration of basalts of oceanic crust and interaction of hot and acidic rainwater with the rocks of greenstone belts and granitoids of continents.

After pronounced cooling of climate in the Early Proterozoic time and neutralization of oceanic water (pH ≈ 7–8) most of the ore elements precipitated from the former hot acidic water solutions of the Archaean ocean (Fig. 5.5). In the Early Proterozoic time (about 2.5–2.3 billion years ago), the largest stratified deposits of gold, uranium, polymetals, cobalt, sulfides and carbonates of iron, magnesium, etc. were formed in such a way. Examples of such deposits include (1) conglomerates of Witwatersrand, in which ore elements (gold, uranium) began accumulating only 2.5–2.4 billion years ago,

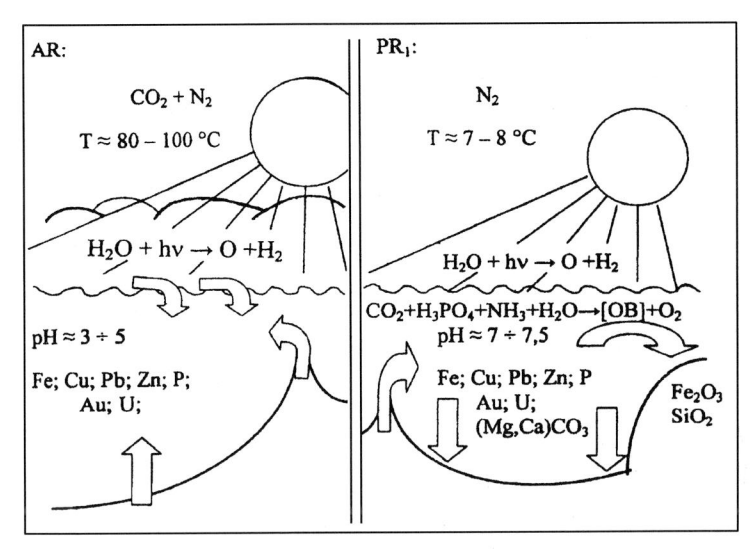

Figure 5.5: The Formation of Sedimentary Ore Deposits during the Early Proterozoic Time due to Cooling and Neutralization of Hot and Acidic Archaean Ocean: (AR) during the Archaean Eon, the Ore Elements, Entering the Ocean from the Rift Zones and Continents, were Dissolved by the Hot and Acidic Ocean Waters; (PR$_1$) during the Early Proterozoic Time, after Oceanic Cooling and Neutralization of its Waters; the Ore Elements, Previously Dissolved in Ocean Water, were Transferred into the Sediments. At the End of Archaean Eon and the Beginning of Early Proterozoic Time, Unique (largest) Iron Deposits were Formed as a Result of Oxidation of Soluble Divalent Iron Hydroxide to the Insoluble Trivalent Iron Oxide.

(2) copper ores of Katanga–Rodesia copper-bearing belts in Africa, (3) gold-bearing conglomerates of Early Proterozoic time on ancient plantforms, and (4) copper-bearing sandstones of Udokan (Siberia).

Climatic changes can also explain the origin of unique iron-ore formations of the Late Archaean and Early Proterozoic time. During most of the Archaean time, the iron content in convective mantle was relatively low, because (at that time) it was concentrated almost entirely in the zones of matter differentiation underlying the convective mantle. By the end of Archaean time, however, the primordeal matter with high concentrations of iron and iron oxides squeezed out of deep Earth's regions began entering the convective mantle. According to our estimates, by the end of Archaean and in the Early Proterozoic time, the average concentration of metallic iron in the mantle could have reached 5.5% and that of divalent iron could have reached 15%. At the oceanic rift zones, the metallic iron rose to the Earth's surface and interacted with oceanic water. Upon contact with water, the hot iron (in the absence of oxygen) was oxidized because of water dissociation and reacted further with carbon dioxide, forming the highly soluble bicarbonate of iron:

$$4Fe + 2H_2O + CO_2 \rightarrow 4FeO + CH_4 + 41.8 \text{ kcal/mole}$$

$$FeO + 2CO_2 + 2H_2O \rightarrow Fe(HCO_3)_2 \tag{5.12}$$

Iron was spread around the entire ocean in the form of $Fe(HCO_3)_2$. In the near-surface layers, due to life activity of cyanobacteria and algae, the divalent iron oxidized to the trivalent state and precipitated:

$$2Fe(HCO_3)_2 + O \rightarrow Fe_2O_3 + 4CO_2 + 2H_2O \tag{5.13}$$

As a result of bacterial metabolism, iron could be transformed to trivalent form, but only to the stoichiometry of magnetite (Slobodkin et al., 1995). Silicon dioxide, which was liberated in the process of hydration of pyroxenes (Eq. (5.3)), was removed from the rift zones jointly with iron. This explains the paragenesis of iron oxides with silicon dioxides in jaspilites of iron ores of Precambrian time (Fig. 5.6).

The massive transfer of iron and other metals from the mantle into hydrosphere could have occurred only if the mantle matter contained significant quantities of these metals and provided that the ocean surface was close to the average level of rift zones at the mid-oceanic ridges or even exceeded this level. Only joint combination of these two conditions could ensure the transfer of iron from the mantle into the hydrosphere and its further accumulation in sediments. In addition, the composition of oceanic crust also played an important role in this process. Thus, the content of iron in the basaltic crust of Precambrian time was considerably lower (about 10 times) than that in serpentines, which formed as a result of hydration of restive areas of mantle matter. Taking into consideration all of the above factors, one could estimate the relative rate of accumulation of iron formations in the Precambrian time, assuming that these formations contained an average of about 50% of iron and that only about 50% of iron was extracted from the oceanic crust rocks.

In addition to the above-described process of formation of iron deposits in the Early Precambrian time as a result of removal of iron from the rift zones, an alternative mechanism could have functioned in the Early Archaean time. During the Early Archaean time, the Earth's core formed only within a narrow equatorial belt of Earth, whereas the rest of its surface was covered with primordeal Earth's matter. The latter contained about 13% of metallic iron and about 23% of its divalent oxide (in silicates). After the beginning of Earth's outgassing and formation of carbon dioxide atmosphere, iron from the

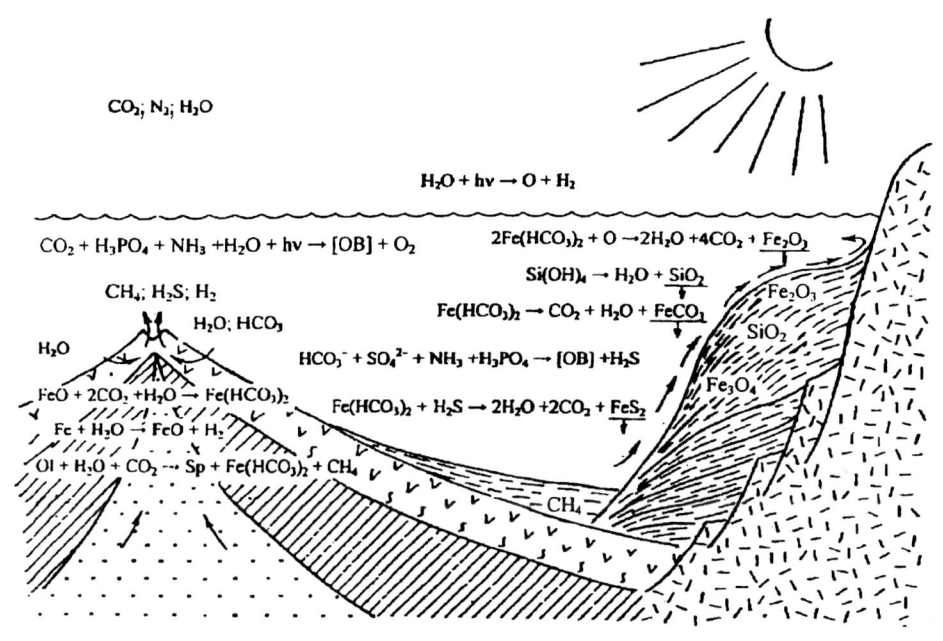

Figure 5.6: Geochemistry of Iron Transfer from the Mantle into Rift Zones and Oceans, and the Formation of Iron Deposits on the Continental Margins in the Early Proterozoic Time.

near-surface layers of these primordeal areas began to be transferred by acidic rainwater (as bicarbonate) into the juvenile sea basins. Upon precipitation, the deposits of Early Archaean time were formed. These ores contain a great deal of siderite (Starostin et al., 2000), forming only after the saturation of seawater with iron bicarbonate: $Fe(HCO_3)_2 \rightarrow FeCO_3 + CO_2 + H_2O$.

The rate of accumulation of iron ores during the Precambrian time is shown in Fig. 5.7. This graph indicates that there were four periods of massive accumulation of iron deposits during the Precambrian time. The earliest deposition of iron ores occurred about 3.8–3.5 billion years ago (Isua Formation in Western Greenland). The second period of intense iron accumulation occurred during the Late Archaean time (3.0–2.6 billion years ago), when the sedimentary–volcanic iron deposits of Kivatin type were formed around the Earth. In Russia, these deposits are represented by the ores of Kostamuksha and other regions of Karelia and Kola Peninsula, by the iron ore complexes of Taratash (Ural Mountains), and by the Starooskol'skiy Series of the Voronezh crystalline massif.

The most intensive period of iron ore accumulation occurred, however, at the end of Early Proterozoic time (from 2.2 to 2.0–1.8 billion years ago).

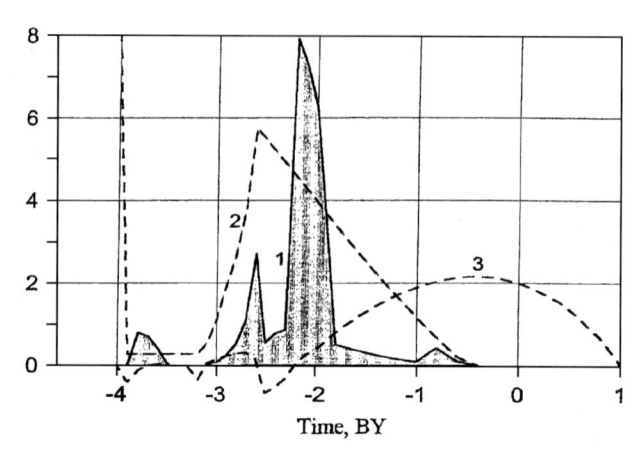

Figure 5.7: The Theoretical Rate of Iron Ore Accumulation during the Precambrian Time. (1) Total Rate of Iron Ore Accumulation, 10^9 ton/year; (2) Concentration of Metallic Iron in the Convective Mantle, %; (3) Position of Oceanic Surface with Respect to Average Level of Crests of Mid-Oceanic Ridges, km.

The iron ore deposits of that time exist on all the continents and many of them accumulated during the same interval of time. The examples of iron formations of that age include (1) deposits of jaspilites of Krivoy Rog (Ukraine), (2) Kursk Magnetic Anomaly (Russia), (3) Karsakpay (Kazakhstan), (4) Hammersley (Western Australia), (5) Lake Superior (Canada and USA), and (6) Guiana (South America). During this period, which constituted only about 5–7% of the total time of geologic development of Earth, at least 70% of the world total resources of iron were formed. According to our estimates, at the time of formation of Early Proterozoic iron ore deposits the rate of iron precipitation reached 3.3 billion tons per year, which is close to the earlier obtained estimates of $(1-3) \times 10^9$ tons per year (Holland, 1989). The total amount of iron precipitated during the Precambrian time, should reach about 3.3×10^{18} tons, which is many orders of magnitude higher than the amount of discovered resources of iron ores (about 3.0×10^{12} tons according to Bykhover, 1994). It is more than 30 times higher than the amount of iron oxides in the sediments of continents (about 0.1×10^{18} tons according to Ronov and Yaroshevskiy, 1978). Possibly, the metasedimentary rocks and granite layer of continental crust contain somewhat greater amount of iron. All this provides indirect evidence that the greater part of precipitated iron sank into the mantle during the Precambrian time through the ancient subduction zones.

A specific feature of this rather unique period of iron ore accumulation is the fact that the process of formation started practically simultaneously (about

2.2 billion years ago) on all continents. In the presented model, the entire chain of events occurred in the following sequence: (1) the oceanic crust was completely "saturated" with water; (2) then, the mid-oceanic ridges were flooded by ocean water (Fig. 5.7); and (3) afterwards, the dissolved hydroxides of iron began entering the ocean water along the rift zones (Fig. 5.6).

By the end of Early Proterozoic time (about 1.8 billion years ago), the massive accumulation of sedimentary iron deposits had stopped almost as abruptly as it started. This was probably caused by the rising level of oceanic surface: the surface level had risen by about 400 m above the peaks of mid-oceanic ridges, i.e. exceeded the thickness of active oceanic layer. In the Mid-Proterozoic and Riphaean times the ocean had a stable stratification with stagnant deep waters (which is corroborated by the wide occurrence of black oil shales). As a result, beginning with that age, the hydroxides of divalent iron (moving along the rift zones) entered the stagnant water and could not be oxidized to the insoluble state. Most likely, the stagnant stratification of water in the World Ocean lasted until the development of new ice age, which spread over the continents of Laurasia and Gondwana at the end of Riphaean time. During the ice ages, the oceanic waters were mixing and, consequently, at the end of Riphaean time the iron oxides arriving from the oceanic rift zones could reach the active layer of ocean. By that time, however, the content of free iron in the mantle decreased considerably (less than 1%) because its larger part was already transferred into the growing Earth's core. As a result, the last Precambrian impulse of iron ore accumulation was the weakest one.

The process of formation of iron ore deposits is based on oxidation of iron as a result of thermal dissociation of oceanic water saturated with carbon dioxide and the hydration of iron-bearing rocks of oceanic crust by this water. In this process, the abiogenic methane was generated according to reaction (5.12), for example. Apparently, the periods of maximal rates of iron transfer into oceanic water can be correlated with the periods of maximal rates of methane generation. This, in turn, should lead to the growth of population of methane-consuming bacteria. Fractionation of carbon isotopes lightens the isotope content of carbon and, consequently, lightens the isotope content of methane and the organic matter of methane-consuming bacteria. This may explain why the organic matter of methane-consuming bacteria is characterized by the extremely low values of $^{13}\delta C_{org}$ (down to -50‰) and the occurrence of local minima in the distribution of $^{13}\delta C_{org}$ at the times of maximal rates of iron ore formation at the end of Archaean and during the Early Proterozoic time. However, in spite of a lower intensity of jaspilite formation in the Late Archaean time, the amplitude of isotope $^{13}\delta C_{org}$ minimum was highest at that time. This can be explained by the existence of dense carbon dioxide atmosphere in the Archaean, whereas in

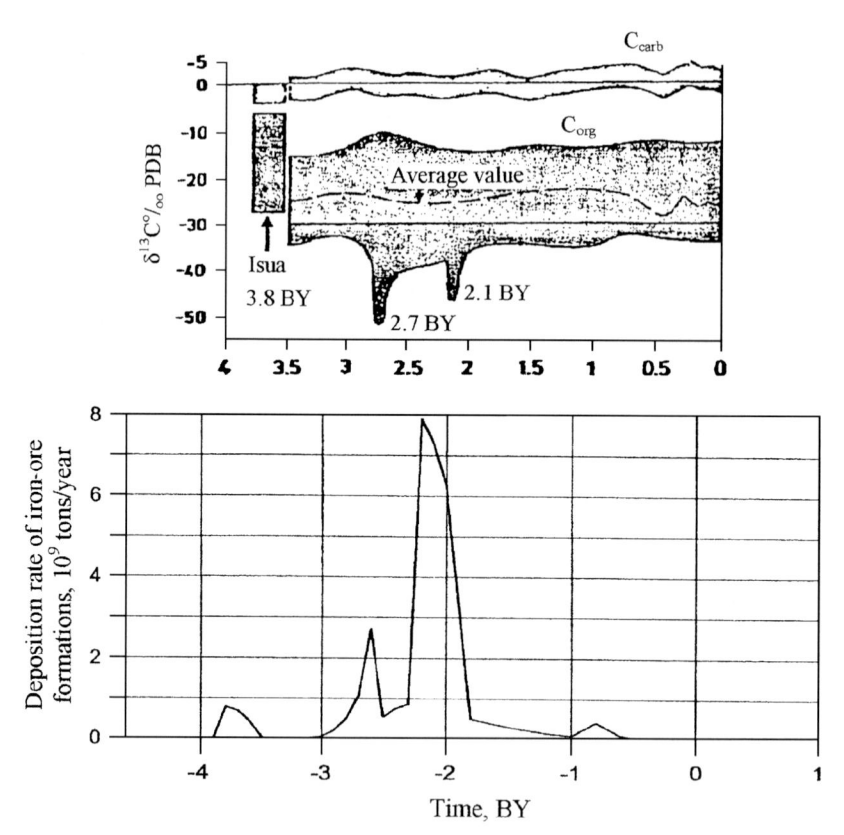

218　*Global Warming and Global Cooling: Evolution of Climate on Earth*

Figure 5.8: Correlation of Carbon Isotopic Shifts in the Organic Matter with the Periods of Iron Ore Accumulation during the Precambrian Time. The Upper Graph Represents the Distributions of $\delta^{13}C_{org}$ and $\delta^{13}C_{carb}$ throughout the Earth's Geologic History (Modified after Shidlowski, 1987); the Lower Graph Represents the Rate of Formation of Iron Deposits during Precambrian Time. Two Local Minima on the Curve of Minimal Values of $\delta^{13}C_{org}$ Clearly Correspond to the Two Periods of Most Rapid Accumulation of Iron Deposits during Precambrian Time.

the Early Proterozoic time, the partial carbon dioxide pressure decreased and the rate of methane generation was reduced correspondingly (Fig. 5.8).

The formation of iron ore deposits during the Early Archaean time should have been accompanied by the generation of methane. At the places of its generation (beyond the layers of sediments), however, methane could not accumulate and entered the atmosphere. In the wet and warm atmosphere of Early Archaean time, under the influence of ultraviolet radiation methane was oxidized according to the following reaction: $CH_4 + H_2O + h\nu \rightarrow CO + 3H_2$, and hydrogen rose into atmosphere. As a result, in the early Archaean time the primitive bacterial forms could not consume the abiogenic methane and, consequently, the isotopic shifts of organic matter were determined by metabolism of bacteria only, without addition of isotopic shifts of methane (Fig. 5.8).

In addition to the above-described unique endogenic (mantle–magmatic) and exogenic (primary–sedimentary) mineral deposits of Early Proterozoic time, one can also identify the metallogenic formations at the subduction zones, which were linked to the calcareous–alkaline and granitoid magmatism. At that time, the first pair-belt metamorphism appeared; pegmatite formations (with muscovite–rare metal, lithium–beryllium, and phlogopite–apatite mineralization) became widespread; and the crystal-bearing, gold–uranium, polymetallic, and pyrite deposits were formed (Sokolov and Krats, 1984).

One can explain the origin of this pronounced impulse of "geosynclinal" metallogenics by using the theory of lithospheric plates. During the Early Proterozoic time, the plate tectonics began "functioning" with formation of the first subduction zones. At the same time, the hydration of oceanic crust was intensified. Therefore, the smelting of continental crust over the subduction zones occurred in the presence of excess water, liberated at these zones as a result of dehydration of oceanic crust. In the Early Proterozoic time, the mantle and, consequently, the oceanic crust were enriched in the primordeal Earth's matter, ascending from the central region of Earth in the process of Earth's core formation. In the subduction zones, most of the siderophilic elements returned to the mantle, whereas lithophilic and, in part, chalcophilic elements and compounds (together with liberated overheated waters) moved upward. Upon combining with continental crust, unique pegmatite and polymetallic deposits were formed.

5.5. Origin of Diamond-bearing Kimberlitic and Similar Rocks

The isotopic composition of carbon in diamonds could not be explained without considering the properties of crust matter (Galimov, 1978). An analogous situation is observed in the high-temperature deep rocks association of carbonatites and kimberlites: the isotope composition of carbon and oxygen indicate that the formation of these rocks is impossible without participation of the crustal carbonic acid of primary-sedimentary origin (Kuleshov, 1986). The origin of these exotic rocks was discussed by Sorokhtin (1981, 1985a, b, c) and Sorokhtin et al. (1996).

According to Sorokhtin et al. (1996), the diamond-bearing kimberlites and similar rocks owe their origin to dragging of heavy (iron-containing) oceanic

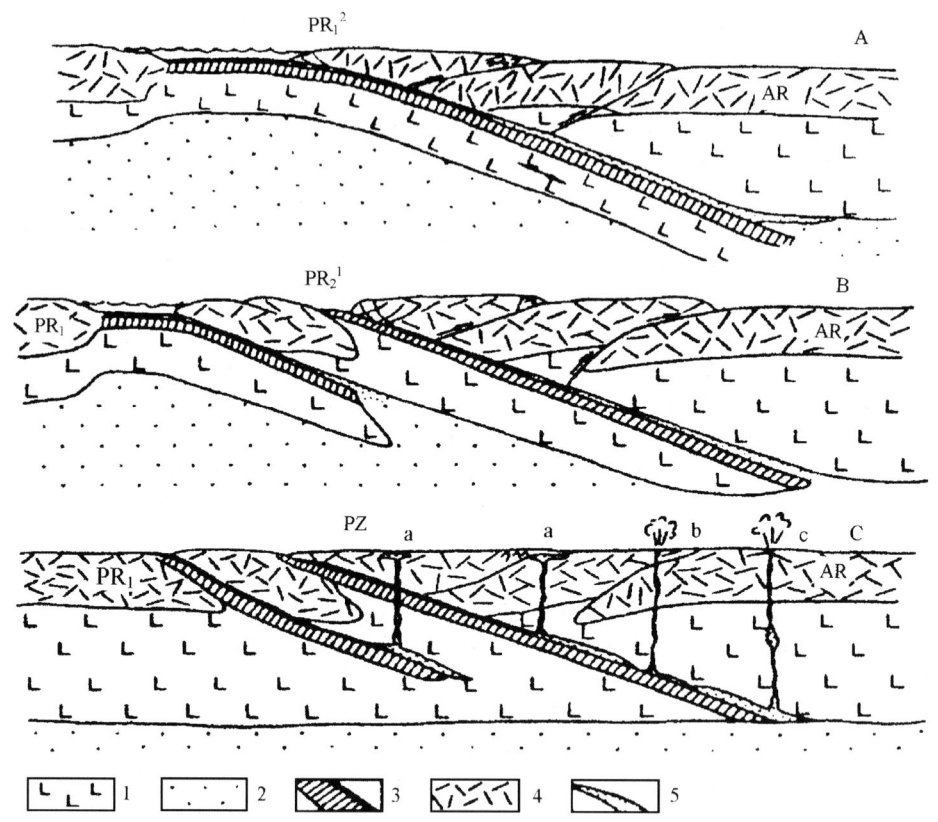

Figure 5.9: The Formation of Deep-Seated Smelts of Alkaline-Ultrabasic, Lamproite, and Kimberlite Compositions (Sorokhtin et al., 1996). (A) At the End of Early Proterozoic Time; (B) at the Boundary between Early and Mid-Proterozoic Times; (C) during the Riphaean and Phanerozoic Times (the Moments of Deep Magma Eruption to the Surface and Formation of: (a) Alkaline-Ultrabasic Intrusions; (b) Mellilite, and (c) Diamond-Bearing Lamproite or Kimberlite Sub-Volcanic Complexes are shown). (1) Lithosphere; (2) Asthenosphere; (3) Early Proterozoic Oceanic Crust with Overlying Heavy Iron-Bearing Sediments; (4) Continental Crust (AR – Archaean Age, PR_1 – Early Proterozoic Age); (5) Deep-Seated Smelts.

sediments of Early Proterozoic time through the ancient subduction zones into deep zones (up to 200–250 km) under the Archaean shields (Fig. 5.9). Due to high density, these heavy sediments fell down into the subduction zones and served there as a lubricant. This probably explains the fact that subduction zones at the end of Early Proterozoic (during Svecofennian Orogeny) and Middle Proterozoic time were deprived of magmatism and calciferous–alkaline volcanism, which is characteristic for island arcs and active margins of continents.

In this model, the formation of smelts at great depths occurred only during the second half of Early Proterozoic time. During the Archaean time, the conditions for generation of magma of considered type did not exist, because of an extremely high tectonic activity and huge heat fluxes, which did not allow the thickness of continental lithospheric plates (together with continental crust) to increase over 60–80 km. The subduction zones did not exist at that time; instead, there were zones of piling up and crushing of relatively thin basaltic oceanic plates. The subduction zones appeared only after the separation of Earth's core at the end of Archaean time, when the thickness of continental lithospheric plates began to increase rapidly. By the end of Early Proterozoic time their thickness reached ≈ 250 km, which created conditions for the formation of deep (diamond-bearing) smelts. This possibility was realized only at the time (2.2 billion years ago) when the heavy iron-ore sediments (of jaspilite type) began accumulating on the ocean floor.

Important role of iron content in the original sediments dragged under the Archaean crust during the Early Proterozoic time is evidenced by carbonatite–magnetite and apatite–magnetite deposits in the intrusions of central type, located in the provinces of proliferation of alkaline-ultrabasic complexes. The magnetite deposits in the Kovdor and Africanda massifs of Kola Peninsula are examples of such iron-bearing intrusions. The iron content in these intrusions reaches 27%, although the overall composition of these rocks (excluding iron) closely resembles that of carbonate–clay and phosphorus-bearing sediments of the oceanic upwelling zones and is completely different from the composition of mantle rocks.

The above-described model allows one to explain most of the specific characteristics of the composition of diamond-bearing and similar rocks, including diamonds themselves and their mineral inclusions. Thus, although kimberlites and lamproites are of deep-seated origin, they were formed from the pelagic sediments. This leads one to conclude that carbon, phosphorus, nitrogen, most of the lithophilic elements (Li, B, F, Cl, K, Ti, Rb, Sr, Y, Zr, Nb, Cs, Ba, Ta, Pb, Th, U), water, and other fluids in the diamond-bearing rocks did not originate in the mantle, but are of primary-sedimentary origin, i.e. they are exogenic. This is evidenced by (1) the high concentration and spectra of rare elements, (2) potassium/sodium and thorium/uranium ratios, (3) isotopic composition of hydrogen, oxygen, sulfur, and strontium in kimberlites, and (4) gas–liquid inclusions in diamonds (H_2O, H_2, CH_4, CO_2, CO, N_2, Ar, C_2H_4, and even ethanol C_2H_5OH, Melton and Giardini, 1974, 1975). Additional evidence is found in the carbon isotope ratios of the diamond crystals, which bear distinct biogenic markers.

The age of kimberlites, based on the strontium and lead isotope ratios in omphacites and inclusions in diamonds, is close to about 2–2.5 billion years (Early Precambrian) (Dowson, 1983), which is in close agreement with our estimates. Other researchers (see Sorokhtin et al., 1996), however, dated the diamond formations as older (about 3–3.4 billion years) using samarium–neodymium and rhenium–osmium ratios in the diamond inclusions.

5.6. Origin of Exogenic Deposits of Minerals

New theories provide new opportunities for exploration of exogenic mineral deposits, for example, bauxites, phosphorites, coals, and gypsum. Although these deposits are found in various climatic zones of Earth, they could only form under certain, often very narrow, intervals of climatic and tectonic conditions. The plate tectonic theory allows one to develop detailed paleogeologic maps of the past geologic periods in order to find the regions with the most favorable climatic conditions and tectonic regimes for generation and accumulation of certain kinds of exogenic mineral deposits. The exploration of evaporites (salt deposits) should be conducted in a region, which experienced arid climate at the time when the inland sea basins could form. For example, the young trenches of the Red Sea type at the early stages of Atlantic-type oceans are most promising in this case. Such conditions developed during the Jurassic Period along the shores of Atlantic Ocean forming at that time (Fig. 5.10) and during the Miocene Epoch in the Red Sea Basin.

Under arid conditions, salt deposits can accumulate also in the residual basins, which formed at the borders of continents in the process of dragging of island arcs over continents. This occurred, for example, at the eastern part of Russian platform upon dragging of the Ural island arc over it during the Permian time. Based on paleogeologic reconstructions, during the Permian time the central and northern areas of Near-Ural Depression were located at about 25–30° northern latitude, i.e. in the arid belt. These conditions determined the formation of salt deposits (including salts of potassium) in the narrow, inland sea basin in front of the Ural Mountains.

On the other hand, coals formed only under conditions of wet tropical and moderate climates during continuous submergence of continental margins under the sea level. As a rule, such conditions arise (1) at the gradually subsiding young continental rift zones, which transform them later into aulacogens or (2) at the passive continental margins at the early stages of development of Atlantic-type oceans. In addition, rapid subsidence of continental margins can occur under the load of island arcs dragging over them.

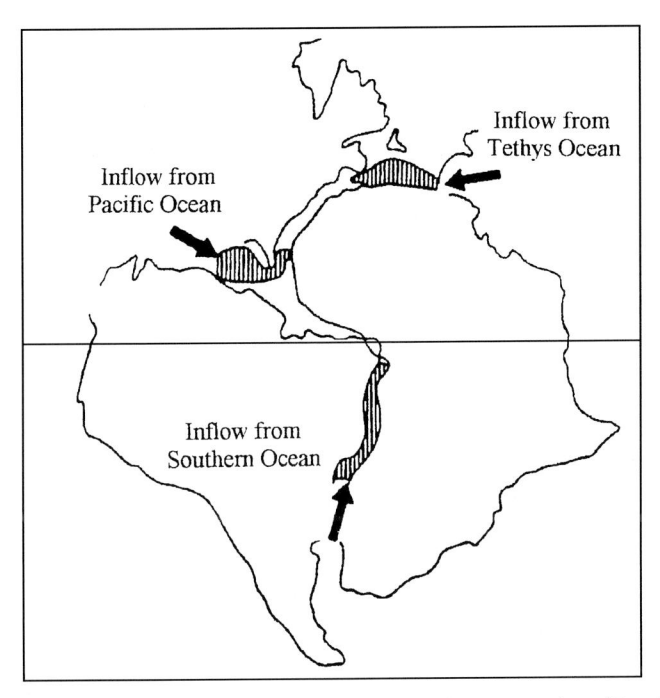

Figure 5.10: The Circum-Atlantic Belt of Salt Deposits (Dashed Areas) (Modified after Berk, 1975).

Such conditions of coal accumulation existed, for example, during the Carboniferous Period in a vast region, stretching over almost the entire Eurasia from England, France, Spain, and through Central Europe and the southern part of Russian Platform, to Kazakhstan and Southwestern Siberia. At that time, this entire region was located in the wet tropical belt and, in addition, experienced a rapid subsidence into the mantle due to closing down of ancient Paleo-Tethys Ocean located to the south.

In exploring for phosphorites, one should focus on the regions of Earth's surface that during the world transgressions (e.g. Vendian, Ordovician–Devonian, and Late Cretaceous transgressions) were located at the nearshore and flooded regions of continents, which were located at the tropical belt, in the zones of uplifting of deep waters or upwelling (Figs. 5.11 and 5.12). These were usually situated at the eastern shores of ancient oceans. The geologists identified four main periods of accumulation of phosphorus: (1) Early- and Mid-Cambrian (Karatau); (2) Early- and Mid-Ordovician (Baltic Basin and Tennessee); (3) Early-Permian (Rocky Mountains); and (4) Late-Jurassic – Cenozoic (Volga and Moroccan basins).

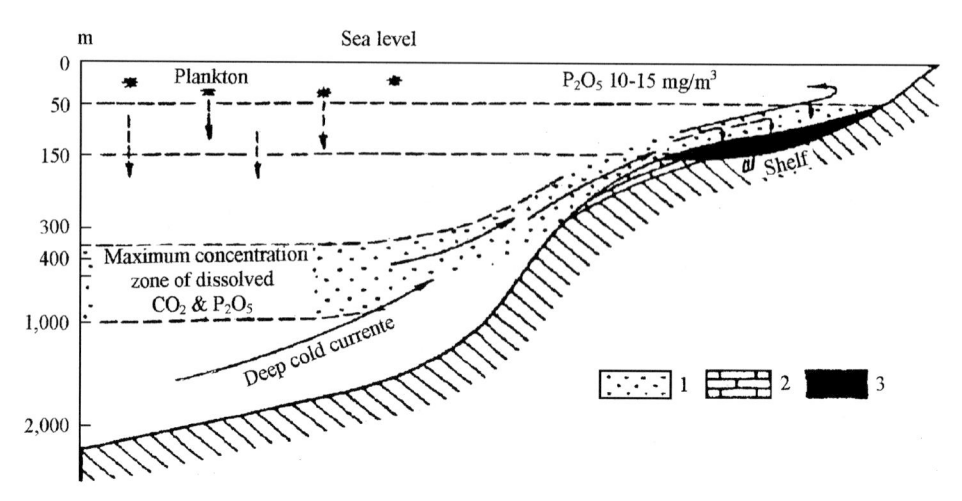

Figure 5.11: The Formation of Phosphorites in the Upwelling Zones on the Eastern Shores of Oceans in the Tropical Regions (Modified after Kazakov; in Sorokhtin and Ushakov, 2002). 1 – Ocean Waters Enriched in Phosphorus Compounds; 2 – Carbonate Deposits on Continental Shelf; and 3 – Phosphorite Deposits.

An analogous approach (using the plate tectonics) can be applied to exploration of bauxites (aluminum ores). The latter formed only in the hot and wet climate of equatorial belt as a result of weathering of basic basalts and argillaceous rocks. Based on the paleoreconstructions, the zones conducive to bauxite formation are located close to the paleo-equator of a corresponding period. The deposits of bauxites of Tikhvin Formation, Urals, Kazakhstan, and South China, for example, many of which are presently located at the relatively high latitudes (about 50–60° of northern latitude), formed at the near-equatorial region during the Early Carboniferous time. Later, due to continental drift, they were dislodged to the regions of higher latitudes.

5.7. Plate Tectonics and Formation of Hydrocarbon Deposits

The plate tectonics theory was successfully applied to the oil and gas exploration. At the end of 1970s (20th century), hypothesis of the existence of powerful mechanism of hydrocarbon generation from organic matter, sucked-in together with the oceanic sediments into the subduction zones, was presented. Figure 5.13 shows the hypothetical mechanism of hydrocarbon

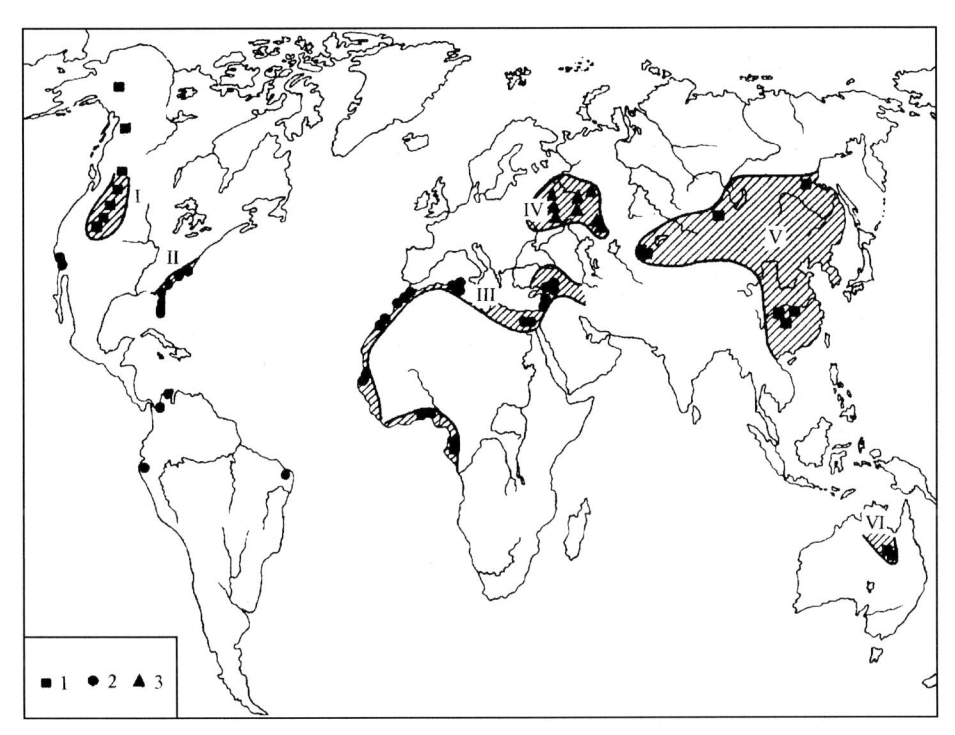

Figure 5.12: The Phosphorite-Bearing World Provinces: (I) Rocky Mountains; (II) Eastern-American Shore Plains; (III) Arabian–African, (IV) Russian Platform, (V) Asian, and (VI) Australian (Modified after Sinyakov, 1987). 1 – Micrograined, 2 – Granular, 3 – Nodular.

accumulation in the island arcs and active margins of continents in the process of subduction of oceanic plates together with pelagic sediments covering them.

To verify this bold hypothesis, one needs to (1) estimate the volume of hydrocarbons generated by this mechanism, (2) prove that pelagic sediments could be sucked-in into the subduction zones, and (3) clarify the mechanisms of formation of marginal or foothill depressions where the main masses of hydrocarbons are concentrated, as a result of migration from the neighboring subduction zones.

The total length of all contemporary subduction zones is about 40,000 km, the average thickness of oceanic sediments is about 500 m, and the average rate of subduction of plates is 7 cm/year. Using this information, one can estimate that about 3 billion tons of sediments are sucked-in annually at the present time. The oceanic sediments contain an average of 0.5% of organic matter, of which only 30% can be transformed into hydrocarbons. This

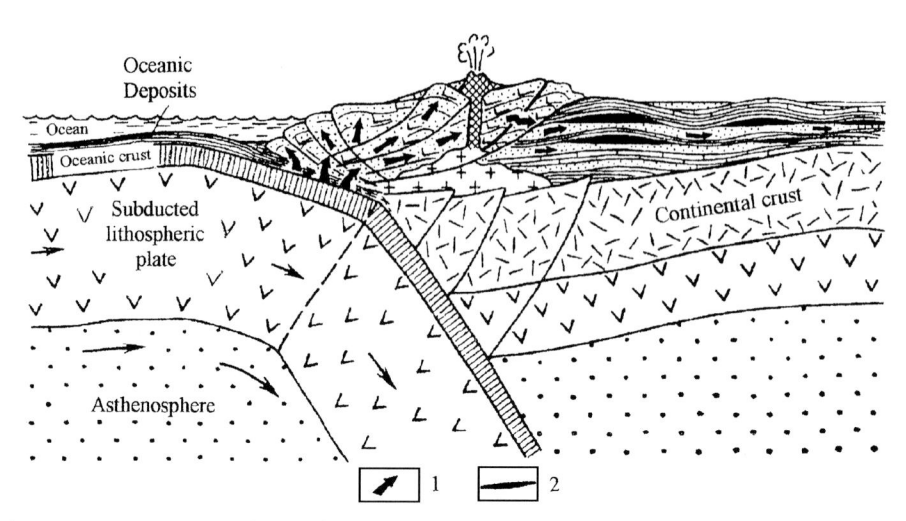

Figure 5.13: Generation of Hydrocarbons in the Zones of Subduction of Oceanic Plates under the Island Arcs and Active Margins of Continents. (1) Paths of Migration of Hydrocarbons from the Underthrust Zones of Plate into the Overthrust Zones of Plate and (2) Hydrocarbon Deposits.

implies that about 5 million tons of hydrocarbons could be generated annually at the subduction zones (Sorokhtin et al., 1974). Although this amount does not seem to be large, during the time of development of highly organized life (i.e. during the last 500–600 million years) about $(2.5–3) \times 10^{15}$ tons of oil and gas could be formed in this manner, i.e. about 1,000 times more than the total mass of these resources identified on the Earth.

The largest accumulations of oil and gas, however, formed in the foothill depressions, in the process of pulling the island arcs and edges of Andean type over the passive margins of continents of Atlantic type with thick layers of sediments, accumulated over these margins during the existence of ocean (Fig. 5.14). Such events occurred frequently in the geologic history of Earth, forming, for example, the Appalachian Mountains, Urals, North American Cordillera, and the larger part of Alpine-Himalayan mobile belt. In the Appalachian Mountains and Ural Mountains the collision process ended 350 and 250 million years ago, respectively; in the Rocky Mountains – about 100 million years ago; and in the Persian Gulf – about 20 million years ago. However, the process of overthrust of the Zagros island arc over the northeastern edge of Arabian platform is going on at the present time, which is evidenced by numerous earthquakes in this region and deformation of the youngest sediments. In the Timor Sea, there is an initial collision stage of the Little Zond Islands island arc with the northern edge of Australian continent.

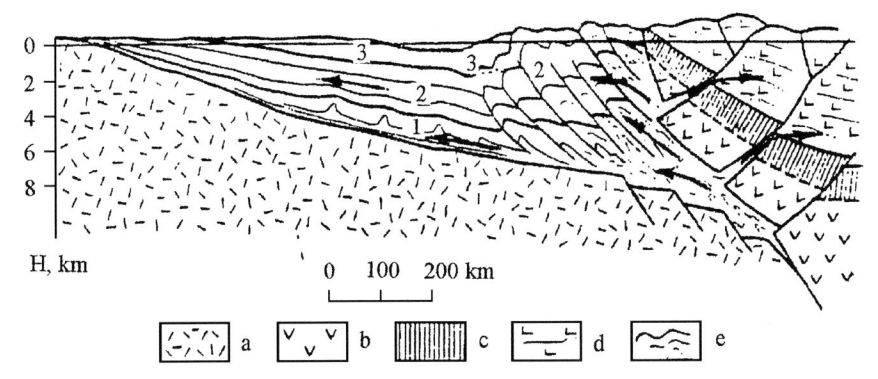

Figure 5.14: Schematic Cross Section of the Overthrust Zone of Island Arc over the Passive Margin of Continental Platform. (a) Precambrian Basement of Continental Platform; (b) Basement of Island Arc; (c) Rocks of Oceanic Crust; (d) Sedimentary-Volcanic Rock Series of Island Arc; (e) Crumpled Rocks of Near-Mountain Depression; (1)–(3) Sedimentary Rocks of Various Ages; Arrows Indicate the Paths of Migration of Hydrocarbons from Subduction Zones.

According to our estimates, the above-described hypothetical mechanism of petroleum formation is extremely powerful. Even if its output is not high, this mechanism can explain the origin of most of the large petroleum provinces of the Earth. For indirect verification of this mechanism, one can compare the distribution of petroleum basins around the World with the location of contemporary and (even more important) ancient subduction zones. After this comparison, one can see that at least 80% of all the World petroleum resources are indeed associated with the subduction zones, which existed during the past geologic periods (Fig. 5.15). Among the numerous examples of this correlation, one can include the unique basins of Persian Gulf, Venezuela, Mid-West of USA, Canada, Alaska, and Indonesia, the classic formations of Appalachian Mountains, Near-Urals Depression, Caucasus Mountains, and Carpathian Mountains, the perspective provinces of Eastern-Siberian platform, and underthrust zones of Verkhoyansk-Kolyma folded belt (Russia).

The proximity of petroleum provinces to the recent and ancient subduction zones (Figs. 5.15 and 5.16) is important evidence in favor of the mechanism of oil accumulation described above.

One also needs to prove that pelagic sediments can be sucked into the subduction zones. At first, this hypothesis was proved theoretically and later verified by drilling. Pelagic deposits are moved toward the subduction zones by the oceanic lithospheric plates. In most cases, there is no crushing of these

228 *Global Warming and Global Cooling: Evolution of Climate on Earth*

Figure 5.15: The Schematic Map of Distribution of Main Petroleum Reserves over the World (Modified after Gavrilov, 1986). (1), (2), and (3) Zones of Subduction of the Paleozoic, Mesozoic, and Cenozoic age, Respectively; (4) Some Intercontinental Rifts; (5) Marginal-Continental Rifts; (6) Petroleum Reserves; (7) Large Deposits of Oil and Gas; (8) Deposits of Bitumen and Heavy Oil; (9) Interplatform Petroleum-Bearing Depressions; (10) Potential Interplatform Petroleum-Bearing Depressions.

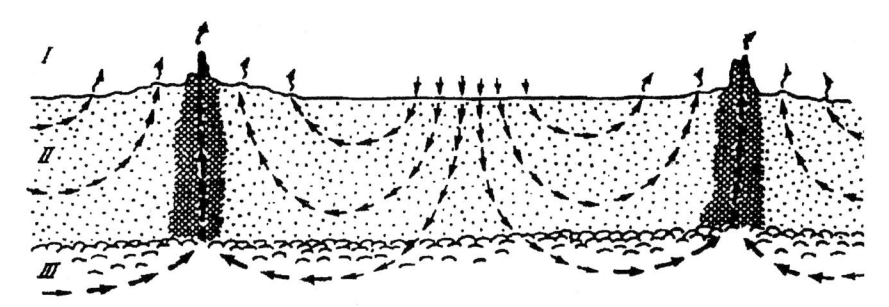

Figure 5.16: Schematic Diagram of Convective Circulation of Sea Water in Porous Deposits of Sedimentary Layer and Basalts of Rift Zone of Gulf of California. (I) Ocean Water; (II) Convective Flows of Ground Water in Sedimentary Series; and (III) Flows of Overheated Fluids in Oceanic Crust. Darkened Areas – Hydrothermal Sulfide Deposits and "Black Smokers".

sediments. There is no excessive accumulation of sediments even within the deep-ocean troughs despite the high depositional rate of about several centimeters per thousand years. At such a rate, most troughs should have been completely filled with sediments within a few dozen million years. Actually, they are devoid of sediments, although some of them existed over hundreds of millions of years as, for example, the Peru–Chile Trough. This indicates the existence of an efficient mechanism of removal of the bottom deposits in the deep-oceanic troughs. This mechanism involves sucking-in sediments into the plate subduction zones. These sediments acted as "lubricating agent".

The amount of sediments getting into the space between the rubbing plates depends on the speed of plate movement and the viscosity of sediments sucked into this space. Sorokhtin and Lobkovskiy (1976) showed that the sediments may be sucked-in under the island arcs without being crushed only if their thickness is below a certain cutoff value, which depends on the plate subduction rate and sediment viscosity. Sorokhtin and Lobkovskiy (1976) calculated that about 500–520 tons of pelagic sediments might be sucked-in without being stripped and crushed under the Kuril, Japan, and Tonga arcs and only about 400–430 tons in the Peru–Chile, Aleutian and Java troughs. The thickness of sediments in deep-water troughs near the Kuril Islands, Japan, and central part of Java does not exceed 300–300 m, whereas in front of the Peru–Chile trough it is less than 100 m.

A completely different situation exists in the Alaska Bay, east of Aleutian arc and northern Java trough, where the subduction rate is low (about 2–3 cm/year) and the thickness of sediments exceeds 500–700 m (sometimes reaching 1,000 m). Similar situation exists near the Atlantic Plate under the Lesser Antilles subduction zone. The cutoff limit value for this region is

250 m, whereas the actual sediment thickness is 500–1,000 m. Consequently, in these regions the plate subduction must have been accompanied by stripping off the deposits and crushing them in front of the island arc of lithospheric margin, forming sedimentary accretion prisms. This can explain the formation of nonvolcanic ridges in the proximity of these subduction zones (Kodiak Island in Aleutian Arc, Lesser Antilles in Atlantic, and Andaman Islands in Indian Ocean).

Dissipation of energy of viscous friction results in heating and, sometimes, partial melting of sediments between the rubbing plates. Thus, the viscosity of sediments within the subduction zone declines drastically (by many orders of magnitude) and the maximal thickness of sediments, which might be preserved, drops significantly. Therefore, the sediments with the density lower than that of the lithosphere cannot be sucked into the subduction zones to a depth greater than 20–30 km. They are squeezed up through the faults and injected (as migmatite granite–gneiss domes or granitoid batholiths) into the island arcs or active continental margins over subduction zones.

The sediments might have been sucked-in under the continental plates only if their density was higher than that of the lithosphere. In such a case, the rate of sucking-in of sediments into subduction zones is higher than the plate subduction rate, because the heavy sediments are moving down into the subduction zones (Monin and Sorokhtin, 1986).

Inasmuch as the present-day rocks are lighter than the lithosphere (average density of 3.3 g/cm^3), they cannot be sucked-in to great depths under the island arcs or active continental margins. During the Precambrian time, however, there were periods when very heavy iron ore formations (e.g. jaspilite with a density of 4–4.5 g/cm^3) had been widely deposited on the oceanic floor and continental shelves. Such sediments could have been easily sucked-in ("dropped down") to deep layers under the continents. They experienced melting and separation of iron compounds from the silicate–carbonate melts and their migration into the mantle.

The sediments accumulating at the continental margins always contain organic matter. In some cases (deltas of large rivers, for example) its concentration reaches several percents, although the concentration of organic matter usually does not exceed 1%. In the process of subsidence of continental margins and their filling up with sediments, the lower layers of sediments are compacted and heated by the upward moving heat flow. As a result, the sediments are lithified (become sedimentary rocks) and their organic matter is gradually transformed into hydrocarbons in the process of thermolysis. This process was thoroughly studied and quantitatively evaluated by I. Gubkin and N. Vassoevich (famous Russian petroleum geologists), who developed the sedimentary–migration theory of petroleum formation.

Using this theory and the principles of plate tectonics, Ushakov (1979) evaluated the conditions of petroleum generation in sedimentary deposits at the passive margins of continents. He showed that "maturation" of oil and gas occurs there 20–30 million years after the formation of continental margin itself. In time, the domain of hydrocarbon generation is extended considerably.

At the passive continental margins, the oil migration occurs only as a result of compaction of underlying sediments and their heating and dehydration. Consequently, the main mass of hydrocarbons occurs in a dispersed state with large oil and gas deposits being rare, whereas giant and unique deposits do not occur at all. To mobilize the oil and gas dispersed in these formations, powerful tectonic forces are needed, which could "squeeze" and "wash out" the hydrocarbons from these formations. This occurs at the second stage of development of marginal depression during the closing down of ancient oceans and thrusting of the island arcs over the former continental margins.

Two important events occur when an island arc closely approaches the continental slope: (1) under the load of island arc, subsidence of continental margin accelerates considerably, which facilitates the subsidence of margin and increases the rate of sediment accumulation in the forming border depression and (2) the pore water together with associated hydrocarbons in the continental shelf belt and now caught under the island arc, start to be squeezed out from the sediments. This process is intensified by the thermal waters (liberated as a result of dehydration of rocks and sediments) rising from the deeper parts of subduction zones. These hot fluids are moving along the sedimentary rocks strata from under the island arc to the regions of least pressure, i.e. toward the continental platform. In this process, the sediments located in front of the moving arc are folded, forming traps, where oil and gas are gradually accumulated.

The emergence of young mountain range signals the end of formation of foothill depressions with associated petroleum basins. The oil and gas accumulating in these depressions have two sources: (1) concentration (mobilization) of dispersed oil (microoil) of the associated source rocks and (2) migration of hydrocarbons from the sediments, which at the time of formation of depressions, were caught under the island arc dragged over the continental margin.

The second source is very important. On assuming that the length of shoreline of the shelf is about 1,000 km with the layer of sediments having the thickness of 15–17 km covering the frontal cornice of island arc (width of 100–120 km) several hundred millions of tons of hydrocarbons could have migrated from the subduction zones toward the depression margin – this occurred, for example, in the Persian Gulf as the Zagross island arc was pulled over the edge of Arabian Platform. However, the real scope of hydrocarbon migration is more modest than the above calculations show. This can be explained by the fact that a considerable part of oil and gas remains in the subduction zones, frequently forming large deposits of petroleum: for example, large oil and gas deposits in the Rocky and Appalachian Mountains and under ophiolite deposits in Cuba, Swiss Alps, and New Zealand.

By the mid-1970s, on drilling through the serpentine ophiolite deposit an oil deposit was discovered in Cretaceous rocks under the overthrust section in Cuba. In 1975, the Pineview oil deposit with reserves of 18.3 million tons of oil was discovered in the Rocky Mountains under the overthrust. According to the estimates of geologists, at the beginning of 1981 the available petroleum reserves in the overthrust belt of the Cordillera Mountain Range were equal to 2.1 billion tons of oil and 2.8 trillion m^3 of gas, which was only twice lower than the total remaining proven reserves in USA. Another example of application of plate tectonics in petroleum geology is the discovery of the large oil deposit "White Tiger", situated in the granites of crystalline basement of Vietnam Shelf. This deposit formed above the underthrust zone of lithospheric plates of Mesozoic age, due to saturation of fractured rocks of crystalline basement by hydrocarbons rising from the subduction zone (Areshev et al., 1996).

According to Sorokhtin and Ushakov (1991) there is an abiogenic mechanism of hydrocarbon generation on the oceanic floor. They believe that the hydration of rocks of oceanic crust by water, containing dissolved carbon dioxide, led to generation of abiogenic methane and hydrogen. Methane formed as a result of oxidation of divalent iron of ultrabasic rocks to the trivalent state and the reduction of carbon dioxide to hydrogen. In this process, hydrogen is liberated due to dissociation of seawater in the presence of divalent iron. Such reactions are exothermal and occur at a temperature of about 400 °C, with generation of considerable amount of energy (Dmitriev et al., 2000; Sorokhtin et al., 2001):

$$4Fe_2SiO_4 + 12Mg_2SiO_4 + 18H_2O + CO_2 \rightarrow 4Mg_6[Si_4O_{10}](OH)_8$$
$$+ 4Fe_2O_3 + CH_4 + H_2 + 144.6 \text{ kcal/mole} \qquad (5.14)$$

$$Fe_2SiO_4 + 3Mg_2SiO_4 + 5H_2O \rightarrow Mg_6[Si_4O_{10}](OH)_8$$
$$+ Fe_2O_3 + H_2 + 21.6 \text{ kcal/mole} \qquad (5.15)$$

$$2FeSiO_3 + 6MgSiO_3 + 5H_2O \rightarrow Mg_6[Si_4O_{10}](OH)_8$$
$$+ Fe_2O_3 + 4SiO_2 + H_2 + \approx 21 \text{ kcal/mole} \qquad (5.16)$$

The rate of methane and hydrogen generation in the oceanic crust reaches 9–10 million tons a year each. The main part of these gases is dissipated in the ocean (by currents, for example) and finally enters the atmosphere. In the presence of bacteria, a noticeable portion of these gases oxidize directly in ocean water according to the following reactions:

$$CH_4 + 2O_2 + (bacteria) \rightarrow CO_2 + 2H_2O \tag{5.17}$$

$$H_2 + O_2 + (bacteria) \rightarrow 2H_2O \tag{5.18}$$

One should emphasize that the oxidation of CH_4 and H_2 (also of H_2S) could occur only in the presence of certain bacteria consuming these gases.

As a rule, methane and hydrogen generated at the hydrothermal rift zones dissipate in the ocean water. In some cases, however, when the slowly separating rift zones are overlain with sediments and in the sedimentary layers at the periphery of oceans (under which the serpentinization of oceanic crust is occurring), the hydrocarbons can accumulate in the sedimentary rocks and form petroleum deposits (Balanyuk et al., 1995). The main factor in transforming CH_4, H_2, and H_2S into complex hydrocarbons is the life activities of bacteria consuming these gases and transforming them into organic matter. In the processes of diagenesis and catagenesis of sediments, the organic matter is transformed into more complex hydrocarbons, although the abiogenic CH_4, H_2, and H_2S can serve as the "feeding base" of such organic hydrocarbons. In addition to the bacterial development of methane into organic matter and more complex hydrocarbons, possibly there is an abiogenic way of synthesis of more complex hydrocarbons due to catalytic reactions in sedimentary rocks (Rudenko and Kulakova, 1986). For example, Al_2O_3 (the major component of shales) can serve as a catalyst. This process can be facilitated by high temperatures in deep rocks and in the subduction zones with a temperature of up to 500 °C.

According to Balanyuk and Dongaryan (1994), a certain part of abiogenic methane at low temperatures of ocean bottom water may form gas-hydrate deposits. The pelagic sedimentary cover is necessary to conserve such deposits. Possibly, the gas-hydrate deposits in thick sedimentary formations of oceanic floor are of abiogenic origin.

Chapter 6

Origin and Development of Life on Earth (Influence of Climate)

6.1. Uniqueness of Earth

Considering the main conditions of Earth's development, one should realize that its evolutionary path was predetermined by the (1) position of Earth in the Solar System, (2) solar luminosity, and (3) mass and chemical content of Earth. If Sun were an unstable star, then the Earth would have been either unbearably hot or cold. If the Sun's mass were considerably higher, then several tens or hundreds of millions of years after its formation it would have exploded and turned into a neutral star or even a black hole. Sun is a quiet stationary star with a medium mass (spectral class G2), which only slightly changes its luminosity over many billions of years. This is especially important because during the last 4 billion years the Earth's life was allowed to develop through a long evolutionary path: from primitive life to its highest forms.

The distance of Earth from the Sun is also optimal. If the distance were smaller, the Earth's surface would have been too hot with an irreversible greenhouse effect (like the one on Venus). If the distance were greater, the Earth would have been frozen and become the "white" planet with a stable ice cover. The formation of massive satellite Moon at the close-to-Earth orbit "pumped" the Earth with tidal energy, accelerated its tectonic development, and span-off the Earth in the direction conducive for a more uniform climate. If the Earth did not have a massive satellite, then the tectonic activity of Earth (like Venus) would have been delayed by 2.5–3 billion years. Under such a scenario, at present time the conditions of Late Archaean time (high surface temperatures and dense carbon dioxide atmosphere) would have been dominant on Earth. Consequently, instead of contemporary highly organized life species, the Earth would have been populated only by primitive bacteria (single-cell prokaryotes).

Considering the Earth's evolution closely linked to the Sun and Moon, one is surprised by the degree of system's balance, which ensured the formation of comfortable conditions for evolution and development of highly organized life on the Earth. Scrutinizing this system, one can observe an optimal mass of the planet, which is able to hold a moderately dense atmosphere on its surface, and also an extremely favorable chemical composition of the Earth. Even insignificant deviations from the initial concentrations of such elements as Fe, FeO, CO_2, H_2O, and N_2, could have led to irreversible and catastrophic consequences for life. If, for example, the original Earth matter contained less water, then the rate of absorption of carbon dioxide would have declined, accumulating in the Earth's atmosphere. As a result, just as during the Archaean time, an irreversible greenhouse effect could have arisen and the Earth would have become a "hot" Venus-like planet. If the content of nitrogen in the Earth were less, then, in the Early Proterozoic time, the Earth would have been a "white", cold planet covered completely with snow. Under the conditions of higher concentration of free (metallic) iron, the free oxygen would not have been able to accumulate in the contemporary (as well as in Proterozoic) atmosphere and, consequently, the kingdom of animals would not have emerged on the Earth. On the other hand, in the case of lower iron concentration, at the present time (or even earlier) an intense emission of endogenic (abiogenic) oxygen should have occurred, with all the life species "burning" in such an atmosphere. In addition, the outgassing of deep-seated oxygen should have led to a strong greenhouse effect, which would have converted the Earth into a "hot" Venus-like planet. This implies that the duration of existence of highly organized and, especially, intelligent life, on any of the planets in the stellar systems of Universe is limited.

It is fortunate that Earth has a comfortable nitrogen–oxygen atmosphere and an abundant land life. Our "lucky" planet of Earth represents an exceptionally rare phenomenon in the Universe. Thus, it is improbable to find another planet populated with intelligent species in our Galaxy.

6.2. The Earth's Climate and Origin of Life

The original Earth, which was formed as a result of accretion of primordial protoplanet matter, was absolutely lifeless. The matter of protoplanet gas-dust cloud formed as a result of explosions of supernovas was completely sterilized by the hard cosmic radiation a long time before the beginning of planetary accretion in the Solar System. In addition, initially the Earth had neither a dense atmosphere nor a hydrosphere, which are favorable for the life development and its preservation. This is explained by the fact that, from the very beginning, the Earth's matter was impoverished in volatile compounds. Even a negligible amount, liberated as a result of impacts and heat explosions of planetesimals, was rapidly absorbed by very porous ground and transferred from the Earth's surface into its deep layers. In addition, during its formation period, the Earth's surface was exposed to an extremely intense irradiation of juvenile Sun, resulting from nuclear reactions. At that time, the Sun, similar to the T-Taurus stars, was at the beginning of its evolution. This intense corpuscular flow (mostly protons and helium nuclei) should have literally blown away the remains of gaseous components from the Earth's surface.

After completion of the first active stage of development of juvenile Sun, its luminosity declined pronouncedly and about 4.6 billion years ago was about 25–30% lower than its contemporary level (Bachall et al., 1982). By that time, however, the initial nitrogen atmosphere formed as a result of degassing of planetesimals fallen on the Earth's surface. Although the equatorial Earth's temperature exceeded the freezing temperature of water, the climatic conditions remained extremely bleak. One side of Earth surface was a cold desert, whereas the other side was continuously and intensively bombarded by the fluxes of severe cosmic radiation.

The unfavorable conditions for life emergence continued until initiation of outgassing of Earth matter. The latter, however, could have occurred only after: (1) the inner temperature of Earth had risen to the level of melting of Earth's matter, (2) formation of the Earth's asthenosphere, and (3) initiation of the mantle convection, i.e. after initiation of most powerful mechanism of gravitational mantle matter differentiation. The formation of asthenosphere and the zonal melting of Earth's matter had led to a pronounced increase in the tidal interaction between Earth and Moon and to a significant overheating of upper layers of mantle at the equatorial belt of Earth. All these important events occurred approximately 4.0–3.9 billion years ago.

At the early stages of Earth's degassing, carbon dioxide started to enter into the atmosphere intensively, and a larger part of water and organic compounds were absorbed by the primordeal regolith of juvenile Earth. Due to high content of iron in the original Earth's matter, the methane began forming in the porous regolith and the Earth atmosphere acquired reducing properties, which were conducive to the origin of life (Galimov, 2001). The high porosity and absorbing capacity of regolith ensured favorable conditions for the (1) formation of complex organic compounds and (2) emergence of life. Probably, life originated in the small pores of primordial regolith after they were filled with degassed and mineralized water (Sorokhtin and Ushakov, 1991). Primary hydrocarbon compounds could have also formed in the process of hydration of iron-containing ultrabasic rocks in the presence of carbon dioxide. The oxides of nitrogen, nitrates, nitrites, ammonia, ammonium sulfates, chlorides, and other numerous compounds of carbon and nitrogen formed due to intense thunderstorm activities in the carbon dioxide–nitrogen atmosphere of Early Archaean time. The compounds of phosphorus probably entered solutions directly from the matter of primordial regolith. At the beginning of Archaean Eon, the conditions necessary for the formation of more complex organic molecules at high atmospheric temperatures were ensured by the pressure of water solutions in the capillaries of porous regolith and the catalytic effect of free and exchangeable cations (Fe, Ni, Cr, Co, etc.) contained in the regolith. The formation of complex proto-organic molecules in the porous regolith was facilitated by the high absorbing capacity and high capillary pressure inside the small regolith pores. The concentration of complex organic molecules there could have reached the level sufficient for the synthesis of more complex organic substances. In the sea basins, these compounds would have been too diluted.

Thus, the life on Earth probably originated in the primordial ground and volcanic ashes at the beginning of Early Archaean time (about 4.0–3.9 billion years ago) at the time of formation of reducing nitrogen–carbon dioxide–methane atmosphere. The emergence of life on Earth had coincided with the first most pronounced tectonic and geochemical milestone in the history of its development, i.e. at the beginning of Earth's core formation (beginning of chemical–gravitational matter differentiation) that had led to the formation of hydrosphere, dense atmosphere, and continental Earth's crust. It is interesting to note (Chapter 1) that the initiation of tectonomagmatic activity on Earth coincided with the beginning of basalt magmatism (formation of basalt "seas") on the Moon.

Experiments showed the possibility of synthesizing complex organic molecules from inorganic compounds upon heating them in the fields of electrical discharges (Oparin, 1957; Miller, 1959; Wilson, 1960; Fox, 1965; Oro, 1965, 1966). Galimov (2001) showed that the origin of life is related to energetic chemical reactions, which reduce the entropy of system. These reactions could occur, for example, with the participation of adenosintriphosphate (ATP), the synthesis of which could have occurred at the early stages of Earth's development. The formation of ATP requires synthesis of adenine (the product of polymerization of cyanic acid – HNC) and of ribose (the product of polymerization of formaldehyde – HCOH). According to Galimov (2001), the synthesis of ATP was necessary for the emergence and development of life on Earth.

In our model, the formation of primary components for the ATP synthesis could have occurred in a natural way. At the beginning of Archaean Eon (see Chapter 1), the Earth was covered mostly by the primordial microporous regolith, which contained up to 13% of free (metallic) iron. After beginning of degassing of Earth (about 4 billion years ago), the first rains saturated the regolith with water with dissolved carbon dioxide. As a result, there was an abundant generation of methane:

$$4Fe + 2H_2O + CO_2 \rightarrow 4FeO + CH_4 + 41.8 \text{ kcal/mole} \qquad (6.1)$$

As the methane entered the atmosphere, the juvenile nitrogen–carbon dioxide–methane atmosphere became reducing in nature.

Formaldehyde formed according to the following reaction:

$$2Fe + H_2O + CO_2 \rightarrow 2FeO + HCOH + 3.05 \text{ kcal/mole} \qquad (6.2)$$

Formaldehyde remained dissolved in water, saturating the regolith, and later was washed out by rainwater into the newly-formed small marine basins, whereas methane was transferred into the atmosphere, resulting in a strongly reducing environment. In the methane-rich reducing atmosphere of the Early Archaean time, the hydrogen cyanide could form due to atmospheric electrical discharges (thunderstorms) according to the following reaction, for example:

$$N_2 + 2CH_4 + Q \rightarrow 2HCN + 3H_2 \qquad (6.3)$$

where Q is the energy of electrical discharges consumed by the reaction.

Thus, at the very beginning of Archaean Eon, conditions conducive to the formation of primary chemical compounds were created. The latter served as the primary components for further synthesis of more complex organic substances and prebiologic compounds. This process was aided by the catalysts (Fe, Cr, Co, Ni, Pt, etc.), which were present in the regolith. The simplest associations of complex organic molecules or primitive (but already containing RNA) compounds could be transferred afterwards into the water of juvenile sea basins of Early Archaean time.

In the process of degassing of Earth and the development of atmosphere, its reducing potential declined gradually due to photodissociation of methane:

$$CH_4 + CO_2 + h\gamma \rightarrow 2HCOH \qquad (6.4)$$

Consequently, with time the atmosphere became almost entirely of nitrogen–carbon dioxide type with small admixture of methane, which was being generated continuously. The methane admixture, however, apparently could have fed the primitive Archaean microorganisms. The further development of life occurred due to high-energetic and low-entropy reactions (Galimov, 2001) and according to the biologic laws of the live matter development, under the influence of directed pressure and environmental "filtering", and afterward in the process of competition of various life forms.

The life emergence on Earth may be explained by the favorable combination of the following factors: (1) composition of primordial Earth's matter;

(2) presence of microporous regolith; (3) atmospheric composition with an appreciable methane concentration; and (4) moderate temperature of juvenile sea basins at the beginning of Early Archaean time. After the first (and probably not too long) period of formation and existence of organic compounds of abiogenic origin, which served as a construction material for the formation of first RNA, the first viruses appeared. Owing to their symbiosis and synthesis of the first amino acids and semi-permeable membranes, the first single-cell life forms (bacteria) were probably formed. Later, because of intense degassing from the bowels of Earth, the carbon dioxide partial pressure rapidly increased. Possibly, this explains the retardation in life evolution during Archaean time, when only the most primitive thermophilic and prokaryotic bacterial forms (archaeobacteria of chalcophilic and siderophilic specialization) dominated among the "life" species. Chemogenic reactions (similar to those used at the present time by the thermophilic bacteria at the hydrothermal springs of mid-oceanic ridges) were probably used as the sources of energy by these prokaryotic bacteria.

Probably, most primitive viruses and single-cell organisms (prokaryotes) appeared during the Early Archaean time. They were isolated from the surrounding environment by a protective semipermeable membrane, but without a separate nucleus. Sometime later, the photosynthesizing single-cell microorganisms (cyanobacteria type), which could oxidize iron, were formed. This is evidenced by the proliferation of iron-bearing formations in the rocks of Early Archaean time (about 3.75 billion years ago), which consist of trivalent iron oxides (e.g. Isua Formation of Western Greenland).

6.3. Influence of Geologic Processes on the Evolution of Earth's Life

The outgassing of Earth, which started at the boundary between the Pre-Archaean and Archaean times, led to the formation of relatively dense and reducing carbon dioxide–nitrogen atmosphere during the Archaean Eon. During that time, the first volcanoes, differentiated magmatic rocks, and first shallow sea basins emerged. By the end of Archaean Eon, these shallow basins combined to form a unified (but still shallow) World Ocean. Due to high atmospheric pressure (2–6 atm) during the Archaean time, the average temperatures of near-surface atmosphere and oceanic water increased to 60–80 °C. Because of carbon dioxide atmosphere, the oceanic water became acidic (pH $\approx$ 3–5).

Despite unfavorable conditions for the life in the Archaean time (approximately 3.6–3.5 billion years ago) the stromatolites began forming. In the Onwerwaht Series of South Africa (3.5–3.3 billion years ago) stromatolites composed of silica formed thin and short-in-extent beds situated between the chert layers in volcanogenic rocks of greenstone belts (Semikhatov et al., 1999). During the Mid-Archaean time, the life forms were characterized by a greater variety, as illustrated by the remains of bacterial life forms in the "Date Tree" formation of South Africa. It is probable that cyanobacteria formed at the same time. The latter could generate small amounts of oxygen. Thus, one can hypothesize that a gradual oxygenation of atmosphere began during the Archaean time. This process was facilitated by the fact that during the Archaean time the free iron content in the convective mantle over the zones of Earth's matter differentiation was insignificant. Thus, the free iron could not bind considerable amount of oxygen.

During the Late Archaean time (about 2.8 billion years ago), the water mass accumulating in the Earth's hydrosphere was increased to a level when separate small sea basins began merging with each other and forming a unified World Ocean, surface of which flooded the peaks of mid-oceanic ridges. At about the same time, due to formation of Earth's core, the free iron and iron-containing silicates (fayalites) started entering the oceanic crust. As a result, the generation of abiogenic methane was intensified, which, in turn, should have led to an increase in the stromatolite deposits in the greenstone belts at the end of Archaean Eon (although their share in the volcanic formations of these belts remained insignificant – Semikhatov et al., 1999). At the same time, however, the consumption of oxygen increased somewhat and its partial pressure decreased somewhat.

After binding of the atmospheric carbon dioxide in the crustal rocks (Early Proterozoic time, about 2.5–2.4 billion years ago), the Earth's atmosphere became composed almost entirely of nitrogen with insignificant admixtures of other gases. Simultaneously, the Earth's surface temperature declined considerably. Pronounced changes in habitat environment at the boundary between Archaean and Proterozoic eons considerably influenced the entire biota of that time. This led to the appearance of new forms of eukaryotic unicell organisms. Possibly, these organisms emerged as a result of biosymbiosis of more primitive prokaryotic bacteria. Eukaryotes could produce oxygen by using energy of sunlight. At this time, micro-algae appeared and plant life started to develop. Revolutionary changes in the oceanic biota occurred at the beginning of Early Proterozoic time with the advent of microorganisms relying on photosynthesis. At this time, blue-green algae proliferated and the growth of stromatolites accelerated markedly (Semikhatov et al., 1999). Also, the intense generation of iron ores and generation of abiogenic methane were interrupted.

At the beginning of Early Proterozoic time, due to pronounced decrease in tectonic activity of Earth, the oceanic trenches deepened, lowering the sea level. As a result, the iron influx to the hydrosphere (and iron ore accumulation) decreased. This led to a noticeable increase in the partial pressure of oxygen about 2.5 billion years ago. This then led to an emergence of new forms of eukaryotic bacteria with an isolated nucleus containing the DNA molecules transmitting the cell genome.

The further evolution of life occurred in accordance with the biologic evolutionary laws, directed mainly by the changing conditions of habitats and the influence of gradual increase in oxygen content of atmosphere. After completion of the main period of iron-ore accumulation and an increase in the partial pressure of oxygen about 1.9–1.8 billion years ago, a new major leap in the evolution of Earth's life occurred: the first multi-cell organisms had emerged (Fedonkin, 2003). During the Mid-Proterozoic time, various species of bacteria and algae appeared (Fig. 6.1). Inasmuch as the metabolism of eukaryotic microorganisms depends on the consumption of small amount of oxygen, they could have proliferated only after an increase in partial pressure of oxygen to the level of 0.01 of its present-day value. At the boundary between the Mid- and Late-Proterozoic times there was an intense proliferation of eukaryotic organisms and phytoplankton. At approximately the same time, the bacterial colonization of the land began as evidenced by the red-colored weathered rocks of the same age.

The total phytoplankton biomass, generating oxygen in the ocean, is limited by the amount of phosphorus compounds dissolved in the ocean water (Shopf, 1982). Its concentration, however, remained in equilibrium with respect to the oceanic crust basalts and was close to the present-day one. This implies that the mass of oceanic phytoplankton changed in time approximately proportionally to the mass of ocean water. Therefore, about 2 billion years ago, the biomass of oceanic organisms was already significant: about one third of its present-day value.

Examining the evolution of life in the Earth's history, one needs to consider the importance of free (metallic) iron, contained in the Precambrian mantle. With the mantle matter, this iron was transferred into the rift zones of mid-oceanic ridges where, upon contact with water, it was converted to the soluble divalent hydroxide, which was further spread around the entire ocean (probably in the form of bicarbonate). Inasmuch as the divalent iron hydroxide actively reacts with oxygen, the larger part of oxygen produced by the Precambrian phytoplankton was consumed as a result of oxidation of iron to trivalent state (to goethite, for example). This explains not only the formation of large deposits of iron ores but also a very low partial pressure of oxygen in the Precambrian atmosphere. After the metallic iron was completely

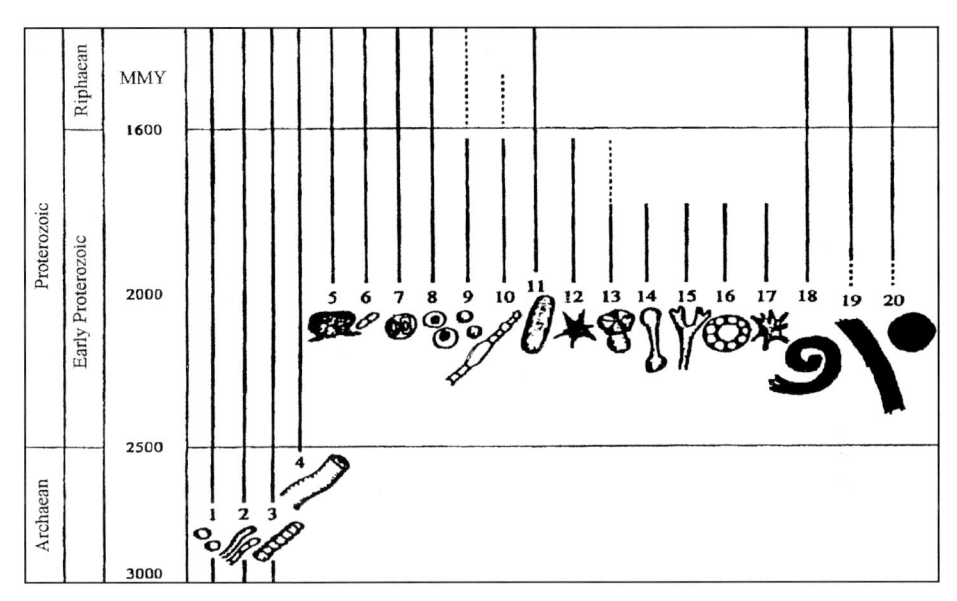

Figure 6.1: Distribution of Main Types of Microfossils in the Archaean and Early Proterozoic Time (According to Semikhatov et al., 1999). Microfossils of the Archaean Time are Represented Mostly by Single Spherical and Thread-like Nanobacteria (1, 2), Trichomes (3), and, Possibly, Cyanobacteria (4). Microfossils of the Early Proterozoic Time Range from Cyanobacteria (5–7), Cocci (8, 9), and Trichomes to Complex Morphologic Forms (11–17) and also Spiral (18), Ribbon-like (19), and Spherical (20) Forms.

transferred from the mantle into the core about 600 million years ago, the major "consumer" of oxygen had disappeared and oxygen started to accumulate in the atmosphere.

With accumulation of oxygen in the atmosphere, the life species began to advance. The first forms of macroorganisms (jellyfish-like animals) emerged during the Wendian time, leaving imprints in many geologic formations. In the Cambrian Period, the first skeletal forms of animals had appeared with almost all of the present-day species of flora and fauna (Fig. 6.2).

The geologic–biologic transitional boundary between the Proterozoic and Phanerozoic eons was clearly recorded in the biologic history of the Earth (biologic explosion). The atmosphere changed from being neutral or slightly oxidizing into oxidizing. Thus, the life forms, the metabolism of which was based on the oxidation of organic matter by oxygen generated by plants (photosynthesis), became more competitive and had a better chance of survival in the natural selection process.

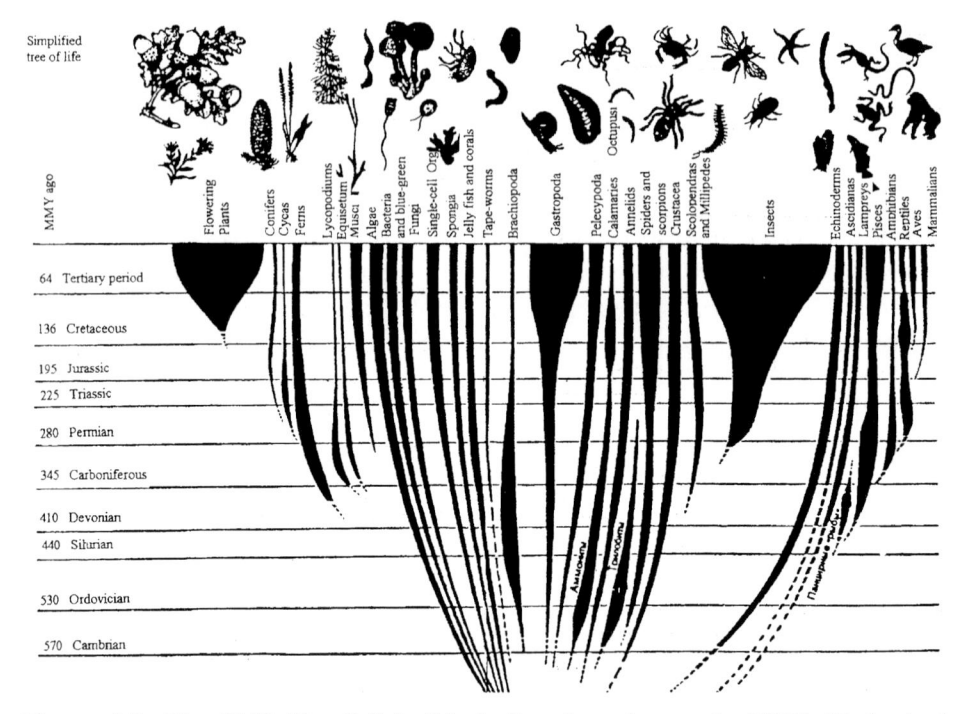

244   *Global Warming and Global Cooling: Evolution of Climate on Earth*

Figure 6.2: The "Life Tree" (Modified after Attenborough, 1979). Biological Explosion at the Boundary between the Proterozoic and Phanerozoic Time.

The life on Earth began flourishing also because of moderately warm climate with slightly oxidizing atmosphere having moderate pressure. In the carbon dioxide atmosphere (having high temperature and pressure) of Archaean time only the most primitive thermophilic forms of bacteria could survive. During the rest of Earth's geologic history, moderate temperatures, favorable for biologic developments, prevailed. Even considerable climatic cooling during the Pleistocene Epoch, with periodic continental glaciations at the polar and near-polar regions of Earth, positively influenced the improvement of highest life forms on Earth, including the development of humans.

6.4. Influence of Continental Drift and Marine Transgressions on Ecology during Phanerozoic Eon

In addition to the gradual increase in the partial pressure of oxygen in atmosphere, the continental drift, oceanic transgressions and regressions, and climatic changes influenced the evolution of life forms significantly. All these factors altered the stationary ecologic niches of biologic species and intensified their competition for survival.

During the Phanerozoic Eon two large-scale global transgressions occurred. The first one occurred from the Ordovician to Devonian time during Caledonean Orogeny (500–350 million years ago) with an amplitude reaching 200–250 m. The second and largest global transgression, caused by the formation of supercontinent Pangaea, occurred during the Late Jurassic–Cretaceous time with amplitude reaching 350–400 m. The accumulation of water in the continental glaciers during the glaciation could have led to global oceanic regressions, causing decrease in sea level by 120–130 m.

Transgressions and regressions of ocean, linked to the eustatic oscillations of oceanic level, should have caused the global variations in Earth's climate in the geologic past. Because the specific heat of water exceeds many times that of atmosphere, any considerable increase in the area of sea surface at the expense of land surface smoothes the seasonal and latitudinal variations in climate. Upon flooding of up to 30% of the area of continental surface during the Mid-Cretaceous time, the smoothing effect of transgressions on the global variations of climate at the moderate and high latitudes was considerable. This was aided by the proliferation of epicontinental seas forming the sea channels, which allowed heat exchange between the moderate and high latitudes. During the regression periods, as the seawater retreated, the seasonal contrasts in temperature increased, and the Earth's climate became more continental.

The main processes influencing the climate and its latitudinal zonation were: (1) changes in the Earth's precession angle due to continental drift and high-latitude glaciations and (2) the nitrogen removal from the atmosphere by bacteria. In addition, the spatial distribution of continents and oceans also influenced the climatic contrasts significantly. Thus, during the ice ages, the land regions, transferred to high latitudes upon movement of lithospheric plates, were first gradually covered by mountain glaciers and, afterwards, by the surface (blanket) glaciers, which acted as global "refrigerators". In addition, the changes in mutual positions of continental massifs altered the circulation routes of oceanic waters, which strongly influenced the Earth climate. The vast contemporary glaciation of Antarctica had started due to general cooling of climate, but reached a maximum only after breaking apart of Australia and its displacement to the North, and the formation of Drake Straight to the south of Tierra del Fuego. After the occurrence of this event, the south circumpolar current was formed, which completely isolated Antarctica from the warmer currents of three neighboring oceans (Scotese, 1997). This isolation of Antarctica also exists today.

In the light of geohistoric interpretation of changes in the global climate of Earth discussed above, one can analyze the nature of most significant ecological milestones of Phanerozoic Eon, during the last 550–600 million years of development of life. Gradual increase in the partial pressure of oxygen in atmosphere allowed the highly-organized life to move to land. This unique event indicated a pronounced transformation of metabolism of oceanic organisms and marked emergence of animal forms possessing lungs (organs perfectly adapted for gas exchange in the air medium).

In addition, the Earth's climate was influenced by the Moon–Sun system, always lowering the precession angle, which, in turn, led to significant cooling. The life activity of nitrogen-consuming bacteria during the Proterozoic and Phanerozoic time also led to gradual atmospheric cooling. On the other hand, the formation of supercontinents led to global warming. As a result of superposition of the above-mentioned factors, a general cooling of climate (10–15 °C) during the Proterozoic and Phanerozoic time occurred, with oscillation periods of about 800 million years and amplitudes of temperature change of about 8–10 °C.

The global tectonic processes also played an important role in the ecologic evolution during the Phanerozoic Eon and earlier. Changes in the spatial distribution, sizes and forms of the continents and oceans during the Phanerozoic time influenced significantly the magnitude and structure of oceanic currents and, consequently, the distribution of biologic productivity, i.e. influenced the formation of ecological systems best adapted to the specific conditions of nature.

Most of the sea species (about 90%) live on the continental shelves and in shallow water near the islands and submarine mounds at a depth of less than 200 m. Thus, during the Phanerozoic Eon the development of marine fauna occurred mostly at shallow depths. The highest variety of shallow marine fauna is now observed at the tropical belt, where a great number of highly specialized species reside. The variety and biomass of marine fauna is decreasing with increasing latitude and, especially, at the middle subtropical latitudes. In the subpolar regions, however, because of higher concentration of dissolved oxygen the bioproductivity starts to increase again. The least productivity of marine fauna is observed at the middle subtropical latitudes. The present-day degree of variation of shallow marine fauna is well correlated with the changes in nutritional resources, which depend on seasonal climate. In addition to this latitudinal factor, there are also the longitudinal factors, which also influence the variety of contemporary marine fauna. In particular, the greater variety of marine fauna is observed in areas with stable nutritional resources. In each latitudinal belt, the highest variety of marine species exists at the archipelagoes and shelves of large oceans.

In the upwelling zones, waters enriched in phosphorus and organic compounds rise along the continental slope to the surface, providing abundant nutrition to the shallow-water organisms. Usually, the upwelling zones exist at the eastern shores of oceans in the tropical belts. At these zones, peculiar oases of life emerge, which flourish among the relatively barren water of surrounding subtropical equatorials. Examples of such oases are the Peruvian and West-African upwelling zones in the Pacific and Atlantic oceans, respectively.

Deep oceanic trenches, forming in the process of rift genesis, provide a significant obstacle to the proliferation of shallow-water fauna. The volcanic arcs, forming over the areas of subsidence of oceanic lithosphere into the mantle, and the inter-plate chains of volcanic islands provide convenient routes for proliferation of sea species, especially when these chains of volcanic islands are latitudinally aligned, or are located within the same climatic belt (the islands of Polynesia or Micronesia, for example). Another important way of species resettlement is the migration of maggot forms of organisms. One, however, cannot assume that the shallow-water species around the world have a common origin. As a result of separated positions of continents, the shallow-water fauna is developing independently in 30 different provinces and is characterized by a low number of species common to all of these provinces. The various estimates show that the number of species of shallow-water fauna is one order of magnitude higher than that of an imaginary single-world shelf province (even having the highest possible variety of species).

The same patterns can be found in the process of resettlement of some species of deep-water fauna. In the biological communities of hydrothermal springs of Pacific Ocean, the large tubular worms and bivalved mussels dominate, whereas in the Atlantic Ocean, the hydrothermal springs are inhabited by small shrimp, which are adapted to consume sulfur bacteria.

Using (1) the above-described patterns of fauna proliferation, (2) data on continental drift, (3) information on eustatic changes in the level of World Ocean, and (4) analyzing the climatic consequences of all the above three factors, it is possible to explain the changes in the number of taxons of the shallow-water fauna during the Phanerozoic Eon. It is also possible to explain, for example, the causes of mass extinction of many animal species at the boundary between the Paleozoic and Mesozoic epochs. During the Paleozoic time, most of the continental massifs were separated and positioned at the tropical and moderate latitudes. All of them had large shelf regions, which led to a significant increase in the variety of shallow-water species in the Ordovician time. This increased variety of species was preserved in the evolutionary process during the larger part of Paleozoic time.

At the transition from the Permian to Triassic time, when all the continents began forming a single supercontinent (Pangaea) and, especially, during the

Early Jurassic time, when Pangaea was completely formed, the Earth's precession angle increased significantly (up to 30–34°), and the Earth's climate became considerably warmer. At the same time, the number of biological provinces and ecologic niches on the Pangaea decreased significantly. In addition, the Permian–Triassic transgression had led to a pronounced decrease of shallow-sea areas. Under such conditions, only the shallow-water species, which could find nutrition at the near-bottom layers, could survive. The species, which survived at the boundary between the Paleozoic and Mesozoic eras, probably were developing under the unstable environmental conditions. On the other hand, the majority of Paleozoic populations, which were developing under the stable environmental conditions (similar to present-day tropical conditions), after formation of Pangaea, appeared to be less adapted to new environment and were destined for extinction. Thus, one can assume that the rapid extinction of many species of sea fauna at the boundary between the Paleozoic and Mesozoic periods was caused by a decrease in the number of ecological niches around the Pangaea supercontinent and by the reduction of bioproductivity of the shallow (shelf) seas on this supercontinent.

During the Early Cretaceous time, Pangaea supercontinent began breaking apart due to very strong uplifting mantle flow, formed under the center of supercontinent. This process was accompanied by a considerable increase in tectonic activity and by the centrifugal parting of separate continents (Fig. 6.3) with an extensive transgression. As a result, the variety of species in kingdom of animals was increased. The variety increased further during the Cenozoic time, following the separation of shelf provinces of different continents from each other and, especially, due to an increased contrast in Earth's climatic zones.

The Cretaceous transgression had led to blossoming of carbonate-consuming fauna and microflora on the shelves and in the epicontinental seas, especially of the cocolithophorid microflora, which formed thick layers of chalk. The same transgression, however, had led to the crisis of coral atolls in the open ocean. Apparently there was an extensive transfer of matter during the Mid-Cretaceous time. This reduced pronouncedly the oceanic sedimentation, but intensified the transfer of calcium carbonate from the open ocean waters to the shallow seas of flooded parts of former land (Bogdanov et al., 1990; Sorokhtin, 1991).

During the Cretaceous time, the position of Earth's continents differed from the present-day one (Fig. 6.4). The shallow epicontinental seas were positioned mostly in the arid zones, where evaporation predominated over precipitation. Thus, such seas acted as natural pumps, sucking water out of the oceans (Fig. 6.5).

The water flowing into the continental seas partially evaporated, increasing the concentrations of dissolved salts, calcium carbonate, and phosphorus

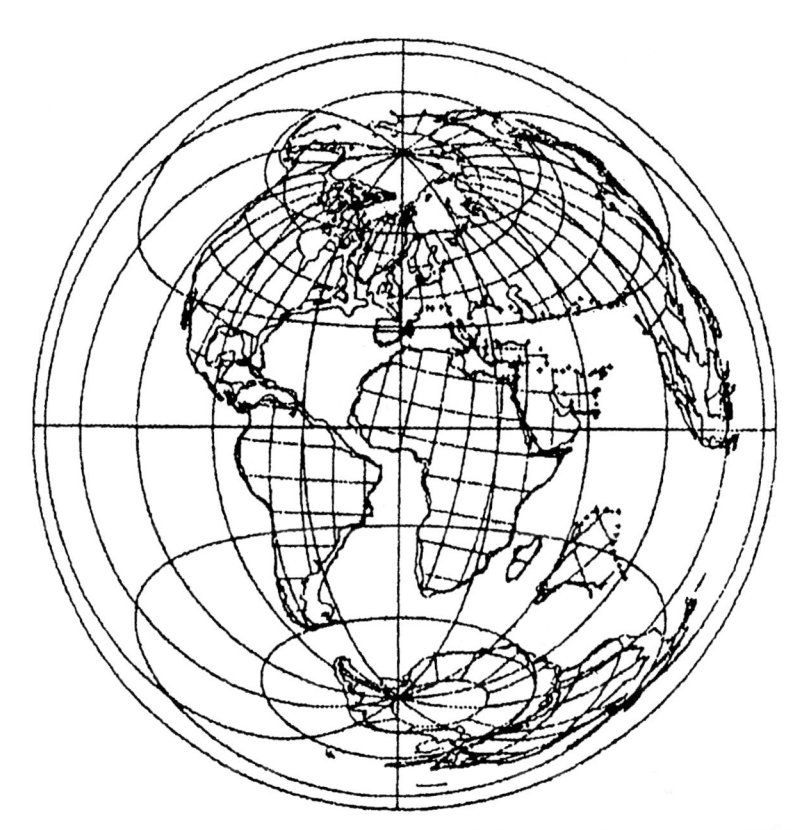

Figure 6.3: Initial Stages of the Pangaea Disintegration about 100 Million Years Ago (Lambert Projection) (Modified after Smith and Brieden, 1977).

compounds. Inasmuch as the epicontinental seas of that time were mostly open, the increase in concentrations was probably not high. Continuous heating and aeration in the wide but shallow sea basins, with the high concentration of $CaCO_3$ and phosphorus compounds led to an intensive development of life (especially, phytoplankton (coccolithophorids) and foraminifera). In shallow waters, there was an intensive development of corals, mussels, and other skeletal animals, which formed thick beds of rudistid–coralline limestones in the middle of Cretaceous Period.

In addition, turbidites were deposited in the epicontinental seas where rivers discharged (Lisitsyn, 1984). The carbonates carried by rivers from continents were precipitated in such seas and did not reach the open ocean. As a result, in the Middle Cretaceous Period the waters of World Ocean were significantly impoverished in calcium carbonate and, probably, in phosphorus compounds.

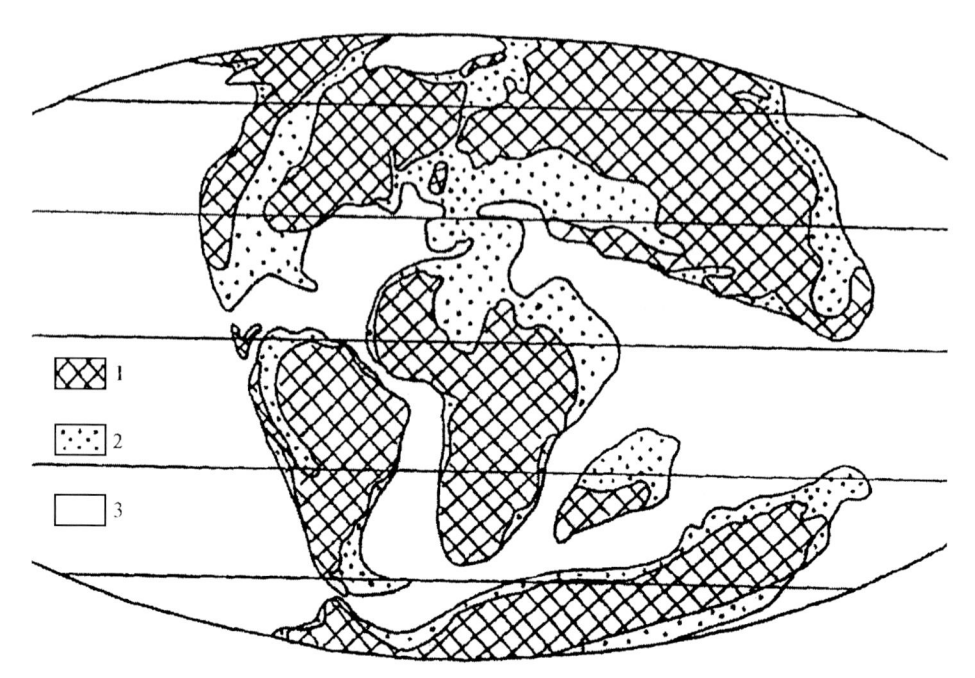

Figure 6.4: Generalized Distribution of the Land and Shallow Epicontinental Seas during the Cretaceous Time (Modified after Kuznetsov, 2003). (1) Land; (2) Shallow Epicontinental Seas; and (3) Deep Oceanic Basins.

There was a shortage of phosphorus compounds and calcium carbonate for construction of bioherms of atolls in the open ocean (Fig. 6.6) during the Apt-Cenomanian time. This led to depression and degradation of reef biota, which used calcium carbonate for building their skeletal and protective structures.

Due to significant scarcity of $CaCO_3$, rudistids, corals, and other skeletal organisms could not form strong structures capable of withstanding the abrasion by oceanic waves, especially during the stormy weather. Under such conditions, the reef construction on oceanic islands could not compensate by their growth for the gradual subsidence of their volcanic foundations. Thus, the erosion of former reefs completely ruined their shallow-water fauna, transforming the former coral atolls and rudistid banks into submarine buttes (guyots) (Fig. 6.7).

During the Apt-Cenomanian time, about 300 flourishing atolls were destroyed in the Pacific Ocean alone. They were transformed later into submarine buttes, with their tops submerged about 1,500 m below sea level. The atoll foundations (volcanic mountains) were formed as a result of membrane

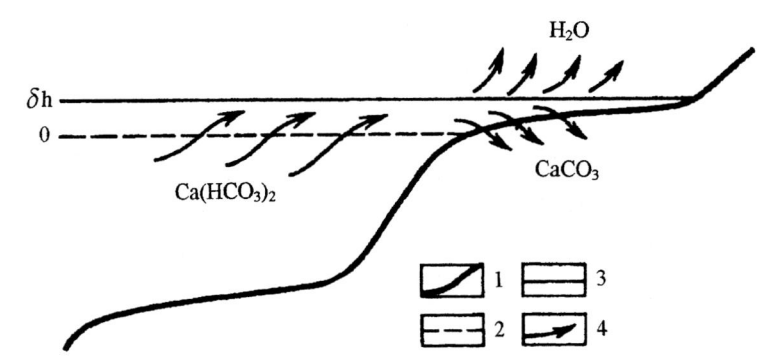

Figure 6.5: The Influence of Marine Transgression on the Redistribution of Areas of Carbonate Accumulation. (1) Hypsometric Curve; (2) Present-Day Sea Level; (3) Sea Level during the Cretaceous Marine Transgression; and (4) Paths of Migration of Carbonates and Water.

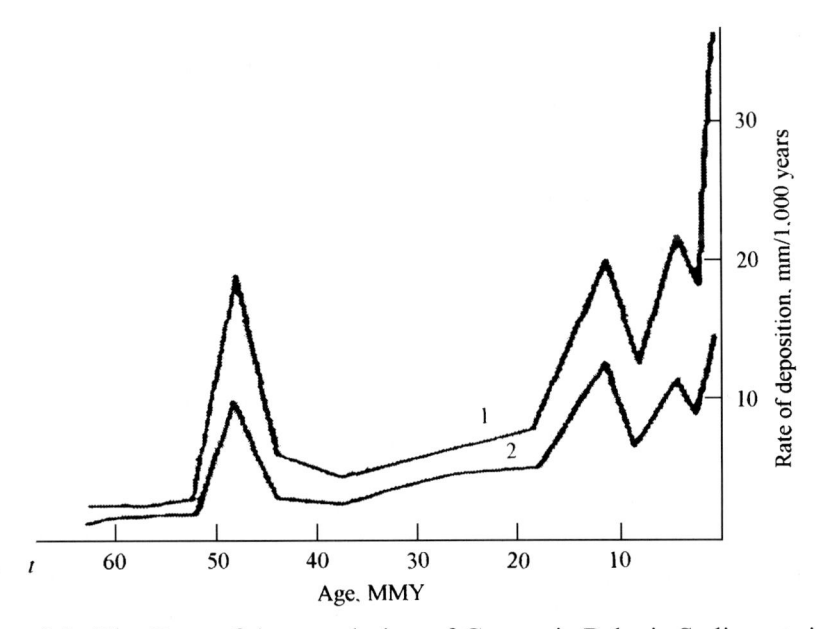

Figure 6.6: The Rate of Accumulation of Cenozoic Pelagic Sediments in the Pacific Ocean (Modified after Davis et al., 1977).

tectonics and deformation of lithospheric plates in the process of their movement on the surface of ellipsoid of Earth's rotation.

Due to continuous stirring of water masses in the oceans during ice ages, the near-bottom water was always saturated with oxygen, which was conducive to the existence and development of benthos. During the warm periods in the

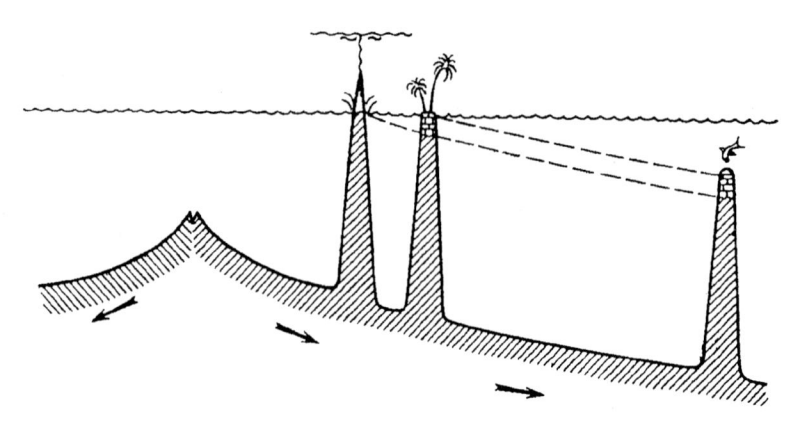

Figure 6.7: Formation of Guyots.

absence of glaciations, when there was no stirring of ocean water, the stagnant warm water devoid of oxygen accumulated in deep basins. Under these conditions, hydrogen sulfide contamination of the water occurred often, which led to accumulation of sapropel muds. In those periods the entire deep-water and benthos fauna (which existed on the ocean floor during the past ice ages) disappeared. After the beginning of new ice age, the benthos formed again due to resettlement of pelagic fauna to the deepwater regions.

Based on deepwater drilling data, the sapropel muds are present in the chalk deposits of Pacific Ocean. This indicates that all of the benthos fauna in this ocean is young (younger than the Mid-Cenozoic time – Kuznetsov, 1991): probably, not older than 55 million years (Nesis, 2001).

The continental drift significantly influenced not only the oceanic but also the land fauna. Thus, the Mesozoic Era and Late Permian time were called the time of reptiles, whereas the Cenozoic Era was the era of mammals. During 200 million years of evolution in the Mesozoic Era, only 20 orders of reptiles were developed, whereas during 65 million years of Cenozoic Era about 30 orders of mammals emerged. This difference can be explained on comparing the conditions necessary for development of reptiles and those for mammals. It is noteworthy that the initial period of intensive development of land reptiles coincided with the time of formation of fourth supercontinent Pangaea. That period (250–200 million years ago) was characterized by significant ocean regression with comparatively mild global climate, which was established at the Late Permian time. Reptiles became abundant in the second half of Mesozoic Era when, after an increase in precession angle as a result of emergence of Pangaea supercontinent, the Earth's climate became warm.

The separation of ecologic provinces within the Gondwana continents probably began somewhat earlier than that of Laurasia. In the middle of

Late Cretaceous time, Africa was distanced significantly from the rest of Gondwana continents (Fig. 6.3). At the same time a larger part of this continent was flooded by shallow seas, which divided it into two or three land regions. South America (separated from Gondwana in the Early Cretaceous time) became completely an isolated ecologic province only in the Late Cretaceous time. It was divided into the two land provinces by a shallow sea at the basin of present-day Amazon River. In addition, during the Late Cretaceous time there were two more isolated land provinces – India and Australia–Antarctica. The latter split into the two independent provinces at the beginning of Cenozoic time (about 40 million years ago).

The main reason for diversification of mammals was existence of 8–10 essentially isolated from each other land provinces, which were formed during the Early Cenozoic time. Merger of several continental fragments during Late Cenozoic time and reduction in the number of ecologic provinces to four [(1) Australian, (2) Central- and South-American, (3) African (to the South of Sahara), and (4) the largest one, which included Eurasia with Hindustan, Northern Africa, and North America] had led to extinction of 13 orders of land mammals. In this selection process, only the orders that in the first half of Cenozoic Era underwent evolution in not-completely isolated ecologic provinces survived. The mammals that before merger developed in the completely isolated provinces were less adapted to changing environment and became extinct.

Based on analysis of the Late Mesozoic and Cenozoic evolution of World Ocean floor, one can conclude that all important events of geologic history (and geochronological division into the eras, periods, and epochs) occurred as a result of merging or splitting of continents in the process of global movement of lithospheric plates.

Paleoreconstructions by Smith and Briden (1977) show that most of the boundaries of changes in biologic communities (which are used to divide the Phanerozoic Eon into periods) coincide in time with the main stages of Earth's tectonics. In the middle of Mesozoic Era, all the continents merged into a unified supercontinent Pangaea (possibly with the exception of Chinese Platform), whereas Cretaceous Period marked the beginning of Pangaea disintegration, which is going on until the present time.

During the Triassic Period, (1) splitting of Europe and Asia had started (but not completed) in the region of present-day Western Siberian plains and (2) separation of North America from Africa and Europe had begun, which led to the formation of young oceanic depression of North Atlantic during the Mid-Jurassic time. At the beginning of Cretaceous Period, the African continent split from South America and Antarctica, which were also separated from each other. At the end of Cretaceous Period and in the beginning of the Cenozoic Era, the contemporary New Zealand Plateau and Lord-Howe underwater ridge split from the unified continent of Antarctica and Australia. These continents were then separated and Australia started drifting toward the Equator. At the same time (at the boundary between the Cretaceous Period and Cenozoic Era) North America, Greenland, and Europe were separated in the Northern hemisphere. As a result, during the Cenozoic Era the Polar Atlantic Ocean was formed. In addition, at the very beginning of Cenozoic Era the Arabian Plate broke away from the African Plate, and the Red Sea and the Aden Bay began forming. At the Mid-Cenozoic time the Hindustan Plate collided with Asia and the largest mountain belt was formed. At the boundary between the Oligocene and Miocene epochs, the Paleoocean Tethys began closing and by the Mid-Miocene time it was closed completely. At the same time, folding in the European part of Alpine–Himalayan foldbelt occurred. It is noteworthy that the process of closing of Tethys was clearly marked by the local increase in temperature of bottom water of World Ocean. During the Late Miocene time, the Caucasus Mountains rose and at the same time the mountains of Central Asia and Himalayas were formed. The process of formation of the gigantic Alpine–Himalayan foldbelt, which separated the Northern Eurasia from its southern regions, is going on at the present time.

The evolution of flora and fauna on continents was determined mostly by the continental drift and the change in Earth's climate. Especially, this should be manifested at the time of merger of separate continents or, vice versa, at the time of disintegration of large continents and separation of their fragments. Alternating events of merging and splitting of continents, with formation of unified and intercontinental oceans, were accompanied by creation of new and closing-down of old ecologic niches, i.e. by drastic changes in environment. Merger and separation of continents could be one of the major causes of species evolution, their preservation and their extinction. As an example of preservation, one can mention the endemic life forms in Australia and South America.

Changes in the global Earth's map were reflected in the paleoclimate and in the evolution of ecologic system of the geologic past. Analyzing the evolutionary problems of flora in the geologic past one needs to take into consideration the continental drift and evaluate the ecologic equilibrium in each one of the regions, which merged or separated during the continental drift (Meyen, 1971, 1981).

The fluctuations in global climate during the Quaternary time, with periodic occurrence of ice ages and interglacial periods, considerably influenced the human evolution, established the times of migration of people, and determined the racial composition of Earth's human population. For example, the human settlement of America across the Bering Strait probably occurred over the land during the lowering of the World Ocean level caused by the last Vurmian glaciation, which lasted approximately from 60 to 12 thousand years ago. The size of this glaciation reached its maximum about 20,000 years ago (Imbry and Imbry, 1988), when (20–12 thousand years ago) the humans settled in the New World.

Thus, during over 3.5 billion years the entire Earth's biosphere developed as a single unified system in close interaction with its geologic development. The studies of evolution of global biosphere can be successful only if they are conducted based on systems approach: (1) considering its strong dependence on geologic scenarios of ancient periods, (2) taking into account the past tectonic and geochemical boundaries in the Earth's evolution, (3) evaluating the ancient climates of Earth, (4) studying the influence of biota on pressure and composition of atmosphere, (5) examining the continental drift, and (6) considering the opening of new and closing of old oceans.

Together with strong direct influence of geologic evolutionary processes on the development of life on Earth, there is an influence of life development on geologic processes. The influence of nitrogen-consuming bacteria on the atmospheric pressure can be cited as an example. Beginning with the Mesozoic Era, life forms played a leading role in controlling the oxygen concentration in the Earth's atmosphere. This not only determined the climatic evolution on the Earth, but also determined the development of highly organized life species. Organic life is important in the sedimentogenesis of carbonates, phosphorites, coals, oil and gas, and pelagic sediments. In addition, it also plays an important role in the process of rock weathering and, consequently, in the circulation of crust matter and formation of mineral deposits (Sorokhtin et al., 2001).

6.5. Future Development of Atmosphere and the Death of Earth

The cooling of Earth climate, caused by the bacterial consumption of atmospheric nitrogen, will be continuing in the near future. Therefore, one should not expect any essential climatic warming in the next 100–200 million years. The contemporary global warming, which started in 17th century (i.e. long before the industrial revolution), probably is temporary and determined by the fluctuations in solar activity. The data on paleotemperature in Sargasso Sea for the last 3,000 years (Keigwin, 1996) indicate that contemporary global warming is temporary, developed on the background of general long-term climatic cooling. Geologic data support this conclusion: 200–100 million years ago, the blanket glaciations almost did not exist on Earth and the average temperatures reached almost 20 °C, but declined to 15 °C by the present time. Thus, a new ice age had begun. At the beginning of Cenozoic Era, the Antarctica was covered with ice and regular periodic glaciations occurred on the continents of North America, Europe, and Asia during Quaternary time.

Despite a gradual increase in solar activity, a slow climatic cooling will continue in the future until reaching a new cool climate state of equilibrium. This future climatic level, which, in our opinion, is determined mainly by the metabolism of nitrogen-consuming bacteria, may not be very favorable for flourishing of highly organized life on Earth.

According to our estimates, in 250 million years the average Earth's surface temperature at the time of development of glaciations will decline to 0 °C; however, during the interglacial periods the average temperature will remain positive (5–8 °C). Due to general decrease in tectonic activity of Earth, at the same time the level of World Ocean will decline by about 200 m. As a result, all the contemporary sea shelves will become uncovered. Despite these changes, at the low and moderate latitudes the environmental conditions for development of highly organized life will remain satisfactory. Approximately in 400 million years, the average temperatures on the Earth's surface will drop to about −5 °C, and the oceanic level will decline by about 400 m with respect to its contemporary elevation. In this situation, all the northern and southern continents at moderate latitudes will be covered by glaciers. Even at the equatorial belt, the elevated regions of continents will be covered by ice.

This cooling trend, however, will not last forever. In about 40 million years, the temperature reduction due to life activities of nitrogen-consuming bacteria will be compensated by an increase in temperature as a result of increase in Sun's luminosity. Then, in about 600 million years, this balance will be disrupted abruptly because of intense degassing of abiogenic oxygen (about 2.1×10^{16} g/year) liberated as a result of core matter generation. If all of this liberated oxygen would enter the atmosphere, then the partial pressure of oxygen would increase at a rate of about 4 atm per million years. Actually, this rate should be about two orders of magnitude lower (about 0.2 atm per million years). This means that after 200 million years upon initiation of oxygen degassing from the mantle, its partial pressure in atmosphere will reach the level of about 4 atm. As a result, the average surface temperature on Earth would rise (due to greenhouse effect) almost to 76 °C. After 200 million years more (in one billion years from now), the partial pressure of oxygen will exceed 14 atm, with the Earth's surface temperatures rising to 110 °C.

Thus, at the very beginning of intense endogenic oxygen degassing all of the land life will literally "burn" in such an atmosphere. Only in the oceans (because of a low solubility of oxygen in water) the highest life forms will exist for some time until they are "boiled" in hot water. After the oceans start boiling (in about 1.5 billion years from now), the irreversible greenhouse effect (with the temperatures of about 550 °C) will develop and even the primitive thermophilic prokaryotes will not survive.

Thus, the most probable estimate of Earth's life longevity is about 4.6 billion years (past 4 billion years plus 0.6 billion years in the future). The duration of existence of highly organized life is even shorter: 1.3 billion years in the oceans (from 600 million years in the past to 700 million years in the future) and about 1 billion years on land (from 400 million years in the past to 600 million years in the future). The theoretical longevity of human species is about 600 million years, but in reality it should be considerably less.

In the future, the greatest peril for Earth should be expected from Sun. Upon exhaustion of nuclear fuel (hydrogen, helium, carbon, etc.) the stars similar to Sun are gradually expanding at the expense of shifting the zone of nuclear reactions from the central regions of star to its periphery. This causes not only an increase in the radius and surface temperature of the star but also in its luminosity (Zel'dovich and Novikov, 1971; Aller, 1976). Thus, after 4.7 billion years of existence, the Sun's luminosity would increase by 30% (Bachell et al., 1982). In the future, the Sun's luminosity will be increasing even more rapidly, which will amplify immensely the strong greenhouse effect caused by the intense generation of abiogenic oxygen.

Chapter 7

Global Forces of Nature Driving the Earth's Climate

7.1. Introduction

Understanding of global forces of nature driving the Earth's climate is crucial for establishing adequate relationship between people and nature and for developing and implementing a sound course of actions aimed at survival and welfare of human race. The latter is especially important in the light of present-day public debates on causes and ways of mitigation of current global atmospheric warming.

After the Kyoto Protocol had been announced in 1997, many researchers around the world criticized its provisions (that imposed drastic restrictions on anthropogenic carbon dioxide emission, especially in developed countries) as meaningless and catastrophic. Logical and quantitative comparison analyses presented in the publications of Robinson et al. (1998), Soon et al. (2001), Bluemle et al. (2001), Baliunas (2002), Sorokhtin (2001a, b), Sorokhtin and Ushakov (2002), Gerhard (2004), and Khilyuk and Chilingar (2003, 2004, 2006) showed that the theory of currently observed global atmospheric warming as a result of increasing anthropogenic carbon dioxide (and other greenhouse gases) emission is a myth. The estimates of the effect of anthropogenic influence on the Earth's climates were presented in Chapter 3. In the worst-case scenario, the effect of anthropogenic carbon dioxide emission on the average surface temperature does not exceed 0.01 °C.

According to the classification proposed by Gerhard et al. (2001), the human intervention constitutes the climatic driver of fourth order. In their classification, the climatic drivers were categorized using two parameters: the range of temperature changes that can be caused by a driver of interest and the length of time during which the driver's effect lasts. Classification of Gerhard et al. (2001), which was modified according to our theory of climatic evolution (Chapter 4), is presented below.

First-order climate drivers: Solar luminosity, solar system geometry, and gaseous composition of atmosphere. To the same category one should also add the factors that directly influence these drivers: solar activity, Milankovitch cycles, rates of outgassing of various components of atmosphere and the powerful factor of microbial activity, which directly influences the gaseous composition of atmosphere (Chapter 4). The temperature range effect by the first-order drivers is estimated at tens of °C and lasts for billions of years.

Second-order climate drivers: Global distribution of continents and oceans on the Earth's surface. This influential driver can explain 15–20 °C of variations over hundreds of millions of years (Frakes, 1979; Lang et al., 1999; Gerhard et al., 2001). The distribution of continental masses significantly influences the dynamic properties of Earth body (the precession angle, for example), and mass and heat transfer around the Earth's surface.

Third-order climate drivers: Orbital and solar variability, large-scale oceanic tidal cycles, and variations in the structure of oceanic currents. The drivers of third order may cause temperature changes of 5–15 °C over the time interval of hundreds to hundreds of thousands of years (Frakes, 1979; Hoyt and Schatten, 1997; Gerhard et al., 2001; Pekarek, 2001).

Fourth-order climate drivers: Volcanoes, weathering, regional tectonics, El Nino, La Nina oscillations, short ocean tidal cycles, solar storms and flares, meteorite impacts, and human intervention. There are many possible causes (natural and anthropogenic) of small temperature changes (up to 5 °C). For example, 18-, 90-, and 180-year cycles driven by ocean tides have been recognized by Keeling and Whorf (1997, 2000). In particular, these cycles change the heat transfer between atmosphere and ocean.

In spite of sketchy and conditional principles of the classification presented here, it allows one to focus on the most influential factors, which determine the temperature variations of highest range and longest duration. After proposing any classification of climate drivers, one should supplement it with the estimates of masses and energies involved in the heat–mass transfers leading to changes in atmospheric temperatures. For example, as the writers show in this book, the entire present-day energy generated by humans is at least four orders of magnitude lower than the energy of solar irradiation delivered to the Earth's surface. Therefore, from the energetic point of view the human intervention is definitely the fourth-order climate control in comparison with the solar luminosity.

The Earth's climate is a generic term for a relatively stable long-term state of Earth's atmosphere. The main parameters that describe the Earth's climate quantitatively are the atmospheric temperature and pressure averaged over certain areas and chosen time intervals. They are determined by the energy and matter flows from inside and outside of the terrestrial body, matter transformation over the Earth's surface (at the interface between lithosphere and atmosphere), and parameters of the Earth's atmosphere and the World Ocean.

Solar radiation supplies the major energy influx to the Earth's atmosphere that determines the Earth's heating and cooling, whereas outgassing supplies the major matter influx to the Earth's atmosphere that determines its chemical composition and physical properties.

The matter transformation at the Earth surface and in the World Ocean is caused by evolutionary physical changes, chemical reactions, and activities of live matter. In addition to these factors, the terrestrial body experiences orbital deviations and spatial mass redistribution that influence the Earth's climate considerably.

7.2. Solar Irradiation Reaching the Earth

The total average solar energy flux currently reaching the Earth's surface, S_0 ($= 1.75 \times 10^{24}$ erg/s), is determined by the so-called solar constant s (≈ 1.37 kW/m$^2 = 1.37 \times 10^6$ erg/cm^2 s) (Horrell, 2003). The total heat flux through the Earth's surface due to energy generated in the mantle and crust is estimated at about 4.3×10^{20} erg/s (Sorokhtin and Ushakov, 2002), which is approximately 0.0257% of the total Earth's solar irradiation. The World total energy production in the year 2003 was equal to 1.34×10^{20} erg/s (Key World Energy Statistics, 2004), which is about 0.0077% of the total solar irradiation reaching the Earth's body. Comparison of the above figures clearly shows that the solar irradiation is a dominant source of energy supply to the Earth's atmosphere and hydrosphere. One can easily estimate that the solar irradiation supplies more than 99.95% of total energy driving the World climate. Thus, heating and cooling of the atmosphere is mostly due to variations in insolation of the Earth (Hoyt and Schaten, 1997). The comprehensive data on the Sun as a primary source of energy for the Earth's climate are presented by Pekarek (2001).

The Sun is about 4.65 billion years old. Its luminosity has been gradually increasing throughout the time of its existence and the present-day energy output is about 40% higher than that at the time of its formation (Gribbin, 1991). This energy output is expected to increase by about 15% in the future 1.5 billion years. The Sun radiates the electromagnetic energy throughout the entire electromagnetic spectrum from gamma rays to radio waves. The Earth's atmosphere absorbs and reflects the solar radiation and, consequently, one needs to use the satellite technology for measuring the total solar irradiance (TSI) at the distance of Earth from Sun. These observations are an absolute must for development of a quantitative climatic theory and explanation of the Earth's climatic changes. The data of direct satellite measurements are available since 1978 (Willson, 1997). In lieu of direct satellite measurements before this time, one can use various variables correlated with TSI: for example, the number of Sun spots, which was recorded since 1610, or solar magnetic activity (Baliunas and Saar, 1992; Baliunas and Soon, 1996). The Solar irradiance is strongest at the peak of magnetic activity and the total irradiance varies directly with the intensity of Sun's activity.

"In recent years increasing number of papers have been published..., reporting correlations between climate and solar variations. The statistical confidence for the correlations is so high that there is now little reason to question that there is a physical link; the main puzzle now is the nature of mechanisms involved" (Tinsley, 1997).

The effect of solar irradiation on global atmospheric temperature can be evaluated using the adiabatic model of heat transfer in the Earth's atmosphere (Chapter 3). To analyze the temperature changes attributed to variations in energy and matter flux, one can use the following convenient form of this model:

$$T(h)/T_0 = (S/S_0)^{1/4}(p(h)/p_0)^\alpha \qquad (7.1)$$

where $T(h)$ is the global atmospheric temperature at any given altitude h; T_0 is the present-day global temperature at sea level; $T_0 = 288$ K; S is the total solar energy flux reaching the Earth's surface; S_0 is the total present average solar energy flux ($S_0 = 1.75 \times 10^{24}$ erg/s); $p(h)$ is the global atmospheric pressure at the altitude h; and p_0 is the global average atmospheric pressure at sea level ($p_0 = 1$ atm).

For a rough estimate of global atmospheric temperature change at sea level attributed to variations in insolation S, Eq. (7.1) can be rewritten in the following form:

$$\Delta T = T - T_0 = T_0[(S/S_0)^{1/4} - 1] = 288[(S/S_0)^{1/4} - 1] \qquad (7.2)$$

Data in Table 7.1 computed using Eq. (7.2) allow one to translate the variations in Earth's insolation into corresponding changes in the Earth's global temperature at sea level.

As shown in Table 7.1, one percent increase in current solar radiation reaching the Earth's body translates directly into approximately 0.86 °C increase in the Earth's global temperature. Using Eq. (7.2), one can also find an upper estimate for possible atmospheric temperature increase due to anthropogenic activities. Even if the entire World energy generated by humans

Table 7.1: Global Temperature Changes Attributed to Variations in Earth's Insolation.

S/S_0	0.85	0.90	0.95	0.99	1.00	1.01	1.05	1.10	1.15
ΔT (°C)	−11.5	−7.49	−3.77	−0.86	0.00	0.86	3.77	7.49	11.5

(1.34×10^{20} erg/s) would be utilized only for heating the Earth's atmosphere, the corresponding atmospheric temperature increase would not exceed 0.01 °C at sea level. If, in addition, one takes into consideration that changes in the global atmospheric temperature are closely correlated with the changes in solar activity (Fig. 7.1), then one has to conclude that the solar irradiation is the dominant energy supply driving the Earth's climate (see also Kondratiev, 1992; Hoyt and Schatten, 1997).

7.3. Orbital Deviations and the Earth's Mass Redistribution

Parameters of the elliptic Earth's orbit (orbital eccentricity, obliquity, and precession index) have been changing over geologic time. The terrestrial mass redistribution also results in the changes of orbital parameters, especially of the precession index (Marov, 1986).

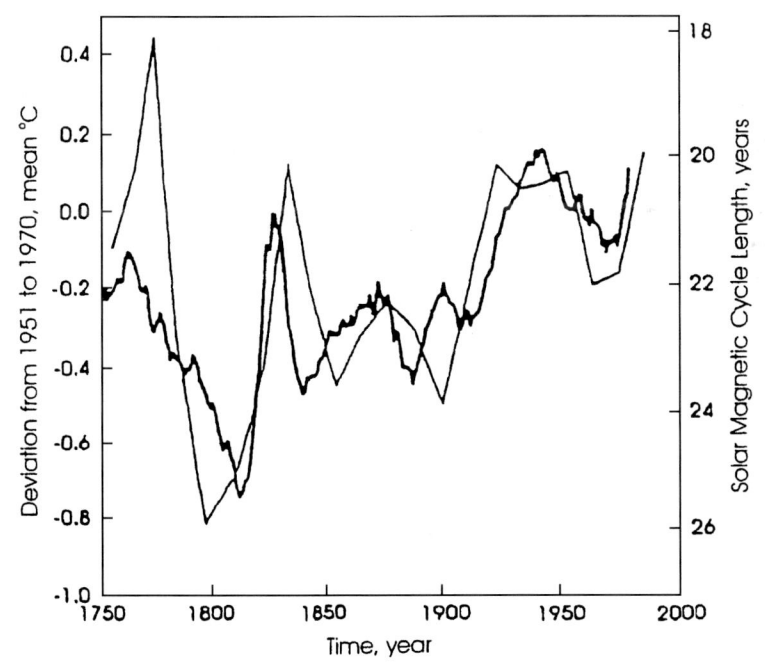

Figure 7.1: Moving 11-Year Average Terrestrial Temperatures (Northern Hemisphere) shown as Deviations in °C from the 1951–1970 Mean Value (Left Vertical Axis and Thick Line) and the Solar Magnetic Cycle Lengths (Right Vertical Axis and Thin Line) (Modified after Robinson et al., 1998, Fig. 3).

Changes in the orbital parameters result in corresponding changes in Earth's insolation. The effect of orbital and rotational changes has been known for a long time (Milankovitch, 1957). The current yearly difference between aphelion (the farthest departure from the Sun) and perihelion (the closest approach to the Sun) is about 3 million miles, which amounts to about a 6% difference in the Earth's insolation from the month of January to the month of July (http://earthobservatory.nasa.gov/Library/Giants/Milankovitch.html).

Paleoreconstructions show (Barron, 1994) that the variations in global average Earth's insolation attributed to the planet's orbital deviations can reach up to 10% of the "long-term" average radiation level (Fig. 7.2). Therefore, maximal orbital deviations result in exactly the same effect on global temperature as 10% change in solar irradiation. Thus, the resulting effect of orbital deviations on global temperature can be evaluated using the adiabatic model of Earth's atmosphere (Chapter 3).

Analyzing Fig. 7.2 and using Eq. (7.2), one can determine that during the last 500,000 years the global temperature deviated 7.5 °C (from the long-term average value) at least 4 times, and more than 20 times this deviation was about 4 °C due to changes in the Earth's orbital parameters. This leads one to an important conclusion on considerable variability of Earth's climate over geologic time due to natural changes in the Earth's insolation caused by orbital deviations and geologic insignificance of recent global warming period with the global temperature increase of 1 °F $\approx$ 0.56 °C during the last century (EPA global warming site, 2001).

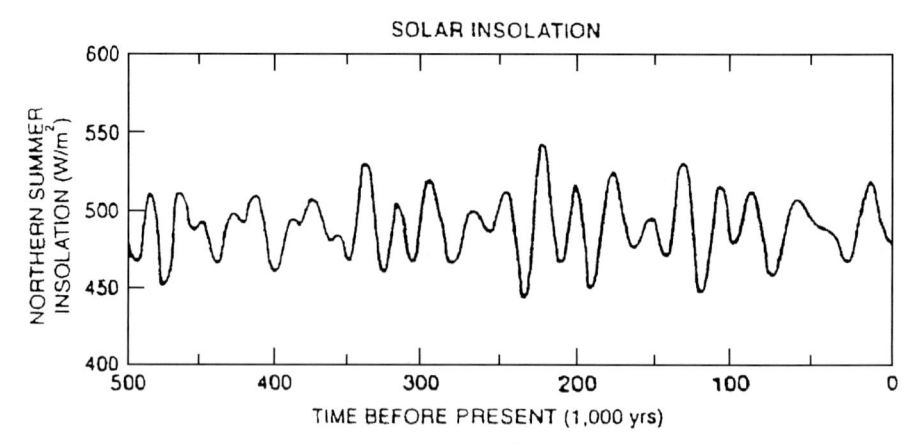

Figure 7.2: Variations in Insolation (W/m^2) due to Deviations of the Earth's Orbital Parameters (Modified after Barron, 1994, Fig. 13).

7.4. The Earth's Outgassing

Atmospheric gases are generated in the inner layers of Earth (mostly in the mantle) over geologic history and are transferred to the atmosphere and hydrosphere by outgassing. Outgassing is a process of upward migration of various gases generated in the mantle and Earth's crust and seeping through the Earth's surface into the atmosphere and the World Ocean (Khilyuk et al., 2000). Various gases (methane, carbon dioxide, water vapor, hydrogen, helium, etc.), formed in the process of chemical reactions at different physicochemical conditions, are continuously migrating upward and have been forming the atmosphere throughout the geologic history. The Earth's atmosphere and hydrosphere were formed about 4 billion years ago by outgassing (Vinogradov, 1967; Holland, 1984; Sorokhtin and Sorokhtin, 2002). This process is going on at the present time.

The rate of outgassing is determined by the rate of tectonic activity (Chapter 2). As a universal measure of the rate of global tectonic activity, one can use the rate of heat flux through the Earth's surface, because its level indicates the magnitude of total energy generated in the mantle. If, for some reason, it is not possible to estimate the value of heat flux, then the rate of oceanic floor spreading (in the spreading zones) can be substituted for it. The rate of spreading is directly translated into the rate of displacement of tectonic plates that presently averages 4.5–5 cm/year (Sorokhtin and Ushakov, 2002).

Main gases generated in the mantle and on the ocean floor are carbon dioxide, methane, and hydrogen. Carbon dioxide (CO_2) dissolves mostly in the oceanic water and transforms to abiogenic methane (CH_4) and carbonates ($CaCO_3$, $MgCO_3$) (Sorokhtin and Sorokhtin, 2002) through the following chemical reactions:

$$4Fe_2SiO_4 + 12\,Mg_2SiO_4 + 18H_2O + CO_2 \rightarrow 4Mg_6[Si_4O_{10}][OH]_8$$
(serpentine)

$$+ 4Fe_2O_3 + CH_4 + 183.47 \text{ kcal/mol of heat} \quad (7.3)$$
(hematite)

$$4Mg_2SiO_4 + 4H_2O + 2CO_2 \rightarrow Mg_6[Si_4O_{10}][OH]_8$$
(olivine) (serpentine)

$$+ 2MgCO_3 + 72.34 \text{ kcal/mol of heat} \quad (7.4)$$

$$2CaAl_2Si_2O_8 + 4H_2O + 2CO_2 \rightarrow Al_4[Si_4O_{10}][OH]_8$$
(anorthite) (kaoline)

$$+ 2CaCO_3 + 110 \text{ kcal/mol of heat} \quad (7.5)$$

These reactions are accompanied by a huge amount of heat release. The released heat contributes considerably to the heat flux through the Earth's surface. Thus, the rate of abiogenic CH_4 generation on the ocean floor in the spreading zones can be used as a measure of global tectonic activity. In turn, the rate of the Earth's tectonic activity can be used as a measure of the Earth's outgassing rate.

Because of high solubility, CO_2 is almost completely dissolved in oceanic water in the form of HCO_3^- ions raising the water acidity. Therefore, one can also use the acidity of oceanic water as a measure of tectonic activity (assuming that the global atmospheric temperature is not decreasing). Thus, the current increase in acidity of ocean water should be interpreted as an indicator of rising tectonic activity at present time. Caldeira and Wickett (2003) discovered the phenomenon of rising oceanic water acidity and traditionally explained it by the increase in anthropogenic carbon dioxide emission to the atmosphere.

CH_4 enters the atmosphere, contributing to the greenhouse effect and also depleting ozone concentration through the following reaction:

$$CH_4 + O_3 + \text{Solar radiation} \rightarrow CO_2 + H_2O + H_2 \qquad (7.6)$$

Due to a high level of current tectonic activity, there is a pronounced increase in the methane gas generation at oceanic floor. Yasamanov (2003) estimates that about 5×10^{15} g/year of CH_4 are currently released to oceanic water at the spreading zones of mid-ocean ridges only. He believes that the increasing concentration of methane leads to significant increase in the atmospheric carbon dioxide content which amplifies the atmospheric greenhouse effect.

Gases accumulating in the atmosphere and the rate of outgassing determine the properties of gaseous atmospheric mixture, in particular, changing the density and thermal capacity of air. One can assume that: the greater the rate of outgassing, the higher the atmospheric pressure. According to the adiabatic theory of heat transfer in the atmosphere, the higher atmospheric pressure leads to an increase in the global atmospheric temperature.

Studying the origin and evolution of Earth's atmosphere, one needs to take into account that the primordial Earth's matter contained only traces of volatile elements (H_2, He) and compounds. All other volatile components (H_2O, CO_2, O_2, HCl, HF), except possibly N_2, were mostly degassed out of the Earth mantle throughout geologic history (Chapter 2). Thus, the main masses of nitrogen, carbon dioxide, oxygen, and water in the atmosphere and hydrosphere of the Earth were degassed out of the mantle.

Outgassing could start only after fusion of the upper layers of the Earth's matter and beginning of the Earth's tectonic activity (about 4 billion years ago). Under plausible assumption that the rate of Earth's degassing is proportional to its tectonic activity and realistic estimates of the original amounts of volatile components in gaseous primordial atmosphere and solid matter of Earth, Sorokhtin and Sorokhtin (2000) modeled degassing of N_2, CO_2, H_2O, and O_2 out of the mantle and their accumulation in atmosphere and hydrosphere. The results of their modeling demonstrated, for example, that the nitrogen of contemporary atmosphere contains 55% of relic gas and 45% of gas of magmatic origin. Geologic evolution of the relative content of nitrogen in the atmosphere (under three different hypotheses of the origin of the dominant amount of N_2) is shown in Fig. 7.3. The most probable "intermediate" evolution of the nitrogen partial pressure in the atmosphere is presented by the curve 2 in Fig. 7.3.

Total mass of CO_2 degassed from the mantle throughout geologic history is estimated at 4.63×10^{23} g, and the CO_2 mass remaining in the mantle is estimated at 4.48×10^{23} g (Sorokhtin and Ushakov, 2002). Therefore, the total mass of CO_2 in the present-day Earth's system is about 9.11×10^{23} g (Sorokhtin and Ushakov, 2002; Khilyuk and Chilingar, 2004). Accumulation

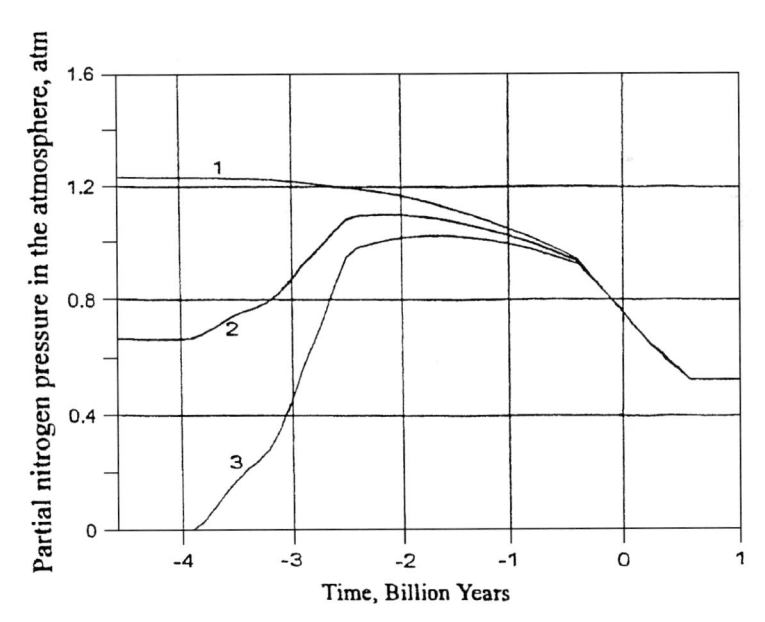

Figure 7.3: Evolution of the Partial Pressure of Nitrogen in Atmosphere. (1) Under Hypothesis that the Entire Earth's Atmosphere Nitrogen was of Primordeal Origin, (2) Most Probable Intermediate Scenario, and (3) Assuming that the Entire Atmospheric Nitrogen Mass was Degassed from the Mantle (Modified after Sorokhtin and Ushakov, 2002).

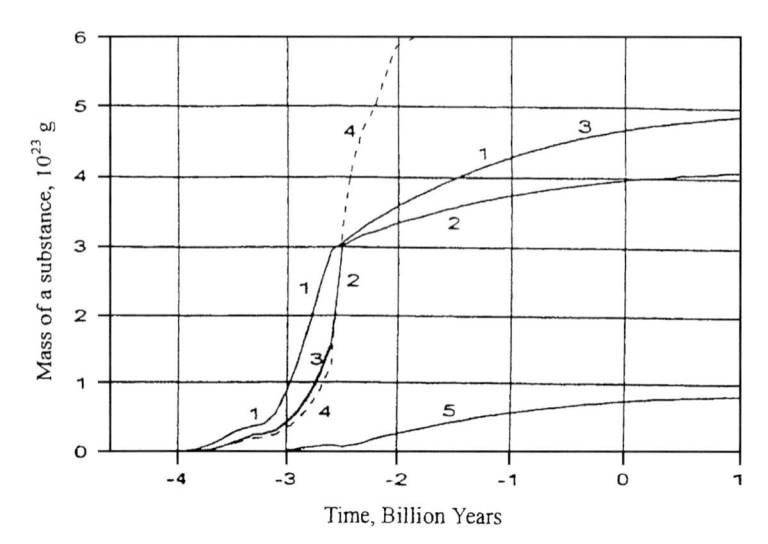

268   Global Warming and Global Cooling: Evolution of Climate on Earth

Figure 7.4: Mass of Carbon Dioxide in the Earth's Crust. (1) Mass of the Carbon Dioxide Degassed from Mantle, (2) Mass of the Carbon Dioxide Accumulated in the Carbonate Rocks of the Earth's Crust, (3) Total Mass of Carbon Dioxide Fixed in Carbonate Rocks, (4) Mass of Water Accumulated in the Earth's Crust, and (5) Mass of Organic Carbon Present in Rocks Recalculated as Carbon Dioxide (Modified after Sorokhtin and Sorokhtin, 2002).

of CO_2 in the upper geospheres (atmosphere, hydrosphere, and the Earth's crust) is shown in Fig. 7.4. Figure 7.5 illustrates the evolution of CO_2 partial pressure in the atmosphere.

To estimate the amount of total anthropogenic CO_2 emission, one can use an excellent compendium of data on estimates of anthropogenic carbon dioxide emission from the Carbon Dioxide Information Analysis Center, Oak Ridge National Laboratory, Tennessee (Marland et al., 2002). The estimates in this compendium are expressed in million metric tons of carbon. The data set comprises the annual releases of CO_2 from 1751 to 2002. They can be roughly sorted out into two groups: (1) before the year of 1900 and (2) after the year of 1900. The data of the first group exhibit linear growth of emission in 18th and 19th centuries, whereas the data from the second group show exponential growth of anthropogenic CO_2 emission in the 20th century. This observation allows one to use a piece-wise approximation for the data of compendium: a linear function for the first group and an exponential function for the second group.

Using the endpoints of the domain intervals for evaluation of the coefficients of approximating equations, the writers obtained the following piece-wise

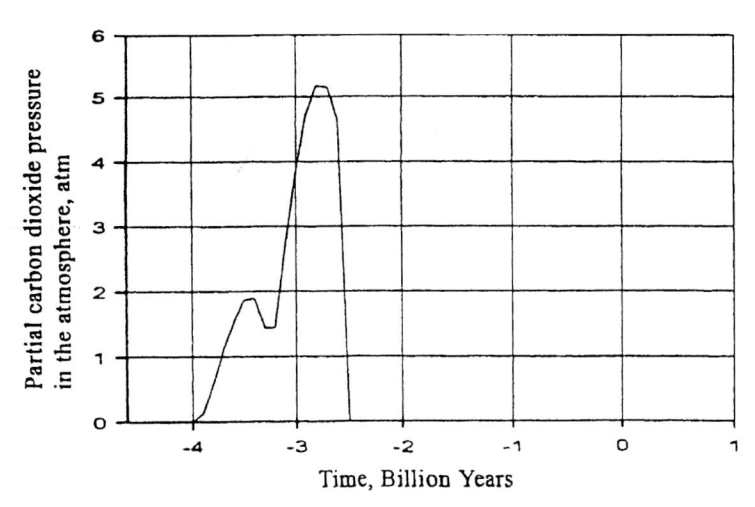

Figure 7.5: Evolution of the Partial Pressure of Carbon Dioxide in the Atmosphere (Modified after Sorokhtin and Sorokhtin, 2002).

function (considering that CO_2 emission prior to the year of 1800 is negligible in comparison with that later):

$$C = 8 + 5.26t \quad \text{if } t \in [0, 100] \text{ and } t = 0 \text{ is taken for the year of 1800,}$$
$$C = 534e^{0.025(t-100)} \quad \text{if } t \in [100, 202]$$

(7.7)

where t is the time (years) and C is the annual carbon dioxide emission rate in 10^6 metric tons of carbon/year.

The writers used this approximating function (Eq. (7.7)) for computing a rough estimate of the total anthropogenic carbon dioxide emission throughout human history. Integrating the first function over the interval [0,100], one obtains $27,100 \times 10^6 = 2.71 \times 10^{10}$ tons. Integration of the second function over the interval [100, 202] results in $253,543 \times 10^6 = 2.53543 \times 10^{11}$ tons. The latter number indicates that the total anthropogenic CO_2 emission in 20th century is about one order of magnitude higher than that in 19th century. Addition of these two numbers allows one to obtain a rough estimate of total anthropogenic carbon dioxide emission throughout the human history (about 2.81×10^{11} metric tons of carbon). Recalculating this amount into the total anthropogenic carbon dioxide emission in grams of CO_2, one obtains an estimate of 1.003×10^{18} g, which constitutes less than 0.00022% of the total amount of CO_2 naturally degassed from the mantle. Comparing these figures, one can conclude that anthropogenic carbon dioxide emission is negligible in global energy–matter transformation processes changing the Earth's climate.

Evaluation of the atmospheric oxygen content evolution throughout the geologic history encounters a great deal of uncertainty, because this process was affected considerably by many fuzzy factors, such as photosynthesis ability of ancient microorganisms. Nevertheless, even qualitative reconstructions based only on geochemical data present a clear picture of continuously increasing oxygen content in the atmosphere. The reconstruction of accumulation of oxygen in sedimentary rocks and in the atmosphere (Sorokhtin and Ushakov, 2002) is presented in Fig. 7.6. Evolution of the atmospheric composition and pressure is shown in Fig. 7.7.

In the light of debates on global warming, one should notice the gradual reduction of atmospheric pressure on the geologic timescale between approximately -0.3 BY and $+0.5$ BY (zero corresponds to the present time). This means that we live in the cooling geologic period and the global warming observed during the past 150 years is just a short episode in geologic history. According to Sorokhtin and Ushakov (2002), the cooling period will continue in the future due to life-activities of the nitrogen-consuming bacteria that transfer the atmospheric nitrogen into soils with subsequent burial in sediments. In approximately 0.6 BY, when endogenic oxygen starts degassing intensively because of formation of the core matter (Fe · FeO) according to the following reaction:

$$2Fe_3O_4 \rightarrow 3Fe \cdot FeO + 5O \qquad (7.8)$$

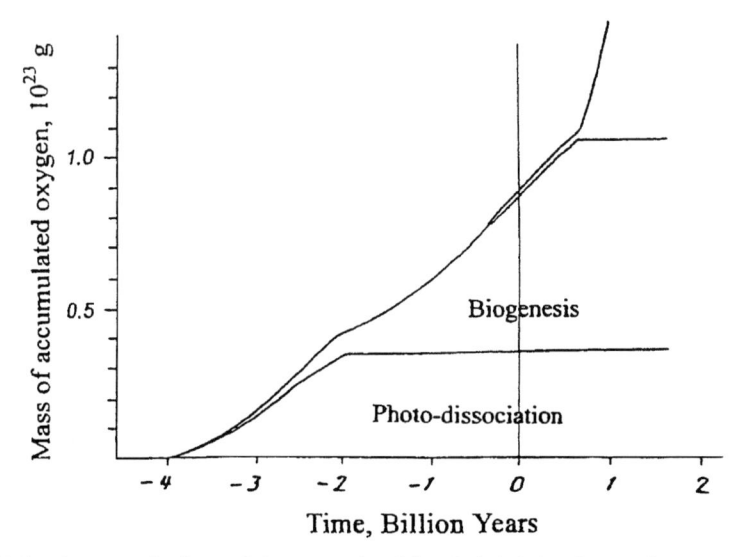

Figure 7.6: Accumulation of Oxygen (and its Origin) in the Sedimentary Rocks and in Atmosphere (Modified after Sorokhtin and Ushakov, 2002, p. 417).

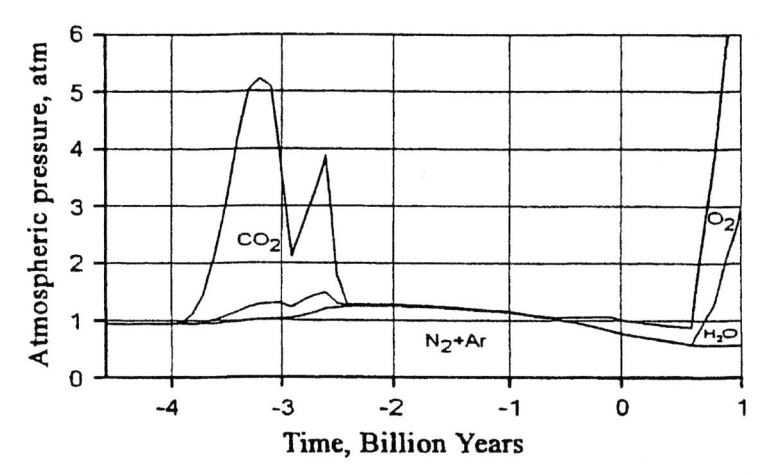

Figure 7.7: Evolution of Composition and Pressure of the Earth's Atmosphere (Modified after Sorokhtin, 2005).

the atmospheric pressure will rapidly increase to over 40 atm. This will result in abrupt increase in the global temperature to over 80 °C (Fig. 7.8). The impact of this hostile environment on highly organized species will be their extinction.

7.5. Global Climatic Cooling due to Increase in CO_2 Content

Increase in CO_2 content leads to global cooling of atmosphere. This, at first sight, paradoxical conclusion can be inferred from the adiabatic theory of heat transfer (Chapter 3). To compare the temperature characteristics of a planet at various compositions of its atmosphere, one can use the following convenient form of atmospheric temperature equation (Sorokhtin et al., 2006):

$$T = b^{\alpha}\{S(1-A) \div [8\sigma\pi^{-1} \times ((\pi/2 - \psi) + \psi \times (1 + \cos\psi)^{-1})]\}^{1/4}(p/p_0)^{\alpha} \tag{7.9}$$

where S is the solar constant at the distance of Earth from Sun; σ is the Stefan–Boltzmann constant; A is the planet's albedo (for the Earth $A \approx 0.3$); ψ is the precession angle; p is the atmospheric pressure; p_0 is the atmospheric pressure at sea level; α is the adiabatic exponent; and the coefficient b is defined by the following expression:

$$b = 1/(1-A)^{1/4\alpha}$$

(for the Earth $\alpha = 0.1905$).

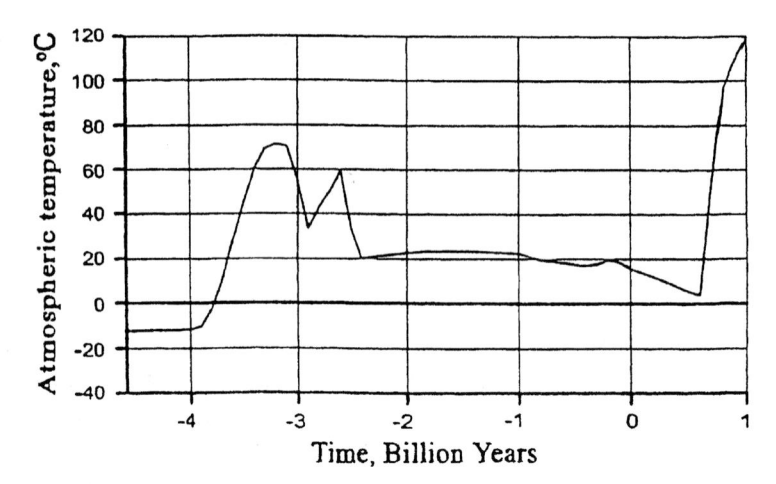

272   *Global Warming and Global Cooling: Evolution of Climate on Earth*

Figure 7.8: Temperature Evolution of the Earth's Atmosphere (Average Atmospheric Temperature at Sea Level) (Modified after Sorokhtin and Ushakov, 2002).

For the present-day nitrogen–oxygen Earth's atmosphere, $b^\alpha = 1.093$. The value of b does not change if the composition of atmosphere changes, whereas the value of b^α changes with the value of adiabatic constant α if the composition of planet's atmosphere changes.

Equation (7.9) can be used for comparison of temperature characteristics of the same planet for various compositions of its atmosphere. Thus, if one assumes that the existing nitrogen–oxygen atmosphere of Earth is replaced entirely by an imaginary carbon dioxide atmosphere with the same pressure of 1 atm and adiabatic exponent $\alpha = 0.1428$, then the value of $b^\alpha = 1.597^{0.1428} = 1.069$ and the near-surface temperature would decline to 281.6 K, i.e. the atmosphere cools by 6.4 °C (instead of warming, as the traditional theory states).

Constructing the distributions of temperature in the carbon dioxide atmosphere, one should take into consideration the fact that for the same pressure the corresponding elevation over sea level is lower than that for the nitrogen–oxygen atmosphere of Earth: $h(CO_2) = h(N_2 + O_2) \times 29/44$, where h is the elevation, and 29 and 44 are the molar weights of nitrogen–oxygen and carbon dioxide atmospheres, respectively. Constructed in such a way temperature distributions are shown in Fig. 7.9. In this figure, the graph of temperature distribution for the carbon dioxide troposphere lies below the graph of distribution for the nitrogen–oxygen atmosphere. Correspondingly, the-near surface temperature for the carbon dioxide troposphere occurs 6.4 °C lower than that for the nitrogen–oxygen atmosphere (not considerably

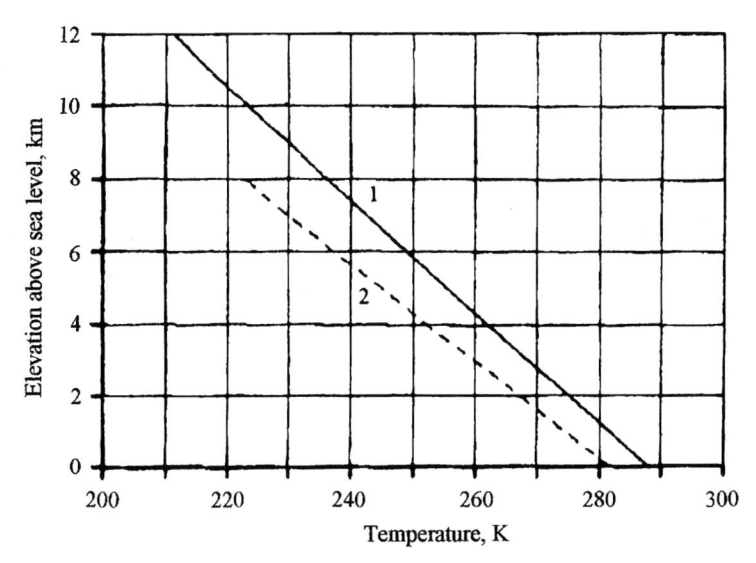

Figure 7.9: Averaged Temperature Distributions in the Earth's Troposphere. (1) Present-Day Nitrogen–Oxygen Atmosphere and (2) Hypothetical Model of Carbon Dioxide Atmosphere.

higher as some scientists continue to believe). Thus, the accumulation of carbon dioxide in great amounts in the atmosphere should lead only to the cooling of climate, whereas insignificant changes in the partial pressure of CO_2 (few hundred ppm) practically would not influence the average temperature of troposphere and the Earth's surface.

By analogy, if one assumes that the existing carbon dioxide atmosphere of Venus is entirely replaced by the nitrogen–oxygen atmosphere at the same pressure of 90.9 atm, then its surface temperature would increase from 735 to 796 K (from 462 to 523 °C). Thus, one can conclude that saturation with carbon dioxide (at the same other conditions) would lead to cooling of the entire planet's atmosphere. The averaged temperature distributions for the existing carbon dioxide and hypothetical nitrogen–oxygen atmosphere are shown in Fig. 7.10.

7.6. Inner Sources of the Earth's Energy

The Earth is a "heat machine". To originate and sustain the large-scale matter–energy transfers and transformations in the inner layers, an astronomic amount of heat has to be continuously generated in the mantle and the core of Earth.

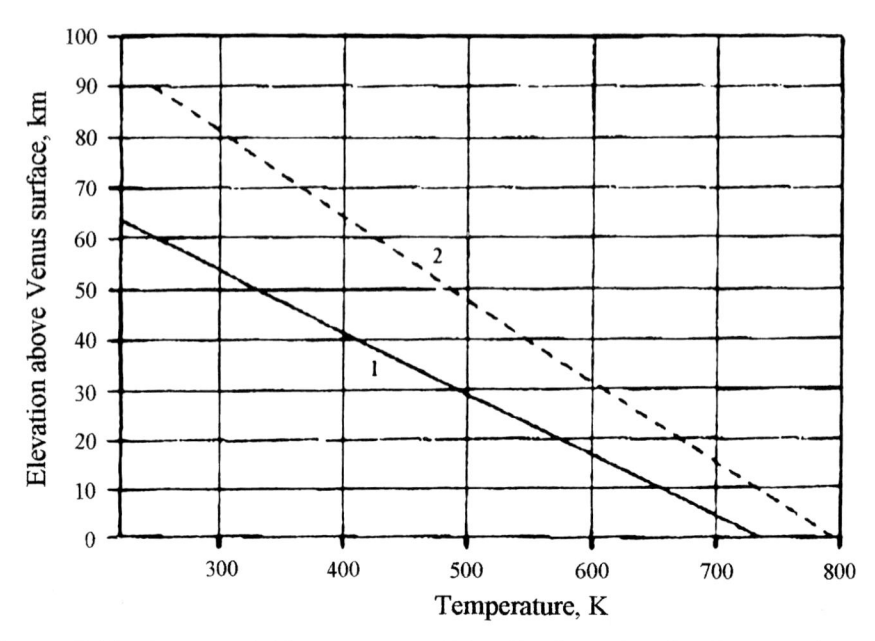

Figure 7.10: Averaged Temperature Distributions in the Troposphere of Venus. (1) Present-Day Carbon Dioxide Troposphere and (2) Hypothetical Model of Nitrogen–Oxygen Atmosphere at the Same Physical Conditions.

To generate gases, the upper silicate mantle matter must be in a plastic (semi-molten) state (Sorokhtin and Ushakov, 2002). The mass of this plastic matter has to be sufficient to develop vertically oriented convective flows in the mantle. The mass of mantle constitutes about 67% of the Earth's total mass of 5.979×10^{27} g (Britannica Student Encyclopedia, 2005). To raise a temperature of this huge mass of mantle matter to the melting point, enormous amount of heat should be generated. What are the energy sources that can generate this heat? The detailed discussion of this question is presented in Chapter 1.

Widespread belief that the necessary energy is supplied by the process of radioactive decay of physically unstable elements cannot pass a simple verification test: the estimated total heat flux through the Earth's surface many times surpasses the heat flux that can be generated in the process of radioactive decay of the entire amount of radioactive matter of Earth. By estimates of Vacquier (1990), the heat from radioactive decay of the entire radioactive matter of our planet can supply at most 25% of the observed total heat flux through the Earth's surface. In addition, the main mass of radioactive matter (about 90% of the Earth's total) is concentrated in the upper layers of continental crust (Khain, 2003). This means that the heat generated by radioactive decay cannot play a leading role in driving the matter–energy transformations in deep layers of the mantle. For example, it is absolutely insufficient to generate energy for heating a huge mass of mantle matter (approximately 4.006×10^{27} g) to the temperatures ranging from 100 °C (at the top of the mantle) to 2,700 °C (at the bottom) and to originate and sustain the mantle matter convection. Therefore, one needs to identify a significantly greater source of energy to drive all physical and chemical transformations in the mantle.

This source of energy was first identified and described by Sorokhtin (1972a, b) and Dubrovskiy and Pankov (1972) as the gravitational matter differentiation that results in heat generation in the process of redistribution of Earth's matter over the terrestrial body throughout geologic history. This redistribution occurs through chemical and physical transformations, driving denser matter toward the center of gravity of Earth. This process is a manifestation of universal principle of minimum potential energy for a closed system: in the Earth's body, a denser matter migrates toward the center of gravity to minimize the Earth's body potential energy.

Gravitational matter differentiation supposedly had started in the Archaean time (about 4 billion years ago) with zonal melting of an original Earth's matter (Sorokhtin and Ushakov, 2002) that was caused by the heat inherited from the protoplanetary "disk" and also accumulated during the accretion period. According to the most likely and popular hypothesis, the Earth's celestial body was formed as a result of homogeneous accretion of cold matter from the protoplanetary cloud (Safronov, 1972). This cold matter was continuously heated by numerous impacts of planetesimals in the process of matter accretion, releasing huge amount of heat. As a result, the temperature in some peripheral zones of the protoplanetary blob was raised above the melting point of iron oxides, forming iron pools in the "subsurface areas" of the protoplanetary "conglomerate".

One can get an idea about the amount of heat generated on these impacts using data from a NASA deep impact probe experiment (http://www.spacetoday.org). The NASA copper-clad projectile was relatively small (mass of about 372 kg). It dissipated energy equivalent to the energy of explosion of 5 tons of TNT on impact with Comet Tempel 1 on July 4, 2005. The impact instantly increased the temperature at the area of collision by several thousand degrees.

The early Earth's matter zonal differentiation during the Archaean Eon resulted in extreme gravitational instability of the planetary body: a high-density peripheral ring layer consisting of molten iron was formed due to zonal melting and was positioned above a lighter material of core. This instability could be resolved in only one way: by squeezing out the core matter through the equatorial belt to the Earth's peripherals and sinking of the heavy molten iron to the center of planetary body, gradually forming an iron core. This process was accompanied by a huge energy release (about 5.5×10^{37} erg; Sorokhtin and Ushakov, 2002), which was used mostly for additional heating of Earth's interior, accelerating the core separation process.

After separation of the Earth's mater into two parts (metallic core and silicate peripherals), the process of squeezing out of heavy iron oxides from the mostly silicate "mantle" into metallic core accelerated due to baric diffusion–differentiation mechanism. This is the most productive mechanism of gravitational matter differentiation. Under high pressures and temperatures at the mantle bottom (about 0.89 Mbar and 2,700 K at a depth of 2,000 km), the atoms of iron and oxygen diffuse through the crystalline grid of solid silicate compounds and form iron oxides, which accumulate in the Earth's core. This process of baric diffusion (or barodiffusion) is accompanied by huge release of heat energy (151.6 cal per one gram of released Fe_2O at the bottom of lower mantle; Sorokhtin and Ushakov, 2002). Therefore, gravitational matter differentiation in the interior of Earth is a major source of Earth's inner heat.

In addition to energy of radioactive decay and gravitational matter differentiation, the Earth's interior is also heated by dissipation of tidal energy caused by gravitational influence of Moon (solar gravitational influence is insignificant and can be neglected). Kinetic tidal energy transforms into tectonic heat of the Earth as a result of inner matter friction in tidal humps, which run around the globe and deform the Earth's body (Khain, 2003).

At the present time, the bulk of tidal energy is released in shallow seas with a considerably smaller part released in deep oceans and the Earth's asthenosphere. Current rate of tidal energy release is estimated at 2.5×10^{19} erg/s (for comparison, the total heat flux through the Earth's surface is presently 4.3×10^{20} erg/s). Two-thirds of this energy is dissipated in the shallow seas due to friction between the bottom currents and seafloor (MacDonald, 1964, 1965). The part of tidal energy dissipated in the mantle does not exceed 6% of the total released tidal energy. This means that the impact of dissipated tidal energy on the Earth's interior heating is insignificant, and that the main sources of tectonic Earth's heat are the gravitational matter differentiation and radioactive decay.

7.7. Smoothing Role of World Ocean

The mantle degassing began in the Archaean time (4 billion years ago). One of the main constituents of degassed mixture was water vapor. After accumulation of sufficient amount of water vapor in the atmosphere (under appropriate atmospheric pressure and temperature conditions), the first aquatic basins were created, which (about 3.6 billion years ago) developed into a unified shallow ocean. As a result, the Earth's global surface temperature began to increase rapidly (Sorokhtin and Sorokhtin, 2002).

Oceans influence the Earth's climate significantly, distributing the heat over the Earth's surface. They effectively absorb, transport, and transfer large amounts of thermal energy, changing the land climates (Bigg, 1996). Huge mass of water in the contemporary World Ocean (approximately 1.37×10^{24} g) smoothes and slows down all processes of heat and mass transfer between the mantle and the atmosphere and accumulates the excess of solar energy. The top 200 m of the World Ocean store 30 times more heat than the atmosphere [http://www.ace.mmu.ac.uk/eae/Climate.html]. Presently observed global atmospheric heating was also accompanied by warming of oceanic water. During the 40-year period from 1969 to 2000, the upper 300 m of the World Ocean warmed on the average by 0.56 °F (NOAA News Online, 2000).

Global oceanic currents transfer heat from the equatorial warmer regions of the ocean to the cold polar waters and thus mix these waters, reducing oceanic and atmospheric temperature differences between various climatic zones. Therefore, the distribution of continents and oceans around the Earth's surface influences the climate significantly (the second-order climate driver in Gerhard's classification: Gerhard et al., 2001).

Contemporary ocean water contains about 90 times more carbon dioxide than the atmosphere. When the global temperature rises, the solubility of carbon dioxide in the ocean water decreases and part of the carbon dioxide content of ocean water is transferred into the atmosphere, creating an illusion that the increased concentration of carbon dioxide is heating the atmosphere as a result of anthropogenic activity (Sorokhtin, 2001a, b; Khilyuk and Chilingar, 2003, 2004). All land plants produce food from photosynthesis that utilizes carbon dioxide and water in the presence of solar radiation. Oceanic plankton consumes carbon dioxide for photosynthesis and building of shells. Absorbed carbon dioxide is replaced by the gas of magmatic origin and by the atmospheric gas (if the global temperature drops).

Interplay of the energy and matter fluxes between the World Ocean and the atmosphere has a pronounced mitigating effect on the Earth's climate. Because of the World Ocean, one can talk about long-term climatic patterns. Otherwise, all the climatic changes would occur much sharper and faster, and it would be impossible for the forces of nature to determine and sustain long-term trends (changes) in the average global temperature that determine the Earth's climate.

7.8. Microbial Activity at Earth's Surface

From all the live species, only bacteria can directly influence the mass and content of the gaseous mixture of the Earth's atmosphere. Considerable changes in concentrations and contents of various gases composing air can alter the Earth's climate. Recent discoveries in microbiology revealed a broad proliferation of methane-generating and methane-consuming bacteria and also (which is even more important) abundance of denitrifying and nitrogen-consuming bacteria. Therefore, the interplay of these "microbial" forces should affect the Earth's climate considerably.

Methane gas entering the atmosphere from such natural sources as (1) wetlands, (2) rice paddies, (3) cattle digestion, (4) termite digestion, and (5) marine organisms, is actually generated by bacteria. There were few studies on methane-generating bacteria in spite of the fact that life activity of these bacteria considerably contribute to the total balance of methane gas in atmosphere. Among important research works on methane-generating bacteria, one should mention the Cornell University study initiated in 2002 with the support of National Science Foundation (www.news.cornell.edu/Chronicle/02/2.14.02/microbial_observatory.html). The purpose of this study was to extract from the wetlands (dominant source of natural methane) and artificially grow the bacteria, and to determine the impact of environmental conditions on the life of bacterial colonies.

Recently, a team of researchers (from Penn State University and University of California, Los Angeles) discovered vastly proliferated methane-consuming archaeobacteria (Orphan et al., 2001, 2002). These bacteria consume most of the methane gas at the bottom of World Ocean, preventing the methane accumulation in atmosphere.

Microbial activities play an important role in the transit of nitrogen from the atmosphere to the Earth's surface and vice versa. Only bacteria are able to consume and accumulate nitrogen directly from atmosphere. They fix it to organic molecules, producing proteins (nitrogen fixation process). All other live species consume and accumulate nitrogen via food chain, from proteins produced by bacterial activity in the process of nitrogen fixation.

Bacteria living on the roots of some plants can fix nitrogen by producing proteins (www.starsandseas.com/SAS%20Ecology). Animals can get necessary nitrogen by eating plants. The remains of plants and animals can be buried in the oceanic sediments, transferring nitrogen into solid matter of Earth. It is not clear yet what is the scope of life activities of such bacteria.

Recently, microbiologists discovered new microbial species consuming nitrogen: spirochetes (corkscrew bacteria) living in termite guts and fresh and saline waters (Lillburn et al., 1999, 2001). It is noteworthy that world total termite biomass is of the same order of magnitude as the biomass of Antarctic krill. The latter accounts for 2–5 times the total human biomass (http://answers.google.com/answers). Huge population and vast proliferation of termites worldwide make spirochetes, living in the guts of termites and generating proteins, a major supplier of nitrogen to the Earth's live matter.

Some soil bacteria, however, consume NO_3^- (nitrates) and convert them into nitrogen, returning the nitrogen gas into atmosphere. This process is called denitrification. Therefore, the increase or reduction of nitrogen content in atmosphere may be determined by the balance between bacterial activities in the processes of nitrogen fixation and denitrification.

7.9. Glacial and Interglacial Periods

Glacial periods are common phenomena throughout the past geologic periods. They occur about once in 100,000 years. During the ice ages, large, thick shields of ice covered the entire continents. This had a pronounced impact on the Earth's climate changing the sea level (total oceanic water volume), climatic belts, and distribution of biota throughout the world.

Traditional explanation of the development of ice ages in the past geologic periods is based on the Milankovitch cycles (http://earthobservatory.nasa.gov/Library/Giants/Milankovitch) – periodic variations in Earth's insolation due to semi-cyclic variations in the Earth's orbital parameters (orbital eccentricity, obliquity, and precession of Earth's axis). These parameters exhibit semi-cyclic changes with periods ranging from 0.1 to 2 million years. Corresponding changes in temperature at the Earth poles can reach up to 15 °C (Sorokhtin, 2005).

These periodic variations in the orbital parameters are well correlated with the occurrence of glacial periods (http://geosciences.ou.edu/). They contradict, however, the irregular, increasingly more frequent occurrence of main ice ages in the geologic past during the last billion years (Chumakov, 2004). Consequently, one needs additional explanations (to oscillating insolation) for the pronounced drop in global atmospheric temperature.

According to the adiabatic theory of heat transfer in the atmosphere, atmospheric temperature drop is determined by the decrease in atmospheric pressure. The latter implies that there are some global mechanisms reducing the partial pressures of the Earth's atmospheric constituents.

One of the proposed hypotheses (http://geosciences.ou.edu/) on the causes of pronounced atmospheric pressure reduction leading to major ice ages is a persistent drop in atmospheric carbon dioxide content during the Earth's geologic past. Possible mechanisms include long-term weathering of silicates and rapid burial of large amounts of organic matter, preventing contained carbon from being oxidized. "This might have occurred about 300 million years ago when large coastal swamps were buried" that led to a significant reduction of the atmospheric carbon dioxide content and consequent temperature drop (http://geosciences.ou.edu/).

One can easily refute this hypothesis by noting that, at that time, the atmospheric CO_2 content was very low (its partial pressure did not exceed 1 matm, and the global atmospheric pressure and temperature were close to their present values (Sorokhtin and Sorokhtin, 2002). Even if the entire content of CO_2 was removed from the atmosphere at that time, this could not have led to considerable temperature drop necessary for the initiation of ice age. Corresponding temperature drop can be conveniently estimated using the adiabatic model. At sea level, the temperature change can be computed using the following formula (Khilyuk and Chilingar, 2003):

$$\Delta T \approx T\alpha\Delta p \tag{7.10}$$

where Δp is the atmospheric pressure change in atm; T and ΔT are the global temperature and temperature change at sea level, respectively; and α is the adiabatic exponent. After substitution of $T = 288\,K$, $\alpha = 0.1905$, and $\Delta p = 10^{-3}$ atm into Eq. (7.10), one obtains $\Delta T \approx 5.49 \times 10^{-2}\,°C$. This insignificant global atmospheric temperature drop (approximately $0.055\,°C$) obviously could not initiate an ice age.

The main constituent of Earth's atmosphere during the last 2.5 billion years was (and continues being) nitrogen. The present partial pressure of nitrogen is about 0.755 atm. Three hundred million years ago, the nitrogen content was higher than the present level. Its partial pressure was approximately 0.95 atm, at the total atmospheric pressure of about 1.1 atm (Sorokhtin, 2005). This means that only pronounced persistent drop in the atmospheric nitrogen content can explain a nonperiodic ice age occurrence. In other words, certain global mechanisms must exist in the Earth's system transforming gaseous nitrogen of the atmosphere into solid matter, which is buried in the sediments.

One of the most plausible explanations is the high nitrogen consumption by the nitrogen-consuming bacteria. In this way, the gaseous nitrogen of the atmosphere can be transferred into the sediments, reducing the partial pressure of nitrogen (and, thus, the total atmospheric pressure) and lowering global temperature of Earth. Considering this process to be irreversible, Sorokhtin (2005) presented the hypothesis on bacterial nature of ice ages. Having used simple exponential model of nitrogen degassing and having estimated the total content of nitrogen in Earth's system, he developed plausible scenarios of evolution of partial atmospheric nitrogen pressure and corresponding global temperature changes throughout the geologic history (Figs. 7.3 and 7.8). In these scenarios, all the glacial periods that occurred since 900 million years ago were attributed to nitrogen-consuming bacteria.

The bacterial hypothesis gains additional support after discovery of nitrogen fixation by symbiotic and free-living (in fresh and saline water) spirochetes (Lillburn et al., 1999, 2001). This process of nitrogen transfer from the atmosphere to Earth's crust, however, is reversible. For example, some soil bacteria consume nitrates and convert them into gaseous nitrogen (www.starsandseas.com/SAS%20Ecology). In this process of denitrification, nitrogen returns back into the atmosphere. To verify the bacterial hypothesis, one needs to estimate the balance between "nitrogen-consuming" and "nitrogen-generating" bacterial activities and prove that the consumption prevails. This problem, which is very important in the context of evaluation of causes and trends of global climatic changes, is yet to be solved.

At present, the research on the bacterial transport of chemical elements and compounds through the Earth's system is in the embryonic stage. Characterizing the level of knowledge on methane-generating bacteria, for example, Segelkin (2002) noted: "They've been at it for millions of years, but practically nothing is known about wetland bacteria that turn organic matter into the greenhouse gas, methane" (www.news.cornell.edu/Chronicle/02/2.14.02/microbial_observatory.html). Microbiologists should culture and grow the bacteria, which generate and consume atmospheric gases, and study the impact of various environmental conditions on growth and decay of microbial colonies and their proliferation.

7.10. Global Warming or Global Cooling?

Do we live in the time of global warming or global cooling? The answer to this question depends on the time span of observed atmospheric changes. The latest 150 years were (and the near future will probably be) a period of global warming (Fig. 7.11). The major causes of currently observed global warming are: rising solar irradiation and increasing tectonic activity. If one

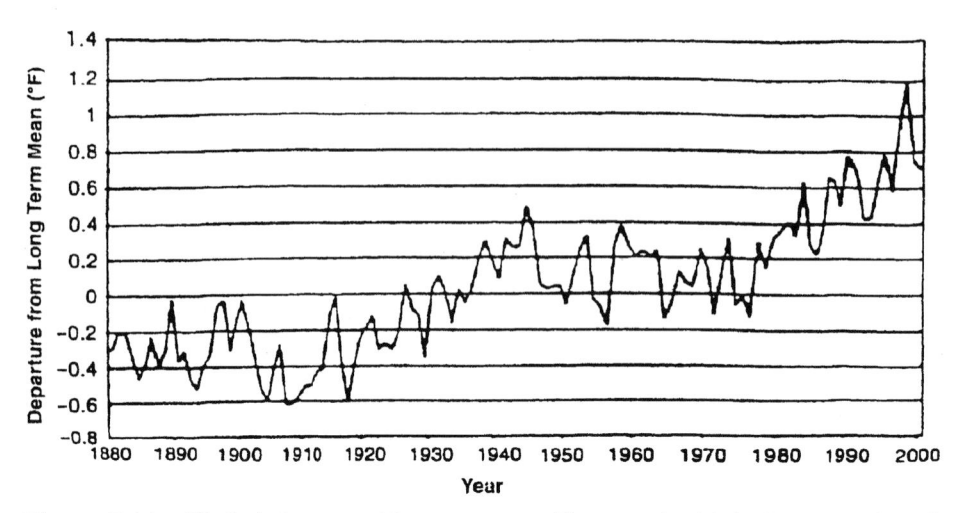

Figure 7.11: Global Average Temperature Changes in 20th Century. (Modified after EPA Global Warming Site: Climate. U.S. National Climatic Data Center, 2001.)

considers the last three millennia, then one can observe a cooling trend in the Earth's climate (Keigwin, 1996; Hoyt and Schatten, 1997; Sorokhtin and Sorokhtin, 2002; Sorokhtin and Ushakov, 2002; Gerhard 2004; Khilyuk and Chilingar, 2004, 2006). During that period, the global temperature deviations were 3 °C with a clear trend of decreasing global temperature by about 2 °C (Figs. 7.12 and 7.13).

This trend started at least 7,000 years ago. Friedman (2005) studied the oxygen isotope composition of Gulf of Aqaba beachrock (carbonate cement). He showed that the temperature of ambient seawater decreased for approximately half of the Holocene. According to Friedman (2005), the average Red Sea water temperature decreased from 33 to 17 °C over an interval of approximately 4,500 years – between the ages of 7.07 ± 0.38 and 2.65 ± 0.23 ka. Thus, the general tendency in the Red Sea area (and over the entire Earth's surface) has been toward global cooling. Combining the Friedman's findings with those of Keigwin (1996) (see Fig. 7.12), one can conclude that the latest cooling geologic time comprises most of the Holocene. Inasmuch as the deduced temperature change occurred during the time of relatively stable concentration of carbon dioxide in the atmosphere, the changes in its content were not driving the Earth's global temperature (Jenkins, 2001).

The cooling trend will probably last in the future (Fig. 7.8). This means that we live in the *cooling geologic time* and the global warming observed during the last approximately 150 years is just a short episode in the geologic

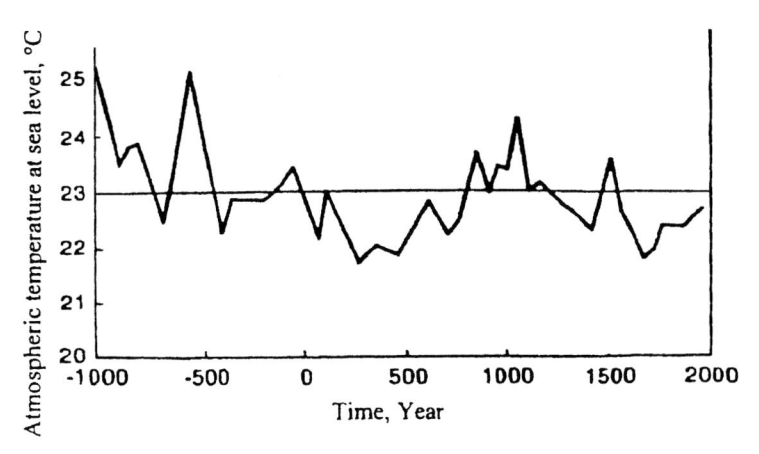

Figure 7.12: Surface Temperature in the Sargasso Sea (with Time Resolution of about 50 years) as Determined by Isotope Ratios of Remains of Marine Organisms Buried in the Sediments of Seafloor. Horizontal Line Represents the Average Temperature over a 3,000-year Period (Modified after Keigwin, 1996).

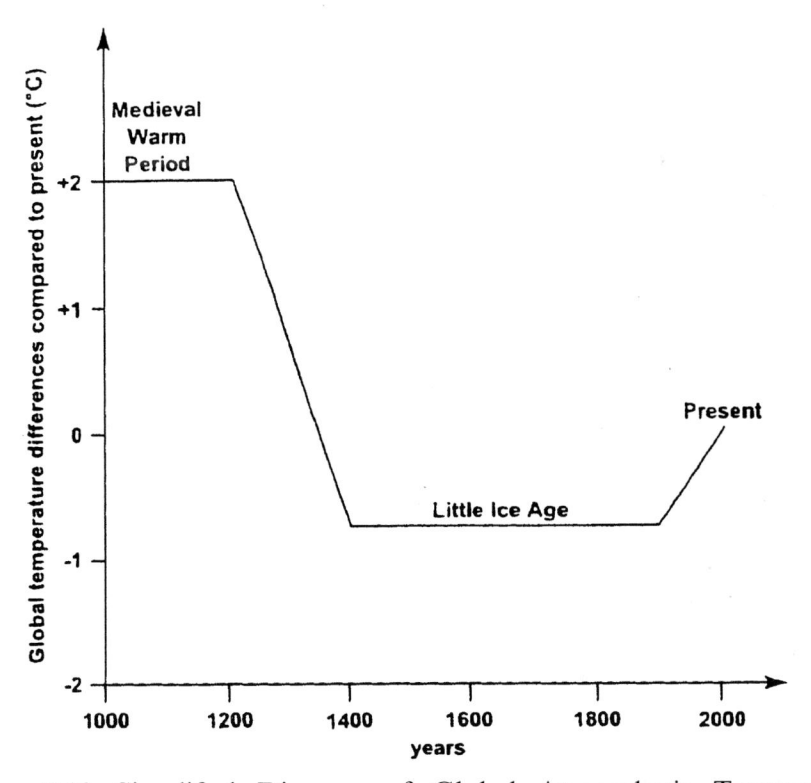

Figure 7.13: Simplified Diagram of Global Atmospheric Temperature Change over the Last 1,000 Years (Modified after Gerhard, 2004).

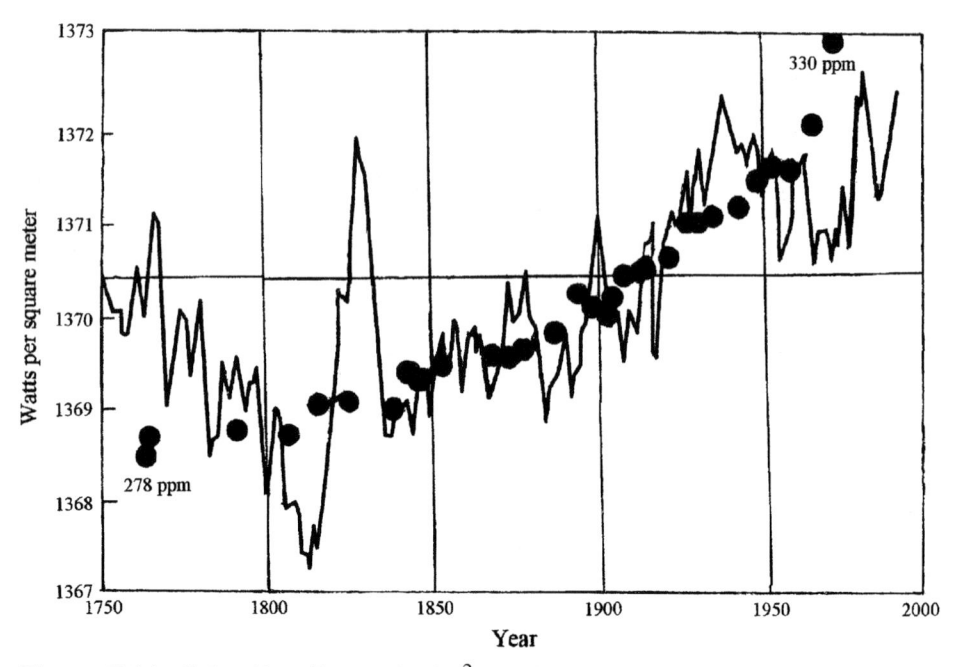

Figure 7.14: Solar Irradiance (W/m²) and CO_2 Concentrations in Atmosphere (ppmv) in Northern Hemisphere (Modified after Hoyt and Schatten, 1997).

history. The cooling period will probably last in the future due to life-activities of the nitrogen-consuming bacteria that transfer the atmospheric nitrogen into soils with subsequent burial in the sediments.

Finally, if we move our horizon two–three billion years forward, we should expect the atmospheric overheating to up to 300 °C due to catastrophic increase of the relative content of oxygen in atmosphere. This oxygen will be degassed out of the iron oxides as a result of chemical reactions accompanying gravitational matter differentiation in the Earth's inner layers.

7.11. Conclusions

The authors identified and described the global forces of nature driving the Earth's climate: (1) solar irradiation as a dominant energy supplier to the atmosphere (and hydrosphere); (2) outgassing as a dominant gaseous matter supplier to the atmosphere; and (3) microbial activities at the interface of lithosphere and atmosphere. The scope and extent of these processes are ≈ 4 orders of magnitude greater than the corresponding anthropogenic impacts (such as emission of the greenhouse gases) on Earth's climate.

Inspection of the global atmospheric temperature changes during the last 1,000 years (Fig. 7.13) shows that the global average temperature dropped by about 2 °C over the last millennium. This means that we live in the cooling geologic period (which comprises most of the Holocene), and the global warming observed during the latest 150 years is just an insignificant episode in the geologic history. The current global warming is most likely a combined effect of increased solar and tectonic activities and cannot be attributed to the increased anthropogenic impact on the atmosphere.

The anthropogenic impact on the global atmospheric temperature is negligible (Sorokhtin, 2001a, b; Khilyuk and Chilingar, 2003, 2004, 2006). Considering the relationship between the carbon dioxide concentration and global atmospheric temperature, one should understand that (according to the adiabatic theory) the saturation of atmosphere with CO_2 could lead only to its cooling (but not to warming). One should also keep in mind the fact that peaks in the Sun irradiation always precede the peaks of CO_2 concentration in the atmosphere (Fig. 7.14), which supports the hypothesis that temperature changes cause the corresponding changes in CO_2 concentration in the atmosphere (Sorokhtin, 2001a, b; Khilyuk and Chilingar, 2003).

The global natural processes drive the Earth's climate: "Climate will change, either warmer or colder, over many scales of time, with or without human interference" (Gerhard, 2004). Any attempts to mitigate undesirable climatic changes using restrictive regulations are condemned to failure, because the global natural forces are at least 4 orders of magnitude greater than the available human controls. In addition, application of these controls will lead to catastrophic economic consequences. Estimates show (http://www.JunkScience.com) that since its inception in February 2005, the Kyoto Protocol has cost about $10 billion a month, supposedly averting about 0.0005 °C of warming by the year 2050. Thus, the Kyoto Protocol is a good example of how to achieve the minimum results with the maximum efforts (and sacrifices). Impact of available human controls is negligible in comparison with the global forces of nature driving the Earth's climate. Thus, the attempts to alter the occurring global climatic changes (and drastic measures prescribed by the Kyoto Protocol) have to be abandoned as meaningless and harmful. Instead, moral and professional obligation of all responsible scientists and politicians is to minimize potential human misery resulting from oncoming global climatic changes.

REFERENCES

Aliven, H., & Arrhenius, G. (1979). *Evolution of Solar System*. Mir, Moscow, 511 pp.
Aller, L. (1976). *Atoms, Stars and Nebulas*. Mir, Moscow, 352 pp.
Alpher, R. A., & Herman, R. (1988). Reflections on early work on "big bang" cosmology. *Physics Today*, **August** (8), 24–34.
Al'tshuller, L. V., Simakov, G. V., & Trunin, R. F. (1968). On the chemical composition of the Earth's core. *Physics of Earth*, **1**, 3–6.
Anatol'eva, A. I. (1978). *The Milestones of Evolution of Red-bed Formations*. Nauka, Novosibirsk, 190 pp.
Attenborough, D. (1979). *Life on Earth*. Oxford University Press, London, 399 pp.
Balanyuk, I. E., & Dongaryan, L. Sh. (1994). The role of hydrothermal methane in formation of gas-hydrate deposits. In: *Geology, Geophysics, and Development of Petroleum Deposits*, Vol. 3, pp. 22–28.
Balanyuk, I. E., Matvienkov, V. V., & Sedov, A. P. et al. (1995). Serpentization of the upper mantle rocks of the oceans as a source of hydrocarbon formation. In: *Geologic Studies and Development of the Earth's Resources*. Russian Federation Committee on geology and development of Earth's resources, Moscow, pp. 34–40.
Baliunas, S. L. (2002). The Kyoto Protocol and global warming. *Imprimis*, **31**(3), 1–7; http://www.hillsdale.edu/.
Baliunas, S. L., & Saar, S. (1992). Unfolding mysteries of stellar cycles. *Astronomy*, **May**, 42–47.
Baliunas, S. L., & Soon, W. (1996). The sun–climate connection. *Sky and Telescope*, **12**, 38–41.
Barron, E. J. (1994). *Climatic Variations in Earth History. Understanding Global Climate Change*: *Earth Science and Human Impact*, Global change instruction program #108, USAR, 23 pp.
Bigg, G. R. (1996). *The Oceans and Climate*. Cambridge University Press, UK, 266 pp.
Bluemle, J. P., Sable, J., & Karlen, W. (2001). Rate and magnitude of past global climate changes. In: L. C. Gerhard, W. E. Harrison, & B. M. Hanson (Eds). *Geological Perspective of Global Climate Change*, Vol. 47. AAPG Studies in Geology, AAPG, Tulsa, OK, pp. 193–212.
Boehler, R. (1993). Temperatures in the Earth's core from melting-point measurements of iron at high static pressures. *Nature*, **363** (10), 534–536.
Boehler, R. (1996). Experimental constrains on melting conditions relevant to core formation. *Geochimica et Cosmochimica Acta*, **60** (7), 1109–1112.
Bullen, K. E. (1973). Cores of terrestrial planets. *Nature*, **243**, 68–70.
Bykhover, N. A. (1984). *The Distribution of World Mineral Deposits over the Epochs of Formation of Ores*. Nedra, Moscow, 586 pp.
Caldeira, K., & Wickett, M. (2003). Anthropogenic CO_2 and ocean pH. *Nature*, **425** (6956), 365–366.

Chumakov, N. M. (1978). *Precambrian Tillites and Tilloids*. Nauka, Moscow, 202 pp.
Chumakov, N. M. (2001). General tendency of the Earth's climate changes during the past 3 billion years. *Reports of the Russian Academy of Sciences*, **381** (5), 652–655.
Chumakov, N. M. (2004). Glacial and interglacial climate in Precambrian. In: *Climate during Global Biosphere Changes*. Nauka, Moscow, 299 pp.
Darwin, G. G. (1965). *The Tides and Similar Phenomena in the Solar System*. Nauka, Moscow, 252 pp.
Dmitriev, L. V., Bazylev, B. A., & Borisov, M. B. (2000). The formation of hydrogen and methane as a result of serpentization of mantle hyperbasic rocks of ocean and the origin of oil. *Russian Journal on Earth's Sciences*, **1** (1), 1–16.
Dormand, J., & Woolfson, W. (1989). *The Origin of the Solar System*. Prentice Hall, NY, 399 pp.
Dowson, J. (1983). *Kimberlite and Xenolite Intrusions*. Mir, Moscow, 300 pp.
Dubrovskiy, V. A., & Pankov, V. A. (1972). On the structure of Earth's core. *Izvestiya AN SSSR, Physics of Earth*, **7**, 48–54.
EPA global warming site (2001). Climate: http://www.epa.gov/globalwarming/climate/index.htm.
Exley, R. A., Mattey, D. P., Cague, D. A., & Pillinger, C. T. (1986). Carbon isotope systematics of a mantle "hotspot": A comparison of Loihi Seamount and MORB glasses. *Earth and Planetary Science Letters*, **78**, 189–199.
Fore, G. (1989). *Foundations of Isotopic Geology*. Mir, Moscow, 590 pp.
Fox, S. W. (1965). A theory of macromolecular and cellular origins. *Nature*, **205**, 328–340.
Frakes, L. A. (1979). *Climates through Geologic Time*. Elsevier, Amsterdam, 310 pp.
Friedman, G. M. (2005). Climate significance of Holocene benchrock sites along shorelines of the Red Sea. *American Association of Petroleum Geologists Bulletin*, **89** (7), 849–852.
Fryer, B. J. (1977). Rare Earth evidence in iron formations for changing Precambrian oxidation states. *Geochimica et Cosmochimica Acta*, **41**, 361–367.
Galimov, E. M. (1978). The origin of diamonds in the light of new data on isotope carbon composition of diamond. In: *Transactions of VII All Union Symposium on the Stable Isotopes in Geochemistry*. GEOCHE, Moscow, pp. 110–113.
Galimov, E. M. (2001). *The Phenomenon of Life: Between Equilibrium and Linearity. Origin and Principles of Evolution*. URSS Editorial, Moscow, 256 pp.
Gamow, G. (1948). The origin of elements and separation of galaxies. *Physical Review*, **74**, 505–511.
Gans, R. F. (1972). Viscosity of Earth's core. *Journal of Geophysical Research*, **77**, 210.
Garrels, R., & McKenzie, F. (1974). *The Evolution of Sedimentary Rocks*. Mir, Moscow, 272 pp.
Gavrilov, V. P. (1986). *The Origin of Oil*. Nedra, Moscow, 176 pp.
Gerhard, L. C. (2004). Climate change: Conflict of observational science, theory, and politics. *American Association of Petroleum Geologists Bulletin*, **88** (9), 1211–1220.

Gerhard, L. C., Harrison, W. E., & Handon, B. M. (2001). Introduction and overview. In: L. C. Gerhard, W. E. Harrison, & B. M. Hanson (Eds). *Geological Perspective of Global Climate Change.* AAPG Studies in Geology, **47**, 3–15.

Goldich, S. S. (1975). The age of Precambrian stratified iron-ore formations. In: *Precambrian Iron-Ore Formations of the World.* Mir, Moscow, pp. 286–297.

Goldreih, P. (1975). The history of Moon orbit. In: *The Tides and Resonance in the Solar System.* Mir, Moscow, pp. 97–129.

Hale, C. J. (1987). Paleomgnetic data suggest link between Archaean–Proterozoic boundary and innercore nucleation. *Nature*, **329** (6136), 233–236.

Holland, H. D. (1984). *The Chemical Evolution of the Atmosphere and Oceans.* Princeton University Press, Princeton, 552 pp.

Horrell, M. A. (2003). Finding the solar constant and average temperature of a planet; http://www.imsa.edu/edu/geophysics/atmosphere/energy/blackbody2.html.

Hoyt, D. V., & Schatten, K. H. (1997). *The Role of the Sun in Climate Changes.* Oxford University Press, New York, 279 pp.

Jenkins, D. A. L. (2001). Potential impact and effect of climatic changes. In: L. C. Gerhard, W. E. Harrison, & B. M. Hanson (Eds). *Geological Perspective of Global Climate Change*, Vol. 47. AAPG Studies in Geology, pp. 337–359.

Joley, D. (1929). *The History of Earth's Surface. The Contemporary Problems of Natural Sciences*, Vol. 43, 190 pp.

Kaula, W. M. (1971). *Introduction to the Physics of Planets of Earth Group.* Mir, Moscow, 536 pp.

Kaula, W. M., & Harris, A. W. (1973). Dynamically plausible hypotheses of lunar origin. *Nature*, **245**, 367–369.

Keeling, C. D., & Whorf, T. P. (1997). Possible forcing of global temperature by oceanic tides. *Proceedings of the National Academy of Sciences*, **94**, 8321.

Keigwin, L. D. (1996). The little ice age and medieval warm period in the Sargasso Sea. *Science*, **274**, 1504–1508.

Khain, V. E. (2001). *Tectonics of Continents and Oceans.* Nauchnyy Mir, Moscow, 606 pp.

Khain, V. E. (2003). *Main Problems of Contemporary Geology.* Nauchnyy Mir, Moscow, 346 pp.

Khain, V. E., & Bozhko, N. A. (1988). *Historical Geotectonics, Precambrian Time.* Nedra, Moscow, 382 pp.

Khilyuk, L. F., Chilingar, G. V., Endres, B., & Robertson, J. (2000). *Gas Migration.* Gulf Publishing Company, Houston, 389 pp.

Khilyuk, L. F., & Chilingar, G. V. (2003). Global warming: Are we confusing cause and effect?. *Energy Sources*, **25**, 357–370.

Khilyuk, L. F., & Chilingar, G. V. (2004). Global warming and long-term climatic changes: A progress report. *Environmental Geology*, **46**, 970–979.

Khilyuk, L. F., & Chilingar, G. V. (2006). On global forces of nature driving the Earth's climate. Are humans involved?. *Environmental Geology*, **50**, 899–910.

Kondratiev, K. Ya. (1992). *Global Climate.* Nauka, Moscow, 359 pp.

Kuleshov, V. N. (1986). *The Isotopic Composition and Origin of Deep-Seated (Plutonic) Carbonatites*. Nauka, Moscow, 126 pp.
Kyoto Protocol to the United Nations Framework Convention on Climate Change, 1997; http://www.epa.gov/globalwarming/publications/reference/index.html.
Lang, C., Leuenberger, M., Schwander, J., & Johnson, S. (1999). 16°C rapid temperature variation in central Greenland 70,000 years ago. *Science*, **286**, 934–937.
Landscheidt, T. (2002). *New Little Ice Age Instead of Global Warming*. Schroeter Institute for Research in Cycles of Solar activity, Germany. thlandscheidt@t-online. de: 20 pp.
Lillburn, T. G., Kim, K. S., Ostrom, N. E., Byzek, K. R., Leadbetter, J. R., & Breznak, J. A. (2001). Nitrogen fixation by symbiotic and free-living spirochetes. *Science*, **292** (5526), 2495–2498.
Lillburn, T. G., Schmidt, T. M., & Breznak, J. A. (1999). Philogenetic diversity of termite gut spirochetes. *Environmental Microbiology*, **1** (4), 331–345.
Lisitsyn, A. P. (1984). The turbidite sedimentation, changes in the oceanic level, interruptions and pelagic accumulation of sediments – Global patterns. In: *Paleooceanology. Transactions of 27th International Geologic Congress*, Vol. 3. Nauka, Moscow, pp. 3–21.
Lisitsyn, A. P. (1993). Hydrothermal systems of the World Ocean – The main supplier of endogenic matter. In: *Hydrothermal Systems and Sedimentary Formations of Mid-Oceanic Ridges of Atlantic Ocean*. Nauka, Moscow, pp. 147–247.
Lisitsyn, A. P., Bogdanov, Yu. A., & Gurvich, E. G. (1990). *Hydrothermal Formations of the Oceanic Rift Zones*. Nauka, Moscow, 256 pp.
Luts, B. G. (1985). *Magmatism of Moving Belts of Early Earth*. Nedra, Moscow, 216 pp.
MacDonald, G. J. F. (1964). Tidal friction. In: *Tides and Resonances in the Solar System Reviews of Geophysics*, **2** (3), 467–541.
MacDonald, G. J. F. (1965). Geophysical deductions from observations of heat flow. In: *Terrestrial Heat Flow Chapter*. Academic Press, Washington, pp. 191–210.
MacDonald, D. J. F. (1975). Tidal friction. In: *Tides and Resonances in the Solar System*. Mir, Moscow, pp. 9–96.
Marland, G., Boden, T., & Andres, R. J. (2002). *Global CO_2 Emission from Fossil-Fuel Burning, Cement Manufacture, and Gas Flaring: 1751–2002*. Information Analysis Center, Oak Ridge National Laboratory, Tennessee. http://cdiac.ornl.gov/ftp/ndp030/global.1751 2002.ems.
Marov, M. Ya. (1986). *Planets of Solar System*. Nauka, Moscow, 320 pp.
Melton, C. E., & Giardini, A. A. (1974). The composition and significance of gas released from natural diamonds from Africa and Brazil. *American Mineralogy*, **59** (7–8), 775–782.
Melton, C. E., & Giardini, A. A. (1975). Experimental results and theoretical interpretation of gaseous inclusions found in Arkansas natural diamonds. *American Mineralogy*, **60** (5–6), 413–417.
Menard, G. W. (1966). *Geology of the Pacific Ocean Floor*. Mir, Moscow, 272 pp.
Meyen, S. V. (1971). The contemporary paleobotanics and the theory of evolution. *Priroda*, **2**, 48–57.

Meyen, S. V. (1981). *The Traces of Indian Grasses.* Mysl', Moscow, 159 pp.
Milankovitch, M. (1957). *Astronomische Theorie der Klimaschwankungen ihr Werdegang und Widerhall.* Serbian Academy of Science, Monograph 280, 58 pp.
Miller, S. (1959). Formation of organic compounds on the primitive Earth. In: *The Origin of Life on Earth.* Pergamon Press, London, pp. 123–135.
Mitchell, A. H., & Garson, M. S. (1984). *The Global Tectonic Position of Mineral Deposits.* Mir, Moscow, 496 pp.
Monin, A. S. (1977). *The History of Earth.* Nauka, Leningrad, 288 pp.
Monin, A. S. (1988). *Theoretical Foundations of Geophysical Hydrodynamics.* Hidrometeoizdat, Leningrad, 424 pp.
Monin, A. S. (1999). *Hydrodynamics of Atmosphere, Ocean, and the Earth's Bowels.* Hidrometeoizdat, Leningrad, 524 pp.
Monin, A. S., Seidov, D. G., Sorokhtin, O. G., & Sorokhtin, N. O. (1987a). Numerical modeling of the mantle convection. *Reports AN SSSR,* **295** (1), 58–63.
Monin, A. S., Seidov, D. G., Sorokhtin, O. G., & Sorokhtin, N. O. (1987b). Numerical experiments on the shapes of mantle convection. *Reports AN SSSR,* **295** (5), 1080–1083.
Monin, A. S., & Sorokhtin, O. G. (1981). On volumetric gravitational differentiation of Earth. *Reports AN SSSR,* **259** (5), 1076–1079.
Monin, A. S., & Sorokhtin, O. G. (1984). The evolution of oceans and geochemistry of continents. In: *Paleooceanology. Transactions of 27th International Geologic Congress,* Vol. K.03. Nauka, Moscow, pp. 22–35.
Monin, A. S., & Sorokhtin, O. G. (1986). On sucking in of sediments to great depths under the continents. *Reports AN SSSR,* **286** (3), 583–586.
Moralev, V. M., & Glukhovskiy, M. Z. (1985). Partial melting of metabasic rocks and the evolution of Precambrian lithosphere. *Reports AN SSSR,* **284** (2), 427–431.
Morelli, A., & Dziewonski, A. (1987). Topography of the core – Mantle boundary and lateral homogeneity of the liquid core. *Nature,* **235** (6106), 678–683.
Murbat, S. (1980). The boundary conditions of evolution of Archaean crust according the age and isotopic data. In: *The Early History of the Earth.* Mir, Moscow, pp. 358–366.
Naimark, L. M., & Sorokhtin, O. G. (1977a). The density distribution in the Earth's model with the herzolite mantle composition and iron oxide core. In: *Tectonics of Lithospheric Plates.* The Institute of Oceanology of AN SSSR, Moscow, pp. 28–41.
Naimark, L. M., & Sorokhtin, O. G. (1977b). The energy of gravitational differentiation of the Earth. In: *Tectonics of Lithospheric Plates.* Institute of Oceanology of AN SSSR, Moscow, pp. 42–56.
Nesis, K. N. (2001). When did the life appear in deep ocean?. *Priroda,* **4**, 68–69.
NOAA News Online (2000). *Story 399.* http://www.noaanews.noaa.gov/stories/htm.
Ohtani, E., & Ringwood, A. (1984). Composition of the core I. Solubility of oxygen in molten iron at high temperatures. *Earth and Planetary Science Letters,* **71** (10), 85–93.

Ohtsuki, K., & Ida, S. (1998). Planetary rotation by accretion of planetesimals with nonuniform spatial distribution formed by the planet's gravitational perturbation. *Icarus*, **131**, 393–420.

Oparin, A. I. (1957). *The Origin of Life*. Izd. AN SSSR, Moscow, 343 pp.

Oppo, D. W., McManus, J. F., & Cullen, J. L. (1998). Abrupt climate events 5,000,000 to 340,000 years ago: Evidence from subpolar North Atlantic sediments. *Science*, **279**, 1335–1338.

Oro, J. (1965). Investigation of organo-chemical evolution. In: *Current Aspects of Exobiology*. Pergamon Press, London, pp. 13–76.

Oro, J. (1966). Prebiological organic systems. In: *The Origin of Prebiological Systems*. Academic Press, NY, pp. 137–162.

Orphan, V. J., House, C. H., Hinricks, K. U., MaKeegan, K. D., & DeLong, E. F. (2001). Methane-consuming *archaea* revealed by directly coupled isotopic and phylogenetic analysis. *Science*, **293**, 484–487.

Orphan, V. J., House, C. H., Hinrichs, K. U., MaKeegan, K. D., & DeLong, E. F. (2002). Multiple archaeal groups mediate methane oxidation in anoxic cold seep sediments. *Proceedings of the National Academy of Sciences of the United States of America*, **99** (11), 7663–7668.

Pitman, W. C., & Heyes, J. D. (1973). Upper Cretaceous spreading rates and the great transgression. *Geological Society of American Abstract*, **5** (7), 2344–2357.

Planet Venus (the Atmosphere, Surface, Inner Structure) (1989). Nauka, Moscow, 482 pp.

Pugin, V. A., & Khitarov, N. I. (1978). *The Experimental Petrology of Deep Magmatism*. Nauka, Moscow, 175 pp.

Reed-Hill, R. E., & Abaschian, R. (1992). *Physical Metallurgy Principles*. PWS-Kent, 355 pp.

Ringwood, A. E. (1977). Composition of the core and implications for the origin of the Earth. *Geochemical Journal*, (11), 111–135.

Ringwood, A. E. (1982). *The Origins of Earth and Moon*. Nedra, Moscow, 175 pp.

Robinson, A. B., Baliunas, S. L., Soon, W., & Robinson, Z. W. (1998). *Environmental Effects of Increased Atmospheric Carbon Dioxide*. 8 pp; info@oism.org; info@marshall.org.

Ronov, A. B. (1993). *The Stratosphere, or the Sedimentary Shell of Earth*. Nauka, Moscow, 144 pp.

Ronov, A. B., & Yaroshevskiy, A. A. (1978). The chemical composition of Earth's crust and its shells. In: *Tectonosphere of Earth*. Nedra, Moscow, pp. 376–402.

Rubey, W. W. (1951). Geologic history of sea water. *Bulletin of Geological Society of America*, **62**, 1111–1147.

Rudenko, A. P., & Kulakova, I. I. (1986). The physicochemical model of abiogenic synthesis of hydrocarbons under natural conditions. *Journal of USSR Mendeleev Chemical Society*, **31** (5), 518–526.

Ruskol, E. L. (1975). *The Origin of Moon*. Nauka, Moscow, 188 pp.

Safronov, V. S. (1972). Accumulation of the planets. In: H. Reeves (Ed). *On the Origin of the Solar System*. Centre Nationale de Recherche Scientifique, Paris, pp. 89–113.

Sakai, H., Marais, D., Ueda, A., & Moore, J. (1984). Concentrations and isotope ratios of carbon, nitrogen, and sulfur in ocean-floor basalts. *Geochimica et Cosmochimica Acta*, **48** (12), 2433–2441.

Salop, L. P. (1973). *General Stratigraphic Scale of Precambrian Time*. Nedra, Leningrad, 310 pp.

Semikhatov, M. A., Raaben, M. E., Sergeev, V. N., Weis, A. F., & Artemova, O. B. (1999). The biotic events and positive isotopic anomaly of carbonate's carbon 2.3-2.06 billion years ago. In: *Stratigraphy. Geologic Correlation*. Nedra, Leningrad, pp. 3–27.

Shidlowski, M. (1987). Application of stable carbon isotopes to early biochemical evolution on Earth. *Annual Review of Earth and Planetary Sciences*, **15**, 47–72.

Shopf, T. (1982). *Paleooceanology*. Mir, Moscow, 311 pp.

Sinyakov, V. I. (1987). *Foundations of Orogenesis*. Nedra, Leningrad, 152 pp.

Slobodkin, A. I., & Eroscheev-Schak, V. A. (1995). The formation of magnetite by the thermophylic anaerobic microorganisms. *Reports AN SSSR*, **345** (5), 649–697.

Smirnov, V. I. (1982). *Geology of Mineral Deposits*. Nedra, Moscow, 669 pp.

Smith, A. G., & Briden, J. C. (1977). *Mesozoic and Cenozoic Paleocontinental Maps*. Cambridge University Press, London, 63 pp.

Sobolev, R. N., Pelymskiy, G. A., & Starostin, V. I. (1997). The development of molybdenum ores in the history of Earth. *Bulletin MOIP, Geology Section*, **72** (5), 65–71.

Sobolev, R. N., Starostin, V. I., & Pelymskiy, G. A. (2000). The types of tin deposits and their distribution in time and space. *MGU, Geology*, **1**, 30–39.

Sokolov, Yu. M., & Krats, O. K. (1984). Metallogenic impulses of endogenic activation of Earth's crust in the Precambrian time. In: *Metallogenics of Early Precambrian Time in the USSR*. Nauka, Leningrad, pp. 4–14.

Soon, W., Baliunas, S. L., Robinson, A. B., & Robinson, Z. W. (2001). *Global Warming – A Guide to the Science*. Risk Controversy Series 1, The Fraser Institute, Vancouver, 62 pp.

Sorokhtin, N. O. (2001). *Evolution of Continental Lithosphere in the Early Precambrian Time* (Using the Eastern Part of Baltic Shield as an Example). PhD Thesis, MGU, Moscow, 368 pp.

Sorokhtin, N. O., & Sorokhtin, O. G. (1997). The level of elevation of continents and possible nature of early Proterozoic ice age. *Reports RAN*, **354** (2), 234–237.

Sorokhtin, O. G. (1971). Possible physicochemical processes of formation of Earth's core. *Reports of RAN*, **198** (6), 1327–1330.

Sorokhtin, O. G. (1972a). The Earth's matter differentiation and development of tectonic processes. *Herald of AN SSSR (Physics of Earth)*, (7), 55–66.

Sorokhtin, O. G. (1972b). Differentiation of the Earth's matter and development of tectonic processes. *Bulletin AN SSSR, Physics of Earth*, **7**, 55–66.

Sorokhtin, O. G. (1974). *The Global Evolution of Earth*. Nauka, Moscow, 184 pp.

Sorokhtin, O. G. (1981). The structure of continental lithospheric plates and origin of kimberlites. In: *Problems of Theoretical Geodynamics and Tectonics of Lithospheric Plates*. Institute of Oceanology of AN SSSR, Moscow, pp. 161–168.

Sorokhtin, O. G. (1985a). *The Tectonics of Lithospheric Plates and Diamond-Bearing Kimberlites.* VIEMS, Moscow, 85 pp.
Sorokhtin, O. G. (1985b). The oceanic riftogenesis and the hypothesis of stretching Earth. In: *The Continental and Oceanic Riftogenesis.* Nauka, Moscow, pp. 121–136.
Sorokhtin, O. G. (1985c). *The Tectonics of Lithospheric Plates and Origin of Diamond-Bearing Kimberlites (General and Regional Geology).* VIEMS, Moscow, 47 pp.
Sorokhtin, O. G. (1988). The origin of Moon and the early stages of development of life. In: *The Life of Earth.* MGU, Moscow, pp. 5–24.
Sorokhtin, O. G. (1990). The greenhouse effect of atmosphere in geologic history of Earth. *Reports AN SSSR,* **315** (3), 587–592.
Sorokhtin, O. G. (1999). The early stages of development of the Earth–Moon system. *Izvestiya RAEN, Earth Sciences Section,* (2), 1–16.
Sorokhtin, O. G. (2001a). Greenhouse effect: Myth and reality. *Herald Russian Academy of Natural Sciences,* **1** (1), 8–21.
Sorokhtin, O. G. (2001b). Temperature distribution in the Earth. *Izvestiya RAN, Physics of Earth,* **3**, 71–78.
Sorokhtin, O. G. (2005). Bacterial nature of the Earth's glacial epochs. *Herald Russian Academy of Sciences,* **75** (12), 1107–1114.
Sorokhtin, O. G., Chilingar, G. V., Khilyuk, L. F., & Gorfunkel, M. V. (2006). Evolution of the Earth's global climate. *Energy Sources,* **28**, 1–19.
Sorokhtin, O. G., Lein, A. Yu., & Balanyuk, I. E. (2001). Thermodynamics of oceanic hydrothermal systems and abiogenic generation of methane. *Oceanology,* **41** (6), 898–909.
Sorokhtin, O. G., & Lobkovskiy (1976). A mechanism of pulling of oceanic sediments into subduction zones. *Isvestiya AN SSSR,* **5**, 3–10.
Sorokhtin, O. G., Mitrophanov, F. P., & Sorokhtin, N. O. (1996). *Origin of Diamonds and the Diamond Perspectives in the Eastern Part of Baltic Shield.* Karelian Scientific Center, RAN, 144 pp.
Sorokhtin, O. G., & Sorokhtin, N. O. (1996). The heat regimes of formation of continental plates during the Early Precambrian time. *Life of Earth,* **29**, 95–113.
Sorokhtin, O. G., & Sorokhtin, N. O. (2002). Origin and evolution of the Earth's atmosphere. *Herald Russian Academy of Natural Sciences,* **2** (3), 6–18.
Sorokhtin, O. G., Starostin, V. S., & Sorokhtin, N. O. (2001). Evolution of Earth and origin of mineral deposits. *Izvestiya Russian Academy of Natural Sciences (Earth Sciences),* **6**, 5–25.
Sorokhtin, O. G., & Ushakov, S. A. (1989a). About the three stages of tectonic development of Earth. *Reports AN SSSR,* **307** (1), 77–83.
Sorokhtin, O. G., & Ushakov, S. A. (1989b). *The Origin of Moon and its Influence on Global Evolution of Earth.* Moscow University, Moscow, 111 pp.
Sorokhtin, O. G., & Ushakov, S. A. (1991). *Global Evolution of Earth.* Moscow University, Moscow, 446 pp.
Sorokhtin, O. G., & Ushakov, S. A. (1998). Influence of ocean on the composition of atmosphere and Earth's climates. *Oceanology,* **38** (6), 928–937.

Sorokhtin, O. G., & Ushakov, S. A. (2002). *Development of Earth.* Moscow University, Moscow, 559 pp.

Sorokhtin, O. G., Ushakov, S. A., & Fedynskiy, V. V. (1974). Dynamics of lithospheric plates and origin of oil deposits. *Reports AN SSSR,* **214** (1), 1407–1410.

Starostin, V. I. (1996). The main geologic-metallogenic periods in evolution of Earth. *Herald MGU, Series 4 (Geology),* **4**, 19–27.

Strakhov, N. M. (1963). *The Types of Lithogenesis and Their Evolution in Earth's History.* Gosgeoltekhizdat, Moscow, 535 pp.

Taylor, S. R., & McLennon, S. M. (1988). *Continental Crust, Its Composition and Evolution.* Mir, Moscow, 384 pp.

Teng, T. (1968). Attenuation of body waves and the Q structure of the mantle. *Journal of Geophysical Research,* **73** (6), 2195–2208.

Tinsley, B. A. (1997). Do effects of global atmospheric electricity on clouds cause climate changes?. *Eos,* **78**, 341–349.

Tugarinov, A. I., & Voitkevitch, G. V. (1970). *Precambrian Geochronology of Continents.* Nedra, Moscow, 434 pp.

Turcotte, D., & Burke, K. (1978). Global sea-level changes and the thermal structure of the Earth. *Earth and Planetary Science Letters,* **41**, 341–346.

Urey, H. C., & Craig, H. (1953). The composition of the stone meteorites and the origin of meteorites. *Geochimica et Cosmochimica Acta,* **4**, 36–82.

Vacquier, V. (1990). The origin of the terrestrial heat-flow. *Geophysical Journal International,* **106** (1), 199–202.

Vinogradov, A. P. (1967). *Introduction to Geochemistry of Ocean.* Nauka, Moscow, 215 pp.

Vinogradov, A. P., & Yaroshevskiy, A. A. (1965). On the physical conditions of zonal melting in geospheres. *Geochemistry,* **7**.

Vinogradov, A. P., & Yaroshevskiy, A. A. (1967). The dynamics of zonal melting and some geochemical implications. *Geochemistry,* **12**.

Voitkevich, G. V., & Lebed'ko, G. I. (1975). *Mineral Resources and Metallogenics of the Precambrian Time.* Nedra, Moscow, 231 pp.

Willson, R. S. (1997). Total solar irradiance trend during solar cycles 21 and 22. *Science,* **277**, 1963–1965.

Yasamanov, N. A. (2003). Modern global warming: causes and ecological consequences. *Bulletin Dubna International University of Nature, Society, and Man,* **1** (8), 12–21.

Yoder, H. (1979). *The Formation of Basaltic Magma.* Mir, Moscow, 248 pp.

Zel'dovich, Ya. B., & Novikov, I. D. (1971). *The Theory of Gravitation and Evolution of Stars.* Nauka, Moscow, 484 pp.

FURTHER READING

Aliven, H. (1963). On the early history of the Sun and the formation of the solar system. *The Astrophysical Journal,* **137**, 981–996.

Alpher, R. A., Bethe, H. A., & Gamow, G. (1948). The origin of chemical elements. *Physical Review,* **73**, 803–804.

Alpher, R. A., & Herman, R. (1948). On the relative abundance of the elements. *Physical Review*, **74**, 1577–1578.
Al'tshuller, L. V., & Sharipjanov, I. I. (1971). On iron distribution in the Earth and its chemical differentiation. *Bulletin of AN SSSR, Physics of Earth*, **4**, 3–16.
Anderson, D. L. (1977). Composition of the mantle and core. *Annual Review of Earth and Planetary Sciences*, **5**, 179–202.
Anderson, D. L. (1989). *Theory of the Earth*. Blackwell Scientific Publishers, Oxford, 366 pp.
Anderson, K. D., Bradley, R. S., & Pedersen, T. F. (2003). *Paleoclimate, Global Change and the Future*. Springer, Berlin, 221 pp.
Areshev, E. G. (2003). *Potential of Oil Deposits of Peripheral Seas of Far East and South-Eastern Asia*. AVANTI, Moscow, 288 pp.
Archagenl'sk Diamond-Bearing Province (Geology, Petrography, Geochemistry, and Mineralogy). MGU, Moscow, 524 pp.
Balanyuk, I. E. (1981). Migration of hydrocarbons from the subduction zones into the marginal depression. In: *The Problems of Theoretical Geodynamics and Tectonics of Lithospheric Plates*. Institute of Oceanology of AN SSSR, Moscow, pp. 168–173.
Barazangi, M., & Dorman, J. (1968). World seismicity map of ESSA coast and geodetic survey of epicenters data for 1961–1967. *Bulletin of Seismological Society of America*, **59**, 369–379.
Belousov, V. V. (1954). *Main Problems of Geotectonics*. Gosgeoltekhizdat, Moscow, 607 pp.
Berner, R. (1994). 3Geocarb IIA revised model of atmospheric CO_2 over Phanerozoic time. *American Journal of Science*, **291**, 56–91.
Berger, A. L. J., Imbrie, J. D., Hayes, G. J., & Salzman, B. (Eds). (1984). *Milankovitch and Climate*. D. Reidal, Dordrecht, Netherlands, 895 pp.
Berzon, I. S., Kogan, S. D., & Pasechnik, I. P. (1968). On construction of thin-layered model of the region of transition from the mantle to the core of Earth. *Reports AN SSSR*, **178** (1), 86–89.
Birch, F. (1958). Differentiation of the mantle. *Geological Society of America Bulletin*, **69** (4), 483–486.
Birch, F. (1965). Energetics of core formation. *Journal of Geophysical Research*, **70** (24).
Blackett, P. M. S. (1956). *Lectures on Rock Magnetism*. Weizman Science Press, Jerusalem, 131 pp.
Blackett, P. M. S. (1961). Comparison of ancient climates with ancient latitudes from rock magnetic measurements. *Proceedings of the Royal Society A*, **263**, 236–248.
Blackett, P. M. S., Clegg, J. A., & Stubbs, P. H. S. (1960). An analysis of rock magnetic data. *Proceedings of the Royal Society A*, **256**, 291–322.
Bogdanov, Yu. A. (1996). Hydrothermal sulfide mineralization in the oceanic rifts. *Oceanology*, **36** (2), 277–287.
Borisyak, A. A. (1922). The origins of continents and oceans. *Priroda*, **1–2**, 13–32.

Borukaev, Ch. B. (1985). *The Structure of Precambrian and the Plate Tectonics.* Nauka, Novosibirsk, 190 pp.

Braterman, P. S., Cairns-Smith, A. G., & Sloper, R. W. (1983). Photooxidation of hydrated Fe – Significance for banded iron formations. *Nature,* **303** (5913), 163–164.

Broeker, W. S. (1997). Thermohaline circulation, the Achilles heel of our climate sytem: Will man-made CO_2 upset the current balance?. *Science,* **278,** 1582–1588.

Bullard, E. C., Everett, J. E., & Smith, A. G. (1965). The fit of continents around Atlantic. In: Symposium on Continental Drift. *Philosophical Transactions of the Royal Society A,* **258,** 41–51.

Bullen, K. E. (1940). The problem of the Earth's density variation. *Bulletin of Seismological Society of America,* **30,** 235–250.

Bullen, K. E. (1975). *The Earth's Density.* Chapman and Hall, London, 390 pp.

Burcke, K., & Wilson, T. (1976). Hot spots on the Earth's surface. *Scientific American,* **235** (2), 46–57.

Chilingar, G. V. (1956). Relationship between Ca/Mg ratio and geologic age. *Bulletin of the American Association of Petroleum Geologists,* **40,** 2256–2266.

Chilingarian, G. V., & Wolf, K. H. (Ed.) (1988). *Diagenesis I.* Elsevier, Amsterdam, 590 pp.

Chumakov, N. M. (1992). Climatic paradox of late Precambrian time. *Priroda,* **4,** 34–41.

Coleman, R. G. (1979). *Ophiolites.* Mir, Moscow, 262 pp.

Condy, K. (1983). *The Archaean Greenstone Belts.* Mir, Moscow, 390 pp.

Cook, F. A., Albangh, D. S., Brown, L. D., Kaufman, S., & Oliver, J. E. (1979). Thin-skinning tectonics in the crystalline Southern Applacians. In: COCORP seismic profiling of the Blue Ridge and Inner Piedmont. *Geology,* **7,** 563–567.

Crowell, J. C. (1999). *Pre-Mesozoic Ice Ages: Their Bearing on Understanding the Climate System.* Geological Society of America, Memoir 192, 106 pp.

Dewey, J., & Bird, J. (1970). Plate tectonics and geoclines. *Tectonophysics,* **10** (5/6), 625–638.

Dickinson, W. R. (1974). Subduction and oil migration. *Geology,* **2** (9), 563–567.

Dmitriev, L. V. (1973). Geochemistry and Petrology of Original Rocks of Mid-Oceanic Ridges. PhD thesis. GEOChE, Moscow, 45 pp.

Dmitriev, L. V., Sharas'kin, A. Ya., & Farafonov, M. M. (1969). Original rocks of rift zones of Indo-Oceanic ridge and some features of their geochemistry. *Reports AN SSSR,* **185** (2), 444–446.

Dmitrievskiy, A. N., Balanyuk, I. E., Sorokhtin, O. G., & Dongarian, L. Sh. (2002). Serpentines of oceanic crust as a source of formation of hydrocarbons. *Geology of Oil and Gas,* **3,** 37–41.

Dolgushin, L. D., & Osipova, G. B. (1989). *Glaciers.* Mysl', Moscow, 448 pp.

Dzivonskiy, A. M., Hales, A. L., & Lapwood, E. R. (1975). Parametrically simple Earth's models consistent with geological data. *Physics of the Earth and Planetary Interiors,* **10,** 12–48.

Elssaser, W. M. (1963). Early history of the Earth. In: *Earth Science and Meteoritics.* North Holland, Amsterdam, pp. 1–30.

Emiliani, C. (1992). *Planet Earth. Cosmology, Geology, and the Evolution of Life and Environment.* University of Cambridge, NY, 718 pp.

Erickson, K. A., McCarthy, T. S., & Trustwell, J. F. (1975). Limestone formation and dolomitization in a lower Proterozoic succession from South Africa. *Journal of Sedimentary Petrology*, **45**, 604–614.

Fedorov, S. A., Bagdasarova, A. M., & Kuzin, I. P. (1969). *Earthquakes and Subsurface Structure of Southern Part of Kuryl Islands' Arc.* Nauka, Moscow, 245 pp.

Ferhugen, G., Turner, F., Weiss, L., Warhafting, K., & Fife, W. (1974). *Earth. Introduction to General Geology.* Mir, Moscow, 847 pp.

Fisher, A. G. (1982). Long-term climate oscillations recorded in stratigraphy. In: *Climate in Earth's History.* Nat. Res. Council, Nat. Academy Press, pp. 97–104.

Fisher, D. (1990). *The Birth of Earth.* Mir, Moscow, 264 pp.

Fisher, O. (1989). *Physics of the Earth's Crust.* Oxford University, London.

Forsyth, D., & Uyeda, S. (1975). On the relative importance of the driving forces of plate motion. *Geophysical Journal of the Royal Astronomical Society*, **43**, 163–200.

Frolova, T. I., & Burikova, I. A. (1992). Andesite volcanism in the history of Earth. *Herald of MGU (Geologic Series)*, **4**, 3–17.

Galimov, E. M. (1984a). $^{13}C/^{12}C$ of diamonds. The vertical zonation of diamond-formation in the lithosphere. In: *Transactions of 27th International Geologic Congress. Vol. 11, Section C11, Geochemistry and Cosmochemistry*, Nauka, Moscow, pp. 110–123.

Galimov, E. M. (1984). On origin and evolution of ocean based on the data on changes in $^{18}O/^{16}O$ in the sedimentary shell of Earth throughout geologic time. *Reports AN SSSR*, **284** (5), 977–981.

Gast, P. W. (1968). Upper mantle chemistry and evolution of the Earth's crust. In: *The History of Earth's Crust.* Princeton University Press, Princeton, pp. 15–27.

Gast, P. W. (1975). Chemical composition of Earth, Moon, and the chondrite meteorites. In: *The Nature of Solid Earth.* Mir, Moscow, pp. 23–41.

Gavrilov, V. P. (1978). *The "Black Gold" of the Planet.* Nedra, Moscow, 190 pp.

Geodynamic map of the Circum-Pacific region (1985). *Free-Air Gravity (to Degree 10).* Tulsa, Oklahoma.

Gerstenkorn, H. (1967). On the controversy over the effect of tidal friction upon history of the Earth-Moon system. *Icarus*, **7** (2), 160–167.

Glickson, A. (1980). Stratigraphy and evolution of primary and secondary greenstone complexes; data on shields of Southern hemisphere. In: *The Early History of Earth.* Mir, Moscow, pp. 264–285.

Gorodnitskiy, A. M., & Sorokhtin, O. G. (1979a). The map of thickness of lithospheric plates. In: *Geophysics of Ocean, Vol. 2, Geodynamics.* Nauka, Moscow, pp. 181–183.

Gorodnitskiy, A. M., & Sorokhtin, O. G. (1979b). The map of theoretical values of heat flux through the oceanic bottom. In: *The Problems of Theoretical Geodynamics and Tectonics of Lithospheric Plates.* Institute of Oceanology of AN SSSR, pp. 122–128.

Gorodnitskiy, A. M., Zonenshine, L. P., & Mirlin, E. G. (1978). *Reconstruction of the Positions of Continents in Phanerozoic Time*. Nauka, Moscow, 121 pp.

Green, D. H., & Ringwood, A. E. (1967). The genesis of basaltic magmas. *Contributions to Mineralogy and Petrology*, **15**, 103–190.

Gubkin, I. M. (1975). *The Study of Oil*. Nauka, Moscow, 1975.

Gutenberg, B. (1959). *Physics of the Earth's Interior*. Academic Press, New York.

Harris, A. W., & Kaula, W. M. (1975). A co-accretional model of satellite formation. *Icarus*, **24**, 516–524.

Heitzler, J. R., Dickson, G. O., Herron, E. M., Pitman, W. S., & Le Pichom, K. (1968). Geomagnetic field reversals and motions of the ocean floor and continents. *Journal of Geophysical Research*, **73** (6), 2119–2136.

Heitzler, J. R., & Hayes, D. E. (1977). Magnetic boundaries in the North Atlantic Ocean. *Science*, **157**, 185–187.

Hess, H. H. (1946). Drowned ancient islands of the Pacific basin. *American Journal of Science*, **244** (11), 772–791.

Hess, H. H. (1962). History of ocean basins. In: *Petrologic Studies. A Volume to Honor A.F. Buddington*. Geol. Soc. Am., pp. 599–620.

Hutchinson, R. W., & Blackwell, G. D. (1988). Time, evolution of Earth's crust, and origin of uranium deposits. In: *Geology, Geochemistry, Mineralogy, and Methods of Evaluation of Uranium Deposits*. Mir, Moscow, pp. 153–175.

Isaaks, B., Oliver, J., & Sykes, L. R. (1968). Seismology and the new global tectonics. *Journal of Geophysical Research*, **73** (18), 5855–5899.

Kapitsa, A. P. (1996). Contradictions in the theory of formation of ozone holes. *The Optics of Atmosphere and Ocean*, **9** (9), 1164–1167.

Kaufman, W. (1982). *The Planets and Moons*. Mir, Moscow, 216 pp.

Keeling, C. D., Charles, D., & Whorf, T. P. (2000). The 1,800 year oceanic tidal cycle: A possible cause of rapid climate change. *Proceedings of the National Academy of Sciences*, **97**, 3814–3819.

Keonjyan, V. P., & Monin, A. S. (1976). Calculation of evolution of the inner layers of planets. *Bulletin of AN SSSR, The Physics of Earth*, **4**, 3–13.

Keonjyan, V. P., & Monin, A. S. (1979). The continental drift and the nature of poles' drift. In: *Geophysics of Ocean, V.2, Geodynamics*. Mir, Nauka, pp. 130–142.

Khain, V. E. (1971). *Regional Tectonics (North and South America, Antarctica, Africa)*. Nedra, Moscow, 548 pp.

Khain, V. E. (1973). On new global tectonics. In: *Problems of Global Tectonics*. Nauka, Moscow, pp. 5–26.

Khain, V. E., Koronovskiy, N. V., & Yasamanov, N. A. (1997). *Historical Geology*. MGU, Moscow, 448 pp.

Khain, V. E., & Lomidze, M. G. (1995). *Geotectonics and Foundations of Geodynamics*. MGU, Moscow, 480 pp.

Khilyuk, L. F., Chilingar, G. V., & Rieke, H. H. (2005). *Probability in Petroleum and Environmental Engineering*. Gulf Publishing Company, Houston, 275 pp.

Khilyuk, L. F., Katz, S. A., Chilingarian, G. V., & Aminzadeh, F. (1994). Numerical criterion and sensitivity analysis for time dependent formation pressure in a sealed layer. *Journal of Petroleum Science and Engineering*, **12**, 137–145.

Khramov, A. N. (1976). Paleomagnetic characteristics of rocks. In: *Physical Properties of Rocks and Mineral Deposits.* Nedra, Moscow, pp. 189–193.

Klein, C., & Bricker, O. (1977). Some aspects of the sedimentary and diagenetic environment of Proterozoic banded iron-formations. *Economic Geology*, **72**, 1457–1470.

Knauth, L. P., & Epstein, S. (1976). Hydrogen and oxygen isotope ratios in nodular bedded cherts. *Geochimica et Cosmochimica Acta*, **40**, 1095–1108.

Knauth, L. P., & Lowe, D. R. (1978). Oxygen isotopic composition of cherts from the Onverwacht Group (3.4 BY), South Africa, with implications for secular variations in the isotopic composition of cherts. *Geological Society of America, Abstracts and Programs*, **10**, 436 pp.

Kotlyakov, V. M., & Danilov, A. I. (1999). Global changes in Antarctic Continent. *Earth and Universe*, **4**, 3–12.

Kovalev, A. A. (1972). Plate tectonics and some specifics of metallogenic analysis. *Geology of Mineral Deposits*, **5**, 90–96.

Kovalev, A. A. (1974). On the causes of intense oil formation in the Middle East. *Geology of Oil and Gas*, **11**, 69–74.

Kovalev, A. A. (1978). *Mobility and the Geologic Criteria of Exploration.* Nedra, Moscow, 287 pp.

Kovalev, A. A., Bogatikov, O. A., Dmitriev, Yu. E., & Kononova, V. A. (1987). General patterns in the evolution of magmatizm in the Earth's history. In: *Magmatic Rocks*, Vol. 6. Nauka, Moscow, pp. 332–348.

Krats, O. K., Khiltova, V. Ya., & Vrevskiy, A. B. (1981). *The Stages and Types of Evolution of Precambrian Crust of Ancient Shields.* Nauka, Leningrad, 164 pp.

Kropotkin, P. N. (1958). The importance of paleomagnetism for stratigraphy and geotectonics. *Bulletin MOIP, Geology Section*, **38** (4), 57–86.

Kropotkin, P. N. (1961). Paleomagnetism, paleoclimates, and the problems of significant horizontal movements of Earth's crust. *Soviet Geology*, **5**, 16–38.

Kropotkin, P. N. (1973). Dynamics of Earth's crust. In: *The Problems of Global Tectonics.* Nauka, Moscow, pp. 27–59.

Kucheruk, E. V., & Ushakov, C. A. (1985). *The Plate Tectonic and Capacity of Oil Deposits (Geophysical Analysis).* VINITI, Physics of Earth, Moscow, 200 pp.

Kun, T. (1977). *The Structure of Scientific Revolutions.* Progress, Moscow, 301 pp.

Kuskov, O. L., & Fabrichnaya, O. B. (1990). The phase relationships in the system $FeO-MgO-SiO_2$ at the boundary of transition zone–lower mantle. *Geochemistry*, **2**, 266–278.

Lamb, H. H. (1995). *Climate, History, and Modern World.* 2nd Ed..,Routledge, New York, 433 pp.

Larson, R. L., Pitman, W. C., III, Golovchenko, X., & Cande, S. C. (1985). *The Bedrock Geology of the World.* Freeman, New York, 235 pp.

Lein, A. Yu., & Sagalevich, A. M. (2000). The "smokers" of Rainbow field – the region of intense abiogenic methane synthesis. *Priroda*, **8**, 44–53.

Lein, A. Yu., Vogt, P. R., & Krein, K. (1998). Geochemical peculiarities of methane-bearing formations of submarine mud volcano in Norwegian Sea. *Geochemistry*, **3**, 230–249.
Le Pichon, X. (1968). Seafloor spreading and continental drift. *Journal of Geophysical Research*, **73** (12), 3661–3697.
Levinson-Lessing, F. Yu. (1940). *Petrography*. Gos. Izd. Geol. Lit., Leningrad–Moscow, 524 pp.
Lisitsyn, A. P. (1974). The age and composition of basaltic floor of oceans. *Reports AN SSSR*, **217** (3), 561–565.
Lisitsyn, A. P. (1978). *The Processes of Oceanic Sedimentation*. Nauka, Moscow, 392 pp.
Lisitsyn, A. P. (1980). The quantitative distribution of terrigenous material. In: *Geologic History of the Ocean*. Nauka, Moscow, pp. 172–191.
Lisitsyn, A. P., & Vinogradov, M. E. (1982). Global regularities in the distribution of life in the ocean and their reflection on the composition of pelagic sediments. The formation and distribution of biogenic sediments. *Bulletin Akademii Nauk SSSR, Geologic Series*, **4**, 5–24.
Lister, C. R. B. (1972). On the thermal balance of a mid-ocean ridge. *Geophysical Journal of the Royal Astronomical Society*, **26**, 516–520.
Litvin, Yu. A. (1991). *The Physicochemical Investigation of Melting of Deep-Seated Matter of Earth*. Nauka, Moscow, 312 pp.
Lobkovskiy, L. I. (1988). *The Geodynamics of Spreading Zones, Zones of Subduction, and the Two-Stage Plate Tectonics*. Nauka, Moscow, 251 pp.
Lobkovskiy, L. I., Nikol'skiy, V. N., & Karakin, A. V. (1986). Geologic-geophysical implications of serpentinization of oceanic lithosphere. *Bulletin MMOIP, Geology Section*, **61** (4), 3–11.
Loehle, C. (1993). Geologic methane as a source for post-glacial CO_2 increase: The hydrocarbon pump hypothesis. *Geophysical Research Letters*, **20**, 1415–1418.
Loehle, C. (2004). Climate change: Detection and attribution of trends from long-term geologic data. *Ecological Modeling*, **171**, 433–450.
Loehle, C. (2006). Climate change in the context of long-term geologic data. In: A. R. Burk (Ed). *Focus on Ecology Research*. Nova Science Publishers, pp. 1–39.
Loper, D. E. (1975). Torque balance and energy budget for the precessionally driven dynamo. *Physics of the Earth and Planetary Interiors*, **11**, 43–60.
Lyustikh, E. N. (1948). On using the theory of O. Yu. Shmidt in the geotectonics. *Reports AN SSSR*, **59** (6), 1417–1420.
MacDonald, C. A., Abbott, A. T., & Peterson, F. L. (1990). *Volcanoes in the Sea, the Geology of Hawaii*. University of Hawaii Press, Honolulu, 518 pp.
Makogon, Yu. F. (1985). *The Gas Hydrates*. Nedra, Moscow, 232 pp.
Mason, B., & Nelson, W. (1973). *The Moon Rocks*. Mir, Moscow, 165 pp.
McKenzie, D., & Richter, F. (1976). Convection currents in the Earth's mantle. *Scientific American*, **235** (5), 72–89.
Meyson (1971). *Foundations of Geochemistry*. Nedra, Moscow, 312 pp.
Milloy, S. (2005). Global warming heats up in senate. http://www.foxnews.com/printer_friendly_story/0,3566,15.,9835,00.html.

Minster, J. B., Jordan, T. N., Molnar, P., & Haines, E. (1974). Numerical modeling of instantaneous plate tectonics. *Journal of the Royal Astronomical Society*, **36**, 541–554.

Mitchell, A. H., & Garson, M. S. (1981). Mineral deposits and their global tectonic setting. In: *Coal Geology and Coal Technology*. Blackwell Scientific Publications, 384 pp.

Miyasiro, A. (1972). Metamorphism and related magmatism in plate tectonics. *American Journal of Science*, **272**, 629–656.

Miyasiro, A. (1976). *Metamorphism and Metamorphic Belts*. Mir, Moscow, 496 pp.

Monin, A. S., & Sorokhtin, O. G. (1979). Development of oceans on the Earth. In: *Geophysics of Ocean, Vol. 2, Geodynamics*. Nauka, Moscow, pp. 259–261.

Monin, A. S., & Sorokhtin, O. G. (1982). The evolution of Earth under the spatial gravitational differentiation of its interior. *Reports AN SSSR*, **263** (3), 572–575.

Monin, A. S., & Sorokhtin, O. G. (1982). The thermal evolution of Earth under the spatial gravitational differentiation of its interior. *Reports AN SSSR*, **266** (1), 63–67.

Monin, A. S., & Sorokhtin, O. G. (1982). The evolution of oceans and Precambrian metallogenics. *Reports AN SSSR*, **264** (6), 1453–1457.

Monin, A. S., Zonenshine, L. P., & Gorodnitskiy, A. M. (1986). Paleomagnetism and pre-Gondwana drift of continents. *Reports AN SSSR*, **286** (6), 1355–1359.

Moore, P. D., Chandler, B., & Scott, P. (1996). *Global Environmental Change*. Blackwell Science, Oxford, England, 244 pp.

Morgan, W. J. (1968). Rises, trenches, great faults and crustal blocks. *Journal of Geophysical Research*, **73** (6), 1959–1982.

Morgan, W. J. (1971). Convection plumes in the lower mantle. *Nature*, **230**, 42–43.

Morgan, W. J. (1972). Deep mantle convection and plate motions. *American Association of Petroleum Geologists*, **56**, 203–213.

Naumov, G. B., Ryzhenko, B. N., & Khodakovskiy, I. L. (1971). *Handbook on Thermodynamic Characteristics*. Atomizdat, Moscow, 240 pp.

Ohtani, E., Ringwood, A., & Hibberson, W. (1984). Composition of the core II. Effect of high pressure on solubility of FeO in molten iron. *Earth and Planetary Science Letters*, **71** (10), 94–103.

Oliver, J. (1982). Probing the structure of deep continental crust. *Science*, **216**, 689–699.

Ozima, M., & Podosek, F. (1987). *Geochemistry of Noble Gases*. Nedra, Leningtad, 343 pp.

Panella, G. (1972). Paleontological evidence on the Earth's rotational history since early Precambrian. *Astrophysics and Space Science*, **16**, 212–237.

Pan'kov, V. L., & Kalinin, V. A. (1975). The thermodynamic characteristics of rocks and minerals under conditions of Earth's shell. *Bulletin of AN SSSR, Physics of Earth*, (3), 3–15.

Parker, R. L., & Oldenburg, D. W. (1973). Thermal model of ocean ridges. *Nature (Physical Science)*, **242** (122), 137–139.

Parrish, J. T. (Ed.) (1998). *Interpreting Pre-Quaternary Climate from the Geologic Records*. Columbia University Press, NY, 338 pp.

Pearson, P. N., & Palmer, M. R. (1999). Middle Eocene seawater pH and atmospheric carbon dioxide concentrations. *Science*, **284**, 1824–1826.

Peive, A. V. (1969). The oceanic crust of the geologic past. *Geotectonics*, (4), 5–23.

Perry, E. C. Jr., & Tan, F. C. Significance of oxygen and carbon isotope variations in early Precambrian cherts and carbonate rocks of Southern Africa. *Geological Society of America Bulletin*, **883**, 647–664.

Piper, J. D. A. (1976). Paleomagnetic evidence for a Proterozoic supercontinent. *Philosophical Transactions of the Royal Society of London*, **280**, 469–490.

Polderward, A. (1957). Chemistry of Earth's crust. In: *The Earth's Crust*. IL, Moscow, pp. 130–157.

Precambrian Continents (Handbook) (1976). *North and South America*. Nauka, SO, Novosibirsk, 240 pp.

Precambrian Continents (Handbook) (1976). *Australia and Africa*. Nauka, SO, Novosibirsk, 224 pp.

Precambrian Continents (Handbook) (1977). *Ancient Platforms of Eurasia*. Nauka, SO, Novosibirsk, 312 pp.

Precambrian Continents (Handbook) (1977). *Main Characterists of Tectonics*. Nauka, SO, Novosibirsk, 264 pp.

Ranelli, G., & Fischer, B. (1984). Diffusion creep, dislocation creep, and mantle rheology. *Physics of the Earth and Planetary Interiors*, **34**, 77–84.

Raymo, M. E., Ganley, K., Carter, S., Oppo, D. W., & McManus, J. (1998). Millennial – Climate instability during the early Pleistocene epoch. *Nature*, **392**, 699–702.

Reid, G. C. (1997). Solar forcing of global climate change since the mid-17th century. *Climate Change*, **37**, 391–405.

Richter, F. M. (1973). Dynamic models of seafloor spreading. *Reviews of Geophysics and Space Physics*, **11**, 233–287.

Ringwood, A. E. (1962). A model for the upper mantle, 1. *Geophysical Research*, **67**, 857–866.

Ringwood, A. E. (1962). A model for the upper mantle, 2. *Geophysical Research*, **67**, 4473–4477.

Ringwood, A. E. (1966). The chemical composition and the origin of Earth. In: *Advances in Earth Sciences*. MIT Press, Cambridge, pp. 276–356.

Ringwood, A. E. (1975). *Composition and Petrology of the Earth's Mantle*. McGraw-Hill, NY, London, Sydney, 584 pp.

Ringwood, A. E. (1990). The system Fe-FeO revisited. *Physics and Chemistry of Minerals*, **17**, 313–319.

Rona, P. (1986). *Hydrothermal Mineralization of Spreading Zones in Oceans*. Mir, Moscow, 120 pp.

Ruddiman, W. F. (Ed.) (1997). *Tectonic Uplift and Climate Change*. Plenum Press, NY, 535 pp.

Runcorn, S. K. (1962). Convection currents in Earth's mantle. *Nature*, **195**, 1248–1249.
Runcorn, S. K. (1965). Changes in the convection pattern in the Earth's mantle and continental drift: evidence for a cold origin of the Earth. *Philosophical Transactions of the Royal Society Series A*, **258**, 228–251.
Rundkvist, D. V. (1968). Accumulation of metals and evolution of genetic type deposits in the history of development of Earth's crust. In: *Endogenic Ore Deposits*. Nedra, Moscow, pp. 212–225.
Savel'eva, G. N. (1987). *The Gabbroic-Ultrabasic Complexes of Ophiolites of Ural Mountains and Their Analogs in the Contemporary Oceanic Crust*. Nauka, Moscow, 246 pp.
Savostin, L. A., Volokitina, L. P., & Zonenshine, L. P. (1980). Paleobathimetry of Pacific Ocean in the Late Cretaceous time. *Oceanology*, **XX** (5), 871–881.
Sclater, J. G., Jaupart, C., & Galson, D. (1980). The heat flow from oceanic and continental crust and heat loss of the Earth. *Reviews of Geophysics and Space Physics*, **18**, 269–311.
Shidlowski, M. (1980). The Archaean atmosphere and evolution of oxygen resource of Earth. In: *Early History of Earth*. Mir, Moscow, pp. 523–634.
Shklovskiy, I. S. (1975). *Stars: Their Birth, Life, and Death*. Nauka, Moscow, 368 pp.
Shklovskiy, I. S. (1976). *Supernovas and Related Problems*. Nauka, Moscow, 440 pp.
Shmidt, O. Yu. (1946). New theory of Earth's origin. *Priroda*, **7**, 6–16.
Shmidt, O. Yu. (1946). *Four Lectures on Theory of Earth's Origin*. AN SSSR, Moscow, 72 pp.
Shmidt, O. Yu. (1947). On possibility of capture in space mechanics. *Reports AN SSSR*, **58** (2), 213–216.
Shtille, G. (1964). Geotectonic separation of Earth's history. In: *Selected Transactions*. Mir, Moscow, pp. 344–394.
Sillitoe, R. H. (1972). A plate tectonic model for the origin of porphyry copper deposits. *Economic Geology*, **67**, 184–197.
Sillitoe, R. H. (1972). Relation of metal provinces in Western America to subduction of oceanic lithosphere. *Bulletin of Geological Society of America*, **83**, 813–818.
Singer, S. F. (1972). Origin of the Moon by tidal capture and some geophysical consequences. *The Moon*, **5**, 206. .
Smirnov, V. I. (1984). Periodicity in the formation of ore deposits in the geologic history. In: *Metallogenics and Ore Deposits*. Nauka, Moscow, pp. 3–10.
Sorokhtin, O. G. (1973). Dependence of topography of mid-oceanic ridges on the rate of movement of lithospheric plates. *Reports AN SSSR*, **208** (6), 1328–1330.
Sorokhtin, O. G. (1973b). The nature of tectonic activity of Earth and the origin of oceans. PhD thesis, IO AN SSSR, Moscow, 304 pp.
Sorokhtin, O. G. (1976). Tectonics of lithospheric plates and the nature of global transgressions. In: *Problems of Paleohydrology*. Nauka, Moscow, pp. 56–69.

Sorokhtin, O. G. (1977). Content of radioactive elements in Earth and radiogenic energy. In: *Tectonics of Lithospheric Plates (Sources of Tectonic Energy and Plate Dynamics)*. IOAN SSSR, Moscow, pp. 7–27.

Sorokhtin, O. G. (1979a). Possible mechanisms of differentiation of the mantle matter. In: *Geophysics of Ocean, Vol. 2, Geodynamics*. Nauka, Moscow, pp. 88–93.

Sorokhtin, O. G. (1979b). Heat fluxes through the mid-oceanic ridges. In: *Geophysics of Ocean, Vol. 2, Geodynamics*. Nauka, Moscow, pp. 178–181.

Sorokhtin, O. G. (1984). *The Tectonics of Lithospheric Plates. The Contemporary Geologic Theory*. Znanie RSFSR, Moscow, 40 pp.

Sorokhtin, O. G., & Balanyuk, I. E. (1982). On connection of petroleum potential with the subduction zones of lithospheric plates. *Oceanology*, **XXII** (2), 236–245.

Sorokhtin, O. G., Dmitriev, L. V., & Udintsev, G. B. (1971). A possible mechanism of formation of Earth's crust. *Reports AN SSSR*, **199** (2), 319–322.

Sorokhtin, O. G., & Sagalevitch, A. M. (1994). Hydrothermal activity at the oceanic floor. *Institute Oceanology Russian Academy of Sciences*, **131**, 151–162.

Takahashi, I. (1986). Melting of dry peridotite KLB-I up to 14 GPa: Implications on the origin of perdotitic mantle. *Journal of Geophysical Research*, **91** (89), 9367–9382.

Taylor, R. G. (1975). *Origin of Chemical Elements*. Mir, Moscow, 232 pp.

Taylor, S. R. (1964). Trace elements abundance and the chondritic Earth model. *Geochimica et Cosmochimica Acta*, **28**, 1989–1998.

Tilton, G. B., & Read, G. W. (1963). Radioactive heat production in eclogites and some ultramafic rocks. In: *Earth Science And Meteorites*. North Holland, Amsterdam.

Tolstikhin, I. N. (1986). *Isotopic Geochemistry of Helium, Argon, and Rare Gases*. Nauka, Leningrad, 200 pp.

Trifonov, V. G., & Pevnev, A. K. (2001). Present-day movements of Earth's crust according to satellite measurements. In: *Fundamental Problems of General Tectonics*. Nauchnyy Mir, Moscow, pp. 377–401.

Trubitsyn, V. P., Belavina, Yu. F., & Rykov, V. V. (1994). Interaction of mantle convection with the continental and oceanic plates. *Reports RAN*, **334** (3), 368–371.

Turcotte, D. I., & Oxbury, E. R. (1978). Intra-plate volcanism. *Philosophical Transactions of the Royal Society of London*, **288**, 501–579.

Unksov, V. A. (1981). *Plate Tectonics*. Nedra, Leningrad, 288 pp.

Urey, H. C. (1952). *The Planets: Their Origin and Development*. Yale University Press, New Haven, CT, 245 pp.

Urey, H. C. (1959). Primitive planetary atmospheres and the origin of life. In: *The Origin of Life on the Earth, V1*. Pergamon Press, London, pp. 16–22.

Ushakov, S. A. (1968). Viscosity and dynamic processes in the crust and upper mantle. *Herald MGU (Geologic Series)*, **1**, 62–75.

Ushakov, S. A., & Fedynskiy, V. V. (1973). Riftogenesis as a control mechanism of Earth's heat losses. *Reports AN SSSR*, **209** (5), 1183–1185.

Ushakov, S. A., & Khain, V. E. (1965). Composition of Antarctica according to geologic and geophysical data. *Herald MGU (Geologic Series)*, **1**, 3–27.
Vassoevich, N. B. (1986). *Geochemistry of Organic Matter and the Origin of Oil*. Nauka, Moscow, 196 pp.
Vinogradov, A. P. (1962). The average content of chemical elements in the main types of erupted rocks of Earth's crust. *Geochemistry*, **7**, 188–189.
Vinogradov, A. P. (1969). The atmospheres of planets of Solar System. *Herald MGU, Geologic Series*, **4**, 3–14.
Wegener, A. (1984). *The Origin of Continents and Oceans*. Nauka, Leningrad, 285 pp.
Wood, J. A. (1962). Chondrules and the origin of terrestrial planets. *Nature*, **194**, 127–130.
Wood, J. A. (1968). *Meteorites and the Origin of Solar System*. McGraw-Hill, NY, 117 pp.
Wood, J. A. (1970). The lunar soil. *Scientific American*, **233**, 14–23.
Wood, J. A. (1979). *The Solar System*. Prentice Hall, New Jersey, 196 pp.
Wood, J. A. (1999). Origin of the solar system. In: J. K. Beatty (Ed). *The New Solar System*. Sky Publ. Co., Cambridge, MA, pp. 13–22.
Yoder, G. S. (1965). *The Origin of Basaltic Magma*. Mir, Moscow, 240 pp.
Zharikov, V. A. (1976). *The Foundations of Physicochemical Petrology*. MGU, Moscow, 420 pp.
Zharkov, V. N. (1983). *The Inner Structure of Earth and Planets*. Nauka, Moscow, 415 pp.
Zharkov, V. N., & Kalinin, V. A. (1968). *The Equations of Conditions of Solids at High Pressures and Temperatures*. Nauka, Moscow, 312 pp.
Zharkov, V. N., Trubitsyn, V. P., & Samsonenko, L. V. (1971). *Physics of Earth and Planets, Figures and Inner Structure*. Nauka, Moscow, 384 pp.
Zonenshine, L. P., Kuz'min, M. I., & Natapov, L. M. (1990). *Tectonics of Lithospheric Plates on the Territory of SSSR*, Vols. 1 and 2. Nedra, Moscow, pp. 328, 334.

Subject Index

abiogenic
   methane 93, 120, 123, 125, 128, 198, 217, 233, 266
   oxygen 118, 256
accretion 13
adiabatic
   exponent 144
   temperature distribution 141
albedo 137
Aldan Shield 67, 203–204
Alpine-Himalayan Mountain Belt 103
Amazon Craton 80
Anabar Shield 203, 204
Andes 201
Anorthosite magmatism 24
Antarctic
   ice core measurements 163
   Station Vostok 164, 170
anthropogenic
   CO_2 emission 269
   emission of carbon dioxide 164
   emission of greenhouse gases 161
   energy production 154
   factor 160
   impact 285
anticyclones 155–156
Aqaba beachrock (carbonate cement) 282
Archaean sediments 90
Arkhangel'sk Province 211
Arrhenius Hypothesis 134, 135
asthenosphere 28, 65–66
atolls 250
average surface temperature 143

bacteria 278
bacterial
   metabolism 128
   nature of ice ages 171, 281
Balkan Peninsula 199
Baltic Shield 178, 210
barodiffusion 276
   mechanism 36, 40
   process 39
basalt magmatism 25, 65
basaltic
   magma 204
   plates 207
   smelts 84
bauxites 224
benthos 252
Berson Layer 37
bicarbonate of iron 93, 119, 214
Big Bang 1
biologic explosion 243
biomass of ocean 106
blooming plants 116
Brazilian Craton 80
bulge, equatorial 184
Burakovskiy Intrusion 210
Bushweld granites 209, 210

$\delta^{13}C$ 109
$\delta^{13}C_{carb}$ 218
$\delta^{13}C_{org}$ 122, 218
C_{org} 105
C_{org}/N_{org} ratio 105
calcium carbonate 249
Caledonian Orogeny 103
Canadian Shield 185
capture theory 13

carbon dioxide
　partial pressure 108
　troposphere 272
carbon isotope ratios 123
carbonate 112–113
　accumulation 251
　alkaline magma 69
　rocks 114
chemical
　composition 10
　composition of convective mantle 58
　composition of mantle 55
chlorine 201
chromite ores 199
climate
　drivers 259–260
　evolution 189
climatic cooling 255, 271
cloud blanket 159
cloudy troposphere 159–160
CO_2
　concentration 135
　consumption 162
　emission 162, 269
　fixation 111, 114
　mass 267
　removal 112
coal accumulation 223
Comet Tempel 1 275
compression 212
condensation of water vapor 141, 147
Congo
　Craton 80
　Protoplatform 80
continental
　crust 64, 67
　crust formation 73
　crust mass 74
　drift 190–191, 244
　margins 215
　momentum 183
　shields 53–54
continents 193
convection 147, 158

convective
　cells 52
　flows 37–38
　heat transfer 151
　mantle 55
　mantle currents 38
cooling of climate 167
Cordilleran Mountain Range 232
core 39
　differentiation 33
　growth 33
　separation 27, 205
Coriolis Force 39
crustal evolution 73
Cupola-C 176
cyanic acid 238
cyanobacteria 240–241
cyclones 137, 156

Death of Earth 255
degassing
　of carbon dioxide 110
　of water 89
denitrification 279
density
　distribution 9
　of carbon dioxide 159
deposits of chromites 199
diamond-bearing
　kimberlites 219
　lamproite 220
　rocks 221
diffusion processes 35
Disco Island 57
dissociation of
　carbon dioxide 116
　sulfuric acid 148
dry and transparent troposphere
　155–156

Earth 10, 42
Earth's
　atmosphere 103
　core 26

earthquakes 20
ecologic provinces 252
endogenic
   (mantle–magmatic) 219
   energy 42
   oxygen 270
endothermal boundary 66
energy
   balance 41, 49
   balance model 74
   generation 50
   sources 274
epicontinental seas 248, 250
equatorial
   bulge 182
   bulging 180–181
eukaryotes 241
   microorganisms 119
   microalgae 118
   organisms 242
European Platform 79
eustatic
   changes in ocean level 101, 102
eutectic alloy Fe.FeO 37
evaporites 222
evolution of
   Earth's atmosphere 272
   Earth's climates 173
   ice ages 177
   temperature of continental surface 193
   temperature of World Ocean 192
evolutionary parameter 40, 41
exogenic
   mineral deposits 222
   (primary–sedimentary) mineral deposits 219

fayalite 59
fixation of carbon dioxide 115
forces of nature 259
formaldehyde 128, 238–239
forsterite 200
freons 131

gas solubility in water 114
gas-hydrate deposits 233
generation of
   abiogenic methane 126
   biogenic oxygen 180
   methane in the oceanic crust 129
geodynamic maps 205
geographic latitude 152
geothermal gradients 65
glacial periods 279
glaciation 178, 184, 186
   period 179
   on continents 176
global cooling 281–282
Gondwana 252
granitoid batholiths 230
   intrusions 67
gravitational
   differentiation 43
   differentiation mechanism 26
   energy 28, 43
   matter differentiation 275
Great Dyke of Zimbabwe intrusion 209
greenhouse effect 133, 140
Gulf of Aqaba (Red Sea) beachrock 168
guyots 252
Guiana Shield 80

heat
   content 42, 206
   flow 95
   flux 47, 49, 51, 61, 102, 274
   of formation 130
   transfer 137, 152
Henry's Law 162
Huronian
   glaciation 119, 171
   Ice Age 112
   system 178
hydration of
   oceanic crust rocks 197
   pyroxenes 94, 198

hydrocarbon migration 226, 232, 233
hydrogen 232
 cyanide 128, 239
 sulfide 199
hydrosphere 83, 87
hydrothermal
 deposits 203
 mineralization 210
 processes 197
 rift zones 233
 sources 196
 springs 247

ice ages 177, 279
impacts of planetesimals 275
index of mobility (χ_i) 85
iron
 bearing deposits 92
 concentration 28
 ore 99, 123
 ore accumulation 216
 ores 87, 93, 95, 97, 125
 oxide (FeO) 29, 36, 59
 oxide hypothesis 9
 transfer 94
isostatic equilibrium 182
isotope composition of beachrock 282
"isotope temperature" of Antarctic cover 188
Isua Formation in Western Greenland 215, 240

jaspilite 214, 221

K_2O/Na_2O ratios 70, 91, 92
Kaapvaal Archaean Craton 66
Karelian folding belt 79, 81
Karsakpay (Kazakhstan) deposits 96, 216
Katanga–Rodesia copper-bearing belts 213
Kazakhstan 96
Kempirsay Massif 199, 211
Kepler's Third Law 14

Keran Diastrophism 77
kimberlites 222
Kola Peninsula 210–211
Kolyma Mountain Range 203, 204
Komsomol'skaya Station 176
Krivoy Rog (Ukraine) 96, 216
Kursk Magnetic Anomaly 96, 216
Kyoto Protocol 135, 259, 285

Lake Superior 216
Lakshmi Plateau 68
Laplace Hypothesis 8
latitudinal surface temperature 154
life
 emergence 235, 239
 tree 244
"life span" of supercontinents 211
lithosphere of ultrabasic composition 89
lithospheric plates 61, 63
luminosity of Sun 172
lunar origin theory 14

magmatic tectonic cycles 30
magnetite 60
mammals 253
mantle
 convection 54
 matter flow 77
 temperature 29, 33–34, 66
marine fauna 246
mass of
 carbon dioxide 113, 115, 268
 carbon dioxide degassed 110
 convective mantle 57
 core 57
 glaciers 187
 water 86–87
massif 203, 209
maturation of oil 231
Mauna Loa, Hawaii 135
Maxwell Mountains 68
Megagaea 76, 98
merit factor 23, 47
mesopause 140

metallic iron 59, 87
methane 166, 218, 233, 239, 266, 278
  emission 165
  generation 121–122, 124–125, 128, 198
  oxidation 129
  reduction 129
methane-consuming
  archaeobacteria 278
  bacteria 122, 126, 217
methane-generating bacteria 278, 281
Mezogaea 96
microbial activity 278
microfossils 243
microoil 231
Mid-Oceanic Ridges 90
migration of
  hydrocarbons 226–227
  people 254
Milankovitch cycles 187, 279
mobility indices 86
Model of Taylor and McLennon 74
molybdenum 203
momentum conservation law 14
mono-cell convection 54
Monogaea 32, 34, 76, 77, 80, 81
Moon 12, 17

N_{org} 106
nitrogen 107–108, 138, 280
  fixation 279
  consuming bacteria 180, 189, 278, 280, 284
normalization coefficient K_0 115
North American Platform 79
Nova 1

ocean 276
  surface 90
  water 163
oceanic
  crust 61
  crust composition 63
  currents 277

lithospheric plates 61, 62, 64
  trenches 90
  water 162
oil migration 231
Olong Group, Karelia 210
Onwerwaht Series 241
open dissipative system 137
orbital deviations 263
organic
  carbon 109
  matter 116, 230
  nitrogen (N_{org}) 105
origin of
  life 235, 236
  mineral deposits 195
orogenic belt 79
outgassing 265–266
oxygen 117, 179, 270, 284
  accumulation 116
  generation 118
ozone 140
  destruction 130
  holes 130
  freon reaction 130

Paleo-Tethys Ocean 223
paleomagnetic data 27
paleoreconstruction 79
Pangaea 248–249
Pangaea O 33
Panskaya Intrusion 210
parameter χ 40
partial pressure of
  carbon dioxide 114, 269
  methane 129
  nitrogen 105, 108, 267, 280
  oxygen 118, 120–121, 180
pelagic sediments 227, 229, 251
Peru–Chile Trough 229
petrogenic elements 58
petroleum basins 227
petroleum
  generation 231
  reserves 228

Phosphorite-Bearing World Provinces 225
photodissociation of methane 118, 122, 239
photosynthesis 277
phytoplankton 242
pillow lavas 91
plankton 277
precession angle 181, 183–185, 187, 190–191
precession cycles 180
prokaryotes 240
prokaryotic biota 118
Proto-Moon 8, 11, 15, 17, 20
Proto-Sun 3

radiation 147
radioactive
    decay 274
    isotope contents 45
    lead isotopes 7
radiogenic
    energy 28, 43
    energy generation 46
reducing environment 118
regolith 127
regressions 100, 103, 245
reptiles 252
rift zones 93, 197, 215
Roche Limit 15
Rodinia 32
Russian Platform 80, 97

salt deposits 222
sapropel muds 252
Sargasso Sea 169, 190
sediment recycling 64
sedimentary
    iron deposits 217
    ore deposits 213
serpentine 88, 112
shield 203
Siberian Platform 203–204
siderite 215
silicates 35

skapolites 201
smoothing role of World Ocean 276
sodalites 201
solar
    activity 169
    irradiation 261–263
    magnetic cycle lengths 170
solid solutions 35–36
solidification of olivine basalts 84
solubility of CO_2 in water 114
South Ural Mountains 199
specific axial rotational momentum 19
specific heat of
    air 144
    carbon dioxide 159
spirochetes 279, 281
$^{87}Sr/^{86}Sr$ isotope ratio 91, 92
Starooskol'skiy Series 215
stretching zones 212
stromatolite
    deposits 127
    formations 124
strontium isotopes 71
subduction
    of oceanic plates 225
    zones 63, 69–70, 96, 201, 225–227, 230, 232
Sudbury intrusions 210
Sun 261
Sun's luminosity 257
Sun–Moon tides 185
supercontinent disintegration 77
supernova 1, 2
Svecofennian 79
synergetic approach 137
system 2–3

Taratash (Ural Mountains) 215
tectonic activity 50, 52, 102, 207–208
tectonomagmatic activity 21, 207
temperature
    changes 262
    distribution 139

distributions in the Earth's body 22
of ocean surface 175
on Earth 193
Ten Thousand Smokes Valley
    (New Zealand) 201
termites 279
thermosphere 139
Thetys 103
tidal energy 31, 44, 47, 276
    generation 48–49
tidal interaction 14, 15–16, 47
tide height 21
Tikhvin Formation, Urals 224
tillites 99
transgression 99, 103, 245, 248, 251
Transvaal Formation 178
tropopause 140
turbidites 249
Tury Formation 178

Ukraine 96
Ukrainian Shield 80
uniqueness of Earth 235
upwelling 223
    zones 247
Ural Mountains 204

Venus 148, 157, 273–274
    atmosphere 149–150
    surface 68
    tropopause 148
    troposphere 149, 173
Verkhoyansk-Kolyma
    foldbelt in East Siberia 203, 227
viscosity of core matter 30, 39
Vitim-Patom Mountains 204
volcanic activity 172
Voronezh Crystalline Massif 96
Vostok Station 188

water
    content in the mantle 87
    mass 88
    vapor 138
Western Greenland 57
wet and radiation-absorbing
    troposphere 156
White Tiger 232

Yapetus Ocean 211

zircon 22
zonal differentiation 29–30